modern intermediate algebra
for college students

SECOND EDITION

Vivian Shaw Groza

Susanne M. Shelley

Sacramento City College

Rinehart Press / Holt, Rinehart and Winston

SAN FRANCISCO

Library of Congress Cataloging in Publication Data

Groza, Vivian Shaw.
    Modern intermediate algebra for college students.

    1. Algebra.   I. Shelley, Susanne, joint author.
II. Title.
QA154.2.G76   1974   512.9′042   73-10216
ISBN 0-03-007461-4

PRINTED IN THE UNITED STATES OF AMERICA

4 5 6 7   038   9 8 7 6 5 4 3 2

CONTENTS

Preface   ix

Chapter 0   REVIEW OF ELEMENTARY ALGEBRA   1

0.1   The Language of Algebra   1
0.2   Sets   8
0.3   The Real Number System   12
0.4   Linear Equations   16
0.5   Operations on Polynomials   23
0.6   Fractions   28
0.7   Graphing and Linear Systems   37

Chapter 1   EXPONENTS AND RADICALS   47

1.1   Positive Integral Exponents   48
1.2   Integral Exponents   55
1.3   Square Roots   61
1.4   Cube Roots and nth Roots   70
1.5   Fractional Exponents   74
1.6   Radicals: Simplification and Products   79
1.7   Radicals: Rationalizing   83
1.8   Sums and Differences of Radicals   87
1.9   Further Simplification of Radicals (Optional)   90
1.10   Scientific Notation   93
       Summary   98

Chapter 2   QUADRATIC EQUATIONS   101

2.1   Complex Numbers: Definition, Addition, and Subtraction   102

2.2   Complex Numbers: Products   109
2.3   Complex Numbers: Quotients   113
2.4   A Geometric Model for Complex Numbers (Optional)   116
2.5   Quadratic Equations: Solution by Factoring   121
2.6   Quadratic Equations: Completing the Square   128
2.7   The Quadratic Formula   133
2.8   Sum and Product of Roots   139
2.9   Equations in Quadratic Form   145
2.10   Irrational Equations   148
2.11   Verbal Problems   152
       Summary   161

Chapter 3   FACTORING, INEQUALITIES, ABSOLUTE VALUE
               166

3.1   Factoring by Grouping and Substitution   166
3.2   Synthetic Division   173
3.3   Remainder and Factor Theorems   178
3.4   Solution of Polynomial Equations by Factoring   184
3.5   Inequalities   190
3.6   Linear Inequalities in One Variable   195
3.7   Quadratic Inequalities in One Variable   201
3.8   Absolute Value: Equalities   209
3.9   Absolute Value: Inequalities   215
       Summary   218

Chapter 4   RELATIONS AND FUNCTIONS   221

4.1   Ordered Pairs and Relations   221
4.2   Functions   232
4.3   Linear Functions   238
4.4   Quadratic Functions   246
4.5   The Circle and the Distance Formula   252
4.6   Linear Inequalities: Two Variables   258
4.7   Quadratic Inequalities: Two Variables   264
4.8   Absolute Value Functions and Relations   271
4.9   Applications (Optional)   275
       Summary   281

Chapter 5   LOGARITHMIC AND EXPONENTIAL FUNCTIONS
               285

5.1   Functions and Inverses   285

5.2 Exponential Functions 293
5.3 Logarithmic Functions 298
5.4 The Logarithmic Theorems 302
5.5 Common Logarithms 305
5.6 Linear Interpolation 309
5.7 Computations 312
5.8 Logarithmic and Exponential Equations 314
5.9 Natural Logarithms 317
5.10 Applications 319
     Summary 326

Chapter 6   LINEAR AND QUADRATIC SYSTEMS 330

6.1 Basic Concepts 330
6.2 Parabolas and Systems 337
6.3 Ellipses and Hyperbolas 342
6.4 Graphic Solution of Quadratic Systems 350
6.5 Algebraic Solution: Substitution Method 356
6.6 Algebraic Solution: Addition Method 359
6.7 Algebraic Method for Symmetric Equations (Optional) 363
6.8 Applications 365
     Summary 370

Chapter 7   SYSTEMS OF EQUATIONS, MATRICES, DETERMIN-
             ANTS 372

7.1 Linear Systems 372
7.2 Matrices (Optional) 380
7.3 Solution of Linear Systems Using Matrices (Optional) 388
7.4 Determinants (Optional) 395
7.5 Solution of Linear Systems Using Determinants (Optional)
     404
7.6 Verbal Problems 411
     Summary 415

Chapter 8   SEQUENCES AND SERIES 417

8.1 Basic Concepts 417
8.2 Sigma Notation 421
8.3 Arithmetic Progressions 425
8.4 Geometric Progressions 431

8.5   Mathematical Induction (Optional)   436

8.6   The Binomial Theorem (Optional)   438

Summary   443

Answers   447

Index   531

PREFACE

While maintaining the salient features of the first edition, this second edition of *Modern Intermediate Algebra for College Students* incorporates various suggestions and ideas based on four years of classroom usage.

Chapter 0 is a brief review of elementary algebra. This has been kept to a minimum to permit more time for the more advanced topics of intermediate algebra. More illustrative examples and drill exercises have been provided, however, to improve the quality of the review.

Chapter 1, dealing with exponents and radicals, and Chapter 2, on quadratic equations, essentially repeat the last two chapters of Groza and Shelley's *Modern Elementary Algebra for College Students, 2nd Ed.*, again allowing flexibility in the use of the texts. Integral and rational exponents are introduced at the beginning of Chapter 1 so that both exponential and radical techniques can be demonstrated for the simplification of radicals. The unit on complex numbers has been moved to Chapter 2.

Some additional units, not in the elementary text, have been included in Chapters 1 and 2 of the intermediate text for more extensive coverage.

Chapter 5, on logarithms, now occupies an earlier position in the text to insure more timely and adequate coverage.

The units on relations, functions, and graphing are presented in Chapter 4, while Chapter 6 deals with linear and quadratic systems in two variables. Topics on analytic geometry have been reduced in scope, although there is sufficient coverage to give the student a sound understanding of the concept of an algebraic system.

Chapter 7, on determinants and matrices, and Chapter 8, on sequences and series, have been minimally modified on the basis of classroom experience.

The units on sets and logic have been removed from the Appendix and incorporated into the text as the needs for these concepts arise. The Table on Roots and Powers and the Table of Common Logarithms have been placed on the inside covers of the book for easy access by student and instructor.

As in the first edition, the exercises are arranged into parallel A and B sets, with all answers to the A sets provided in the back of the book. A solutions manual is available for the B exercises. Exercises are graded from drill to more difficult; the most difficult exercises are starred.

More illustrative examples and more exercises have been included in every chapter of the text to increase the learning and comprehension potential of the student.

Motivation and interest are again continually supplied and maintained by introductory explanations, by historical notes when appropriate, and by verbal problems and applications that have been expanded and greatly improved.

The authors would like to express their appreciation to the following persons who reviewed the first edition and provided invaluable suggestions for improving this second edition:

Mary Carter Smith, Laney College, California
Mary Louise Adams, Chabot College, California
P. B. Sampson, East Los Angeles College, California
Lloyd J. Rochon, Chabot College
Henry Harmeling, Jr., North Shore Community College, Massachusetts
Michael J. MacCallum, Long Beach, California.

V. S. G. and S. S.

REVIEW OF ELEMENTARY ALGEBRA

0.1 THE LANGUAGE OF ALGEBRA

DEFINITIONS

Numerals are symbols that name numbers according to some specified system, such as our modern decimal system.

A **variable** is a letter that may designate the name of any number in a given set of numbers. (The letters most commonly used are x, y, and z.)

A **constant** is a letter or numeral which designates the name of exactly one number during a particular discussion. (The letters most commonly used are a, b, and c.)

The **sum** of two numbers is the result obtained by adding the numbers.

The **difference**, or **remainder**, is the result obtained when one number is subtracted from another.

A **term** is one of the numbers that is being added or subtracted.

A **product** is the result obtained when two or more numbers are multiplied.

A **factor** is one of the numbers that is being multiplied.

A **quotient** is the result obtained when one number, the **dividend**, is divided by another number, the **divisor**.

The **square** of a number is the product obtained by using the number as a factor two times.

The **cube** of a number is the product obtained by using the number as a factor three times.

A **square root** of a number is a number whose square is the given number.

A **cube root** of a number is a number whose cube is the given number.

SUMMARY OF THE OPERATION SYMBOLS

In this section, the result of an operation is restricted to belong to the set of numbers of arithmetic—that is, the counting numbers $\{1, 2, 3, \ldots\}$, the number zero 0, and quotients of counting numbers.

	Number Symbols		
Operation	*Two Numerals*	*Numeral and Letter*	*Two Letters*
Addition	$3 + 4$	$x + 4$	$x + y$
Subtraction	$7 - 3$	$x - 4$	$x - y$
Multiplication	$3 \cdot 4$ or $3(4)$	$3x$	xy
Division	$\dfrac{12}{3}$	$\dfrac{x}{3}$	$\dfrac{x}{y}$
Squaring	3^2	x^2	Not applicable
Cubing	5^3	x^3	Not applicable
Square root	$\sqrt{9}$	$\sqrt{x}$	Not applicable
Cube root	$\sqrt[3]{27}$	$\sqrt[3]{x}$	Not applicable

Certain expressions for the algebraic operations occur frequently. Some of these are summarized in the following table:

$a + b$	a plus b	$a - b$	a minus b
	the sum of a and b		the difference between a and b
	a added to b		(when b is subtracted from a)
	a more than b		b subtracted from a
	a greater than b		b less than a
	a increased by b		b smaller than a
	a augmented by b		a decreased by b
			a diminished by b
ab	a times b	$\dfrac{a}{b}$	a divided by b
	the product of a and b		the quotient when a is
	a multiplied by b		divided by b
$2x$	twice x	$\dfrac{x}{2}$	one half of x
	the double of x		
$\dfrac{3x}{5}$	three fifths of x	$\dfrac{x}{3}$	one third of x

GROUPING SYMBOLS

Parentheses	( )
Brackets	[ ]
Braces	{ }
Bar (vinculum)	——

Grouping symbols are used to indicate the order in which the operations are to be performed.

The innermost set of grouping symbols indicates the operation that is to be performed first.

CONVENTION

Unless the grouping symbols indicate otherwise, the operations are to be performed in the following order:

1. Taking square roots and/or cube roots as read from left to right.
2. Squaring and/or cubing as read from left to right.

3. Multiplication and/or division as read from left to right.
4. Addition and/or subtraction as read from left to right.

RELATIONS

SUMMARY OF RELATION SYMBOLS

Symbol	Verbal Translation	Examples
$=$	Equals, is equal to	$3 + 4 = 7$
$\neq$	Does not equal	$3 + 4 \neq 5$
$<$	Is less than	$3 < 7$
$>$	Is greater than	$7 > 3$
$\not<$	Is not less than	$7 \not< 3$
$\not>$	Is not greater than	$3 \not> 7$
$\leq$	Is less than *or* is equal to	$x \leq 7$
$\geq$	Is greater than *or* is equal to	$x \geq 3$

An **equation** is a statement having the form $A = B$ with the understanding that A and B are names of the same number.

THE SUBSTITUTION AXIOM

If $A = B$, then A may replace B or B may replace A in any algebraic expression without changing the number that is being named, or in any algebraic statement without changing the truth or falsity of the statement.

An **inequality** is a statement that can be expressed in any one of the following forms:

$A < B$ Read "A is less than B."
$A > B$ Read "A is greater than B."
$A \leq B$ Read "A is less than or equal to B."
$A \geq B$ Read "A is greater than or equal to B."

EXAMPLE 0.1.1   Simplify $125 - \{5 + [3(7 - 2)^2]\}$.

Solution

$$125 - \{5 + [3(7 - 2)^2]\}$$
$$= 125 - \{5 + [3(5)^2]\}$$
$$= 125 - \{5 + [3(25)]\}$$
$$= 125 - \{5 + 75\}$$
$$= 125 - 80$$
$$= 45$$

EXAMPLE 0.1.2　Simplify $\sqrt{(17 - 15)(17 + 15)}$.

Solution　　　$\sqrt{(17 - 15)(17 + 15)} = \sqrt{(2)(32)}$
$$= \sqrt{64}$$
$$= 8$$

EXAMPLE 0.1.3　Simplify
$$\frac{20 - [10 - 2(7 - 3)]}{20 - [10 + 2(7 - 3)]}$$

Solution

$$\frac{20 - [10 - 2(7 - 3)]}{20 - [10 + 2(7 - 3)]} = \frac{20 - [10 - 2(4)]}{20 - [10 + 2(4)]}$$

$$= \frac{20 - [10 - 8]}{20 - [10 + 8]}$$

$$= \frac{20 - 2}{20 - 18}$$

$$= \frac{18}{2}$$

$$= 9$$

EXAMPLE 0.1.4　Evaluate $3(x + 4) - \dfrac{x}{2}$ for $x = 6$.

Solution　　　$3(x + 4) - \dfrac{x}{2} = 3(6 + 4) - \dfrac{6}{2}$

$$= 3(10) - 3$$
$$= 30 - 3 = 27$$

EXAMPLE 0.1.5　Evaluate $\dfrac{x^3 + y^3}{x + y}$ for $x = 5$ and $y = 3$.

Solution　　　$\dfrac{x^3 + y^3}{x + y} = \dfrac{5^3 + y^3}{5 + y}$

$$= \frac{5^3 + 3^3}{5 + 3}$$

$$= \frac{125 + 27}{5 + 3}$$

$$= \frac{152}{8} = 19$$

EXAMPLE 0.1.6   Express each of the following in symbols:

Expression	*Solution*
Twice the sum of x and 7	$2(x + 7)$
Six more than one half the square of x	$\dfrac{x^2}{2} + 6$
The cube of the difference when 5 is subtracted from x	$(x - 5)^3$
Eight less than the quotient of y divided by 2 is 10	$\dfrac{y}{2} - 8 = 10$
The sum of the squares of x and y is less than the square of the sum of x and y	$x^2 + y^2 < (x + y)^2$
The cube root of 3 times x is greater than or equal to the square root of one third of y	$\sqrt[3]{3x} \geq \sqrt{\dfrac{y}{3}}$

EXERCISES 0.1

Express Exercises 1–18 in symbols.

1. The sum of 5 and N
2. The product of 5 and N
3. The difference when 5 is subtracted from N
4. The square of N
5. The quotient when 5 is divided by N
6. The cube of N
7. The square root of 5
8. The cube root of N
9. Five times the sum of x and 8
10. The remainder when 4 is subtracted from the product of 6 and x
11. The square of the quotient of y divided by 5
12. One half the difference when the square root of y is subtracted from the square of x
13. The sum of the cube of n and the square of the difference when 1 is subtracted from n
14. The quotient when twice the product of x and y is divided by the sum of x and y
15. Ten less than the product of 3 and x is equal to 7.

16. The square of the sum of x and 4 is less than the quotient of x divided by 4.

17. The product of 6 and the remainder when x is subtracted from 9 is greater than or equal to 8.

18. The cube root of the sum of x and y does not equal the sum of the cube root of x and the cube root of y.

19. Write an algebraic expression to indicate that if one begins with a number x, then multiplies it by 3, then adds 6, then multiplies by 4, then subtracts 12 times the original number x, the result is always equal to 24.

20. Write in symbols: If 5 is subtracted from four fifths of a certain number x, then the difference is 3 times the sum of the number x and 2.

Evaluate Exercises 21–45.

21. $2(x - 3)$ for $x = 5$

22. $(x + 2)(x - 3)$ for $x = 5$

23. $x + 2(x - 3)$ for $x = 5$

24. $3x - (7 - x)$ for $x = 4$

25. $(3x - 7) - x$ for $x = 4$

26. $3x^2$ for $x = 2$

27. $(3x)^2$ for $x = 2$

28. $\dfrac{5x}{6 - x}$ for $x = 1$

29. $\dfrac{5x + x^2}{5x + 25}$ for $x = 10$

30. $\dfrac{5x}{5x} + \dfrac{x^2}{25}$ for $x = 10$

31. $x - [x - (x - 2)]$ for $x = 7$

32. $\sqrt{x^2 - 9}$ for $x = 5$

33. $\sqrt[3]{2x^3 + 11}$ for $x = 2$

34. $(x^2 + [x - 2])(x^2 - [x - 2])$ for $x = 4$

35. $x^2 - 2\{4x - [(x - 4)(x + 4)]\}$ for $x = 6$

36. $\dfrac{5x + 7y}{5x - 7y}$ for $x = 7, y = 5$

37. $(x + y)(x - y)$ for $x = 39, y = 38$

38. $\sqrt{x^2 + y^2}$ for $x = 6, y = 8$

39. $x + y + z$ for $x = 3, y = 8, z = 7$

40. xyz for $x = 2, y = 3, z = 5$

41. $x - y - z$ for $x = 25, y = 15, z = 5$

42. $x - (y - z)$ for $x = 25, y = 15, z = 5$

43. $\dfrac{xy}{xz}$ for $x = 7, y = 0, z = 3$

44. $\sqrt[3]{(x + y)^2 - (x - y)^2}$ for $x = 2, y = 1$

45. $\dfrac{x^2 + xy + xz}{x + y + z}$ for $x = 5, y = 6, z = 7$

Determine whether each of the statements in Exercises 46–55 is true or false for the given value of the variables.

46. $5(x - 2) = 5x - 10$ for $x = 6$ **47.** $x^3 = 3x$ for $x = 5$

48. $2x + 5 < 12$ for $x = 2$ **49.** $2x + 5 < 12$ for $x = 5$

50. $3x - 5 > 7$ for $x = 3$ **51.** $3x - 5 > 7$ for $x = 4$

52. $3x - 5 > 7$ for $x = 5$ **53.** $\dfrac{8}{x + 2} = \dfrac{8}{x} + \dfrac{8}{2}$ for $x = 2$

54. $x^2 - 4x \leq 60$ for $x = 10$

55. $2(xy) \neq (2x)(2y)$ for $x = 3,\ y = 5$

0.2 SETS

DEFINITIONS

A **set** is a well-defined collection of objects called *elements* or *members* of the set.

A collection is **well-defined** if it is always possible to determine whether or not a particular object belongs to the set.

Set membership is indicated by the notation $x \in S$, meaning "x is an element, or member, of set S."

For example, if S is the set of counting numbers and if $x = 5$, then $5 \in S$ means that 5 is a counting number.

A **finite set** is a set for which there is a counting number that indicates how many elements are in the set.

An **infinite set** is a set that is not finite but has a never-ending list of elements.

In the **listing method**, a particular set is defined by listing or stating the names of the members enclosed by braces and separated by commas.

The set of **digits**, $D = \{0, 1, 2, 3, 4, 5, 6, 7, 8, 9\}$, is a finite set.

The set of **natural numbers**, $N = \{1, 2, 3, 4, 5, \ldots\}$, is an infinite set.

In the **rule** or **set-builder method**, a particular set is defined by using braces to enclose the description that consists of one or more letters used to designate a member of the set, a vertical bar, $|$, read "such that," and then a statement describing the property or rule for set membership.

EXAMPLE 0.2.1   List the elements in the set $A = \{x \mid x \in N \text{ and } x \text{ is less than } 5\}$.

Solution   The set is read: "The set of all elements x such that x is a natural number and x is less than 5." Therefore, $A = \{1, 2, 3, 4\}$.

The **empty set**, also called the **null set**, is the set that has no members. In symbols, the empty set is denoted by $\varnothing$ or by $\{ \ \}$.

A **universal set**, U, is a set to which the elements of all other sets in a particular discussion must belong.

SUBSETS

The set A is a **subset** of the set B if every element of A is an element of B. In symbols, $A \subset B$. (Some authors prefer to write $A \subseteq B$, reserving the notation $A \subset B$ to mean A is a **proper** subset of B—that is, A is not identical to B. The set B is called an **improper** subset of itself.) For example,

$$\{3, 6, 9\} \subset \{1, 2, 3, 4, 5, 6, 7, 8, 9\}$$

EQUAL SETS

Two sets A and B are said to be **equal**, or **identical**, if and only if

$$A \subset B \quad \text{and} \quad B \subset A$$

In symbols, $A = B$. Thus $\{3, 6, 9\} = \{9, 6, 3\}$. Note that the order in which the elements are written is not important; it is only necessary that the two sets have exactly the same objects in them.

DEFINITIONS

If r and s are natural numbers and if there exists a natural number n such that

$$r = sn$$

then	r is a **multiple** of s
and	s is a **factor** of r
and	r is **divisible** by s
and	s is a **divisor** of r

A natural number n is **even** if and only if there is a natural number k such that $n = 2k$.

A natural number is **odd** if and only if it is not even.

A natural number p is a **prime** if and only if $p \neq 1$ and p has no divisors different from 1 and p.

A natural number n is a **composite** if and only if $n \neq 1$ and n is not a prime.

The **union** of two sets A and B is the set of all elements that are in A, or in B, or in both A and B. In symbols, $A \cup B$ designates the union of A and B.

The **intersection** of two sets A and B is the set of all elements that are in *both* A and B. In symbols, $A \cap B$ designates the intersection of A and B.

In Examples 0.2.2–0.2.6, list the elements of each given set.

EXAMPLE 0.2.2 $\{x \mid x \in N \text{ and } x \text{ is a factor of } 12\}$

Solution $\{1, 2, 3, 4, 6, 12\}$

EXAMPLE 0.2.3 $\{x \mid x \text{ is a prime and } x \text{ is a divisor of } 60\}$

Solution $\{2, 3, 5\}$

EXAMPLE 0.2.4 $\{x \mid x \in D \text{ and } x \text{ is odd}\}$

Solution $\{1, 3, 5, 7, 9\}$

EXAMPLE 0.2.5 $\{y \mid y \in N \text{ and } y \text{ is even}\}$

Solution $\{2, 4, 6, 8, 10, \ldots\}$

EXAMPLE 0.2.6 $\{d \mid d \in D \text{ and } d \text{ is greater than } 9\}$

Solution $\{ \ \}$ or $\varnothing$

EXAMPLE 0.2.7   Given $A = \{2, 4, 6, 8, 10, 12\}$

and $B = \{3, 6, 9, 12\}$
list the elements in (a) $A \cup B$ and (b) $A \cap B$.

Solution   a. $A \cup B = \{2, 3, 4, 6, 8, 9, 10, 12\}$
            b. $A \cap B = \{6, 12\}$

EXAMPLE 0.2.8   Given $A = \{1, 3, 5, 7, 9, \ldots\}$, the set of odd numbers

and $B = \{2, 4, 6, 8, 10, \ldots\}$, the set of even numbers
list the elements in (a) $A \cup B$ and (b) $A \cap B$.

Solution   a. $A \cup B = \{1, 2, 3, 4, 5, \ldots\} = N$
            b. $A \cap B = \{ \ \}$

EXERCISES 0.2

In Exercises 1–16, list the elements of each indicated subset of the set of natural numbers, N. (Note that $x \in N$.)

1. $\{x \mid (8 - x) \in N\}$

2. $\left\{x \mid \dfrac{30}{x} \in N\right\}$

3. $\left\{x \mid \dfrac{x}{30} \in N\right\}$

4. $\{x \mid (x - 15) \in N\}$

5. $\{x \mid \sqrt{x} \in N\}$

6. $\{x \mid \sqrt[3]{x} \in N\}$

7. $\left\{x \mid \dfrac{60}{x} \text{ is a prime}\right\}$

8. $\left\{x \mid x \text{ is prime and } \dfrac{60}{x} \in N\right\}$

9. The set of primes between 30 and 50

10. The set of odd numbers between 30 and 40

11. The set of even numbers between 17 and 25

12. The set of composites between 19 and 27

13. $\{x \mid x \text{ is a factor of } 28\}$ **14.** $\{x \mid x \text{ is a multiple of } 13\}$

15. $\{x \mid x \text{ is a divisor of } 18\}$ **16.** $\{x \mid x \text{ is divisible by } 25\}$

For Exercises 17–21, list (a) $A \cup B$ and (b) $A \cap B$. The universal set is N, the set of natural numbers.

17. $A = \{1, 5, 12, 20\}$
$B = \{5, 15, 20, 25\}$

18. $A = \{x \mid x = 2k, k \in N\}$
$B = \{x \mid x = 2k - 1, k \in N\}$

19. A is the set of composites between 1 and 10.
B is the set of primes between 1 and 10.

20. A is the set of odd numbers between 30 and 40.
B is the set of multiples of 3 between 30 and 40.

21. $A = \{x \mid x \text{ is a factor of } 30\}$
$B = \{x \mid x \text{ is a prime less than } 30\}$

22. List all subsets of $\{2, 3, 5\}$.

23. List all subsets of $\{1, 3, 5, 7\}$.

In Exercises 24–33, let $N =$ *set of natural numbers*
$P =$ *set of primes*
$C =$ *set of composites*
$E =$ *set of even numbers*
$O =$ *set of odd numbers*

Determine whether each statement is true or false.

24. $P \subset N$ **25.** $O \subset P$

26. $C \subset O$ **27.** $43 \in P$

28. $37 \in C$ **29.** $P \cup C = N$

30. $E \cup O = N$ **31.** $P \cap E = \{2\}$

32. $E \cap N = E$ **33.** $E \cup N = N$

0.3 THE REAL NUMBER SYSTEM

DEFINITIONS

The set of **natural numbers** $N = \{1, 2, 3, 4, 5, \ldots\}$.

The set of **integers** $I = \{\ldots, -3, -2, -1, 0, 1, 2, 3, \ldots\}$.

The natural numbers are also called the **counting numbers** or the **positive integers**.

The numbers in the set $\{\ldots, -3, -2, -1\}$ are called the **negative integers**.

The set of **rational numbers** $Q = \{p/q$, where p and q are integers and $q \neq 0\}$. Examples of rational numbers are $\frac{2}{3}$, $-\frac{5}{9}$, 5, 0, and -2.

The set of **irrational numbers** consists of the infinite decimals whose digits are not repeating. Examples of irrational numbers are $\sqrt{2}$, $-\sqrt{5}$, and π.

The set of **real numbers**, R, is the union of the set of rational numbers and the set of irrational numbers.

A line whose points are named by using numbers is called a **number line**.

The **origin** is the point on a number line assigned the name 0 (zero).

The **coordinate** of a point is the number that names the point.

The **graph** of a number is the point whose name is this number.

In this text, the positive direction on a horizontal number line is to the right.

The set of **real numbers**, R, is the set of all the numbers that can be associated with points on a number line (see Figure 0.3.1).

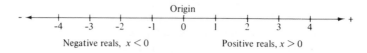

FIGURE 0.3.1. The real number line.

The numbers whose graphs are to the right of zero are called the **positive real numbers,** $x > 0$.

The numbers whose graphs are to the left of zero are called the **negative real numbers,** $x < 0$.

PROPERTIES OF THE SET OF REAL NUMBERS

$(x, y,$ and z real numbers)

Axioms

1. Closure	$x + y$ is a real number xy is a real number	
2. Commutative	$x + y = y + x$ $xy = yx$	(Addition and multiplication are commutative, subtraction and division are not)
3. Associative	$(x + y) + z = x + (y + z)$ $(xy)z = x(yz)$	(Addition and multiplication are associative, subtraction and division are not)
4. Distributive	$x(y + z) = xy + xz$	(Multiplication distributes over addition)
5. Identity	There exists a unique real number 0 such that $x + 0 = 0 + x = x$	(0 is the additive identity)
	There exists a unique real number 1 such that $x \cdot 1 = 1 \cdot x = x$	(1 is the multiplicative identity)
6. Inverse	For each real number x there exists a unique real number $-x$ called the additive inverse of x, such that $x + (-x) = 0$	(Additive inverse)
	For each real number x, except $x = 0$, there exists a unique real number $1/x$ called the *reciprocal* of x, such that $x(1/x) = 1$	(Multiplicative inverse)
7. Trichotomy	Exactly one of the following relations is true:	

$$x < y \quad \text{or} \quad x = y \quad \text{or} \quad x > y$$

8. Completeness	There is a one-to-one correspondence between the set of real numbers and the set of points on the number line.

Definitions

		Examples
1. Subtraction	$x - y = x + (-y)$	$5 - 2 = 5 + (-2)$
2. Division	$\dfrac{x}{y} = x\left(\dfrac{1}{y}\right)$ if $y \neq 0$	$\dfrac{12}{3} = 12\left(\dfrac{1}{3}\right)$

Theorems

1. Closure,
    subtraction $x - y$ is a real number $2 - 5 = -3$ and -3 is real

    division $\dfrac{x}{y}$ is a real number if $y \neq 0$ $-\dfrac{2}{5}$ is a real number

2. Multiplication
    by zero $x \cdot 0 = 0 \cdot x = 0$ $-5 \cdot 0 = 0, \, 0 \cdot 0 = 0$

3. Negative of a
    negative $-(-x) = x$ $-(-5) = 5$

4. Reciprocal of a $\dfrac{1}{1/x} = x$ $\dfrac{1}{1/5} = 5$
    reciprocal

5. Addition $-x + (-y) = -(x + y)$ $-5 + (-2) = -(5 + 2)$
                                                    $= -7$
        $-x + (+y) = -(x - y)$ $-5 + (+2) = -(5 - 2)$
                                                    $= -3$
                                                    $-2 + (+5) = -(2 - 5)$
                                                    $= -(-3) = 3$

6. Multiplication $(-x)(-y) = xy$ $(-5)(-2) = 10$
        $(-x)(+y) = -xy$ $(-5)(+2) = -10$

CONVENTIONS

$$x + y + z = (x + y) + z$$
$$xyz = (xy)z$$
$$x + y - z = x + y + (-z)$$
$$x - y + z = x + (-y) + z$$
$$x - y - z = x + (-y) + (-z)$$

EXERCISES 0.3

Simplify Exercises 1–10 and identify each as a member of one or more of these sets: natural numbers, integers, rationals, real numbers.

1. $2 - 7$

2. $\frac{3}{4} - \frac{1}{2}$

3. $\frac{6}{5}(\frac{1}{2} + \frac{1}{3})$

4. $\sqrt{25 - 16}$

5. $\sqrt{9-4}$ **6.** $\frac{5}{2} - \frac{1}{2}$

7. $\dfrac{9-3}{3-9}$ **8.** $\dfrac{4-8}{6}$

9. $\dfrac{6}{4-7}$ **10.** $(-\frac{3}{5}) - (-\frac{3}{5})$

Name the axiom that is illustrated by each of Exercises 11–20.

11. $-2 + 3 = 3 + (-2)$ **12.** $-2 + [-(-2)] = 0$
13. $(-\frac{1}{2})(-2) = 1$ **14.** $2(x + 5) = 2x + 10$
15. $-2 + (3 + 4) = (-2 + 3) + 4$ **16.** $(-\frac{2}{3})(\frac{1}{2}) = (\frac{1}{2})(-\frac{2}{3})$
17. $(\frac{3}{5})(\frac{1}{1}) = \frac{3}{5}$ **18.** $-10(-\frac{1}{2})(-\frac{1}{3}) = 5(-\frac{1}{3})$
19. $-2(-3 + 4) \in R$ **20.** $-5 + 0 = -5$

Simplify Exercises 21–28 by using one or more identity or inverse axioms.

21. $\frac{1}{4}(4x)$ **22.** $x + 3 - 3$

23. $x - 5 + 5$ **24.** $-2\left(\dfrac{-x}{2}\right)$

25. $-\frac{1}{2}(-2x + 7 - 7)$ **26.** $8\left(\dfrac{x}{8} - 6 + 6\right)$

27. $\dfrac{-3(x + 5)}{-3}$ **28.** $x\left(\dfrac{4x - 9}{x}\right)$

Perform the indicated operations in Exercises 29–56.

29. $(-25)(4)$ **30.** $(-8)(-125)$
31. $(-17) + (-13)$ **32.** $-15 + 10$
33. $-15 + 25$ **34.** $0 - 7$
35. $4 - 7$ **36.** $-2(4 - 7)$

37. $(5 - 8)(8 - 8)$ **38.** $\dfrac{-5}{-1}$

39. $\dfrac{-4 - 6}{2}$ **40.** $\dfrac{18 + (-6)}{-4}$

41. $12 - 3 - 15$ **42.** $-12 - 3 + 15$
43. $(7 - 8)(7 - 9)(7 - 10)$ **44.** $6 - 5 - 4 - 3$
45. $14 - 27 + 13$ **46.** $(5 - 4)(5 - 6)(5 - 8)$

47. $8 - 2 + 2 - 8$

48. $\dfrac{8 - 2}{2 - 8}$

49. $(-5)^2 + (-5)^3$

50. $(-7)^2 - 4(3)(-6)$

51. $\dfrac{8 - (-2)}{-4 - 1}$

52. $\dfrac{-3 - (-3)}{3 - (-3)}$

53. $\dfrac{-2 - 1}{4 - 7}$

54. $(9 - [5 - 7])(9 + [5 - 7])$

55. $\dfrac{(\frac{1}{2} - 1)(\frac{1}{2} - 2)(\frac{1}{2} - 3)}{(1)(2)(3)}$

56. $\dfrac{-5(-5 - 1)(-5 - 2)(5^2)(-2)^3}{(-1)(2)(3)}$

*By using one or more of the axioms or theorems of the set of real numbers,
rearrange the operations in Exercises 57–64 so that the computation can be
performed in the easiest way possible. Then perform the computation.*

57. $(-\frac{2}{3})(-35)(-\frac{3}{2})$

58. $(-57)(-\frac{5}{3}) + (-57)(\frac{2}{3})$

59. $(\frac{3}{4} + \frac{7}{12}) + (-\frac{3}{4})$

60. $-12(\frac{2}{3} - \frac{3}{4})$

61. $\dfrac{(-9)(+5)(+4)(-5)}{-25}$

62. $(-25)(79)(-4)$

63. $(-8)(-\frac{7}{9}) + (-8)(-\frac{2}{9})$

64. $(64 - 75) - (25 - 36)$

0.4   LINEAR EQUATIONS

DEFINITIONS

An **open equation** is an equation that contains a variable and becomes
either true or false when the variable is replaced by the name of a specified
number.

A **solution** or **root** of an open equation in one variable is a number from
a specified set of numbers that makes the equation true when the variable is
replaced by the name of this specified number.

The **solution set** of an open equation is the set of all solutions of the
equation.

To **solve an equation** means to find its solution set.

Two equations are **equivalent** if and only if they have the same solution
set.

A **linear equation** in one variable is an equation equivalent to $x = a$,
where x is a variable and a is a constant.

THE AXIOMS OF EQUALITY (x, y, and z any real numbers)

The reflexive axiom: $x = x$
The symmetric axiom: If $x = y$, then $y = x$.
The transitive axiom: If $x = y$ and $y = z$, then $x = z$.
The substitution axiom: If $x = y$, then x may replace y, or y may replace x, in any algebraic statement without changing the truth or falsity of the statement.

DISTRIBUTIVE AXIOM AND RELATED THEOREMS

Examples

Distributive axiom: $a(b + c) = ab + ac$	$3(x + 5) = 3x + 15$
Theorem: $ba + ca = (b + c)a$	$2x + 5x = (2 + 5)x = 7x$
Theorem: $a(b - c) = ab - ac$	$3(x - 5) = 3x - 15$
Theorem: $ba - ca = (b - c)a$	$2x - 5x = (2 - 5)x = -3x$
Theorem: $-(a + b) = -a + (-b)$	$-(x + 5) = -x - 5$
	$-(x - 5) = -x + 5$

THE EQUIVALENCE THEOREMS

($A = B$ is any open equation; C is any real number.)

Addition Theorem

$$A = B \text{ if and only if } A + C = B + C$$

If the same number is added to each side of an equation, the resulting sums are equal and the two equations are equivalent.

Subtraction Theorem

$$A = B \text{ if and only if } A - C = B - C$$

If the same number is subtracted from each side of an equation, the resulting differences are equal and the two equations are equivalent.

Multiplication Theorem

$$A = B \text{ if and only if } AC = BC \text{ and } C \neq 0$$

If each side of an equation is multiplied by the same nonzero number, then the resulting products are equal and the two equations are equivalent.

Division Theorem

$$A = B \text{ if and only if } \frac{A}{C} = \frac{B}{C} \text{ and } C \neq 0$$

If each side of an equation is divided by the same nonzero number, then the resulting quotients are equal and the two equations are equivalent.

EXAMPLE 0.4.1   Solve for x: $5(2 - x) = 4x - (x + 4)$.

Solution

$$5(2 - x) = 4x - (x + 4)$$
$$10 - 5x = 4x - x - 4 \qquad \text{(Using the distributive axiom to remove parentheses)}$$
$$10 - 5x = 3x - 4 \qquad \text{(Using the distributive axiom to collect like terms)}$$
$$-5x = 3x - 14 \qquad \text{(Subtracting 10 from each side)}$$
$$-8x = -14 \qquad \text{(Subtracting } 3x \text{ from each side)}$$
$$x = \tfrac{14}{8} \qquad \text{(Dividing each side by } -8)$$
$$x = \tfrac{7}{4}$$

Check:
$$5(2 - \tfrac{7}{4}) = 4(\tfrac{7}{4}) - (\tfrac{7}{4} + 4)$$
$$5(\tfrac{1}{4}) = 7 - (\tfrac{23}{4})$$
$$\frac{5}{4} = \frac{28 - 23}{4}$$
$$\tfrac{5}{4} = \tfrac{5}{4}$$

The solution set is $\{\tfrac{7}{4}\}$.

EXAMPLE 0.4.2   Solve for t: $3t - 4(t + 5) = t - (2 - t)$.

Solution

$$3t - 4(t + 5) = t - (2 - t)$$
$$3t - 4t - 20 = t - 2 + t \qquad \text{(Removing parentheses)}$$
$$-t - 20 = 2t - 2 \qquad \text{(Collecting like terms)}$$
$$2t - 2 = -t - 20 \qquad \text{(Exchanging sides)}$$
$$3t - 2 = -20 \qquad \text{(Adding } t \text{ to each side)}$$
$$3t = -18 \qquad \text{(Adding 2 to each side)}$$
$$\frac{3t}{3} = -\frac{18}{3} \qquad \text{(Dividing each side by 3)}$$
$$t = -6$$

Check
$$3(-6) - 4(-6 + 5) = -6 - (2 - (-6))$$
$$-18 - 4(-1) = -6 - (2 + 6)$$
$$-18 - (-4) = -6 - 8$$
$$-18 + 4 = -14$$
$$-14 = -14$$

The solution set is $\{-6\}$.

EXAMPLE 0.4.3   Mixture problem. A piggy bank contains 23 coins, consisting of dimes and quarters. The total value is $3.35. How many coins of each kind are there?

Solution   Let d = the number of dimes; then $23 - d$ = the number of quarters.

Item	Unit Value	Number of Items	Value
Dimes	10 cents	d	$10d$
Quarters	25 cents	$23 - d$	$25(23 - d)$
Mixture		23	335 cents

Equation:
$$10d + 25(23 - d) = 335$$
$$10d + 575 - 25d = 335$$
$$-15d = -240$$
$$d = 16$$

There are 16 dimes and $(23 - 16)$ or 7 quarters.

EXAMPLE 0.4.4   Uniform motion problem. At a certain time two trains start from the same depot and travel in opposite directions. If one travels 35 miles per hour (mph) and the other travels 60 mph, in how many hours will the trains be 285 miles apart?

Solution   [r = rate (mph), t = time (hours), d = distance (miles)]

Sketch

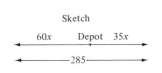

Chart			
Formula	r $\cdot$	t =	d
One train	35	x	$35x$
Other train	60	x	$60x$

FIGURE 0.4.1

Equation:

Distance of one train + distance of other = total distance

$$35x \quad\quad + \quad\quad 60x \quad\quad = 285$$
$$95x = 285$$
$$x = 3$$

Therefore, the trains will be 285 miles apart in 3 hours.

EXAMPLE 0.4.5 How many cubic centimeters of distilled water must be added to a 75 percent acid solution (75 percent acid and 25 percent water by volume) to obtain 100 cubic centimeters of a solution that is 30 percent acid?

Solution Let x = the number of cubic centimeters of water to be added.

Item	Unit Value (% of acid)	Number of Items (number of cc)	Value (amount of acid)
Distilled water	0	x	0
75% acid	0.75	$100 - x$	$0.75(100 - x)$
30% acid	0.30	100	$0.30(100)$

Equation:
$$0.75(100 - x) = 0.30(100)$$
$$75 - 0.75x = 30$$
$$-0.75x = -45$$
$$x = 60 \text{ cubic centimeters}$$

EXERCISES 0.4

Name the axiom that is illustrated by each of Exercises 1–10.

1. $-\frac{1}{2} + \frac{1}{3} = \frac{1}{3} + (-\frac{1}{2})$ **2.** $-\frac{1}{2} + [-(-\frac{1}{2})] = 0$

3. $(-\frac{1}{2})(-2) = 1$

4. $-\frac{2}{3}(3 + (-\frac{1}{5})) = -\frac{2}{3}(3) + -\frac{2}{3}(-\frac{1}{5})$

5. $(-\frac{1}{2} + \frac{1}{3}) + \frac{1}{4} = -\frac{1}{2} + (\frac{1}{3} + \frac{1}{4})$ **6.** $(-\frac{3}{5})(\frac{1}{2}) = \frac{1}{2}(-\frac{3}{5})$

7. $(-\frac{3}{5})(\frac{1}{1}) = -\frac{3}{5}$ **8.** $-10[(-\frac{1}{2})(-\frac{1}{3})] = 5(-\frac{1}{3})$

9. $-2(-4 + 5)$ is a real number.   **10.** $\frac{3}{5} + 0 = \frac{3}{5}$

State which equivalence theorem or theorems (addition, subtraction, multiplication, division) are needed to solve each of Exercises 11–20.

11. $2x - 5 = 8$ **12.** $\dfrac{x + 6}{2} = 7$

13. $\dfrac{x}{2} + 9 = 15$ **14.** $3x + 8 = 2$

15. $5x = 2x + 6$ **16.** $3x = 8x - 20$

17. $4x - 5 = 7 - 2x$ **18.** $6x + 7 = 5x + 2$

19. $\dfrac{5x - (x - 2)}{3} = 6$ **20.** $3(x - 2) - (x + 2) = 4$

In Exercises 21–50, solve each equation and check.

21. $x + 5 = -7$ **22.** $3x + 2 = x + 4$

23. $5 - 2y = 15$ **24.** $3 + 4x - 2 = 4 + 2x$

25. $2(x - 4) = 3x - 16$ **26.** $3(y + 2) + 4 = 5y - 2$

27. $4 - (3x + 8) = 17$ **28.** $5(z - 1) - (1 - 4z) = 30$

29. $5t + 7t - 2 = 5 + 7 - 2t$ **30.** $x + 2 = x + 3$

31. $2(x - 5) = 2x - 10$ **32.** $6 - \dfrac{x + 2}{3} = 4$

33. $4 = 2x + 3$ **34.** $4x - (x - 5) = 4 - (x + 7)$

35. $0 = \dfrac{x + 5}{2}$ **36.** $7x - 5(x - 2) = 20$

37. $9 - 6(2 - x) = 7x$ **38.** $x + 4 = 2x - 3$

39. $z + 1 = z$ **40.** $3 - x = x - 3$

41. $4(x + 3) = 12$ **42.** $2(x - 3) - 3(x + 1) = 0$

43. $y - 2 = 2 - y$ **44.** $4 - 3(x + 2) = 7$

45. $3(x + 2) = 3x + 6$ **46.** $0 = 2(x + 5)$

47. $\dfrac{t + 7}{2} = 5$ **48.** $5 - \dfrac{x + 3}{2} = 6$

49. $2(x - 3) + 3(x + 2) = x + 8$ **50.** $x - 3[x - 3(x - 3)] = 1$

Write an algebraic expression to represent each of Exercises 51–60.

51. Four less than 6 times a certain number x

52. Nine more than one half a certain number x

53. The number of cents in x dimes and twice as many nickels

54. The amount of money invested at 4 percent if x dollars is invested at 5 percent and a total of \$5000 is invested

55. The distance apart two cars are at the end of 5 hours if they travel in opposite directions and if the rate of one is 3 times the rate x of the other

56. The width of a rectangle whose perimeter is 48 and whose length is x

57. The weight of a mixture when x pounds of one kind of rice is mixed with 40 pounds of another type

58. A overtakes B at the end of x hours. If A started 30 minutes after B, how many hours did B travel?

59. The total income from x dollars invested at 5 percent and \$2000 invested at $6\frac{1}{2}$ percent

60. The sum of three consecutive integers if the smallest is x

Solve and check Exercises 61–70.

61. Six more than the double of a number is equal to 5 times the difference obtained when 3 is subtracted from the number. Find the number.

62. A bus and a train leave the same station at the same time, and both travel north. The rate of the train is 15 mph less than twice the rate of the bus. At the end of 3 hours, the train is 90 miles farther north than the bus. Find the rate of each.

63. How many ounces of copper must be added to 18 ounces of an alloy composed of 40 percent copper to make an alloy composed of 50 percent copper?

64. How much paint is required to paint 3 walls of a room if each wall is a rectangle with a length 2 feet less than twice the height? The perimeter of each wall is 50 feet, and 1 gallon of paint covers about 400 square feet.

65. A housewife paid $2.21 for some fruit consisting of peaches costing 29 cents a pound, plums at 25 cents a pound, and bananas at 10 cents a pound. If she bought twice as many pounds of plums as peaches and 1 pound less of bananas than plums, how many pounds of each kind of fruit did she buy?

66. A corner lot has the shape of a right triangle (one angle is 90 degrees). The sides forming the 90-degree angle are along the streets, and one of these sides is 11 feet longer than the other side. To widen the street along the longer side, a strip 10 feet wide is removed from the lot, thus reducing the length of this side by 11 feet. The area of the lot was decreased by 2310 square feet. Find the original length of the longer side along the street.

67. During an average life of 70 years, it is estimated that a person sleeps 3 years more than he works, that he spends one half the years he works on recreation and church functions, 7 years eating and drinking, 5 years traveling, 2 years dressing, and 3 years being ill. How many years does he work?

68. A person traveling on a plane is allowed 44 pounds of baggage free but must pay $3.20 per pound for all excess baggage. A wife had 5 pounds of baggage more than her husband had. Together they paid $67.20 for excess baggage. How much baggage did each have?

69. A rush order is received in Boomtown for some machine parts that must be delivered from Bolttown, 280 miles away. To speed the delivery, a station wagon traveling 65 mph leaves Boomtown at 11 a.m. to intercept the delivery truck traveling 50 mph. If the delivery truck left Bolttown at 10 a.m., at what clock time is the delivery truck intercepted?

70. A dillar, a dollar,
Say, algebra scholar,
Can you tell me my age?

Twice nine years ago,
I have eight years to go.
Use x, mathematical sage.

0.5 OPERATIONS ON POLYNOMIALS

DEFINITIONS

The **nth power** of x, x^n, where the real number x is called the **base** and the natural number n is called the **exponent**, is defined as follows:

$$x^1 = x \quad \text{and} \quad x^n = \underbrace{x \cdot x \cdot \ldots \cdot x}_{(n \text{ factors})}$$

For example, $x^3 = xxx$ and $2^5 = (2)(2)(2)(2)(2) = 32$.

A **monomial** is a constant c; a term of the form cx^n, where x is a variable and n is a natural number; or a product of terms of the form cx^n.

For example, 3, x^2, $5xy$, and $\frac{1}{2}x^3z^2$ are monomials.

A **polynomial** is a monomial or a sum of monomials.

For example, $x^3 - 5x^2 + 6x - 7$ is a polynomial, $2x^5$ is a polynomial, and $x^2 - 5xy + 4y^2$ is a polynomial.

A **binomial** is a polynomial of two terms.

For example, $x + 5$, $2x - 3y$, $x^2 - 9$, and $8x^3 + 27$ are binomials.

A **trinomial** is a polynomial of three terms.

For example, $x^2 - 6x + 5$, $x + y + 5$, and $x^4 - x^2y^2 - 2y^4$ are trinomials.

Any indicated factor of a monomial is called a **coefficient** of the product of the other factors.

For example, in $5xy$, 5 is the coefficient of xy, $5x$ is the coefficient of y, and x is the coefficient of $5y$.

A **numerical coefficient** of a monomial is its constant factor.

For example, 2 is the numerical coefficient of $2x^3$, and -5 is the numerical coefficient of $-5xy$.

The **degree of a monomial** is the number of times a variable occurs as a factor.

For example, the degree of $7x^2$ is 2, the degree of $-4y^3$ is 3, the degree of $6x^2y^3$ is 5, and the degree of $3xy$ is 2.

The **degree of a polynomial** is the greatest degree of any of its terms. For example, the degree of $4x^3 - 5x^2 - 6x + 7$ is 3.

A polynomial is said to be **arranged in descending powers of a variable** when the term of greatest degree in this variable is written first, at the left, the term of next greatest degree is written second, and so on, with the term of least degree written last, at the right.

If the preceding order is reversed, the polynomial is said to be **arranged in ascending powers of a variable**.

For example, $x^3 + 5x^2 - 7x + 4$ is arranged in descending powers of x, and $7 + 3y - y^2$ is arranged in ascending powers of y.

Similar terms or **like terms** are terms whose literal factors are the same.

For example, $3x$ and $5x$ are like terms, $7x^2$ and $-4x^2$ are like terms, and $\frac{1}{2}xy$ and $-6xy$ are like terms. On the other hand, $3x$ and $3x^2$ are not like terms. Also, $5x$ and $5y$ are not like terms.

THE FIRST THEOREM OF EXPONENTS

If x is a real number and m and n are natural numbers, then

$$x^m x^n = x^{m+n}$$

THE SECOND THEOREM OF EXPONENTS

If x is a real number and m and n are natural numbers, then

$$(x^m)^n = x^{mn}$$

SPECIAL PRODUCTS AND FACTORS

Distributive axiom: $AB + AC = A(B + C)$

Simple trinomial: $X^2 + (A + B)X + AB = (X + A)(X + B)$

General trinomial: $ACX^2 + (AD + BC)X + BD = (AX + B)(CX + D)$

Square of binomial: $A^2 + 2AB + B^2 = (A + B)^2$

Cube of binomial: $A^3 + 3A^2B + 3AB^2 + B^3 = (A + B)^3$

Difference of squares: $A^2 - B^2 = (A - B)(A + B)$

Difference of cubes: $A^3 - B^3 = (A - B)(A^2 + AB + B^2)$

Sum of cubes: $A^3 + B^3 = (A + B)(A^2 - AB + B^2)$

DEFINITION

To **completely factor a polynomial over the integers** means to express the polynomial as a product of polynomial factors of lowest degree with integral coefficients.

Polynomials are added or subtracted by using the commutative and associative axioms to rearrange the terms and by using the distributive axiom to combine similar terms.

EXAMPLE 0.5.1   Add $8x^2 - 5x + 7$ to $x^2 + 3x - 4$.

Solution

$$
\begin{aligned}
(8x^2 - 5x + 7) + (x^2 + 3x - 4) &= (8x^2 + x^2) + (-5x + 3x) + (7 - 4) \\
&= (8 + 1)x^2 + (-5 + 3)x + (7 - 4) \\
&= 9x^2 - 2x + 3
\end{aligned}
$$

EXAMPLE 0.5.2   Subtract $4x^2 + xy - 6y^2$ from $2x^2 + 3xy - 9y^2$.

Solution

$(2x^2 + 3xy - 9y^2) - (4x^2 + xy - 6y^2)$ (Recalling that "subtract a from b" is translated as "$b - a$")

$= (2x^2 + 3xy - 9y^2) + (-4x^2 - xy + 6y^2)$ (Using the definition of subtraction)

$= (2x^2 - 4x^2) + (3xy - xy) + (-9y^2 + 6y^2)$

$= (2 - 4)x^2 + (3 - 1)xy + (-9 + 6)y^2$

$= -2x^2 + 2xy - 3y^2$

EXAMPLE 0.5.3   Multiply, writing the result as a single polynomial: $(2x - 5)(3x^2 + x - 4)$.

Solution   (horizontal method)

$$
\begin{aligned}
(2x - 5)(3x^2 + x - 4) &= 2x(3x^2 + x - 4) - 5(3x^2 + x - 4) \\
&= (6x^3 + 2x^2 - 8x) + (-15x^2 - 5x + 20) \\
&= 6x^3 + (2x^2 - 15x^2) + (-8x - 5x) + 20 \\
&= 6x^3 - 13x^2 - 13x + 20
\end{aligned}
$$

EXAMPLE 0.5.4 Multiply $(4x + 3y - 1)(2x - y + 2)$.

Solution (vertical method)

$$
\begin{array}{l}
4x + 3y - 1 \\
2x - y + 2 \\
\hline
8x^2 + 6xy \quad\quad\quad - 2x \\
\quad\quad - 4xy - 3y^2 \quad\quad + y \\
\quad\quad\quad\quad\quad + 8x + 6y - 2 \\
\hline
8x^2 + 2xy - 3y^2 + 6x + 7y - 2
\end{array}
$$

EXAMPLE 0.5.5 Factor completely over the integers

$$6bx^3 + 11bx^2 - 2bx$$

Solution First find the largest common monomial factor, bx.

$$bx(6x^2 + 11x - 2) \quad\quad\quad \text{(By the distributive axiom)}$$

Now use the general trinomial pattern to find the numbers A, B, C, and D so that

$$
\begin{array}{lll}
AC = 6 & BD = -2 & \text{and} \quad AD + BC = 11 \\
6 \cdot 1 = 6 & (-1)2 = -2 & \quad 6 \cdot 2 + (-1)1 = 11
\end{array}
$$

$$bx(6x - 1)(x + 2)$$

$$
\begin{array}{c}
-x \\
12x \\
\hline
11x
\end{array}
$$

Thus $6bx^3 + 11bx^2 - 2bx = bx(6x - 1)(x + 2)$.

EXAMPLE 0.5.6 Factor completely over the integers $3x^5 - 48x$.

Solution

$$
\begin{array}{ll}
3x^5 - 48x = 3x(x^4 - 16) & \text{(Distributive axiom)} \\
\quad\quad\quad = 3x(x^2 - 4)(x^2 + 4) & \text{(Difference of squares)} \\
\quad\quad\quad = 3x(x - 2)(x + 2)(x^2 + 4) & \text{(Difference of squares)}
\end{array}
$$

EXAMPLE 0.5.7 Factor completely over the integers $8x^3 - 125y^3$.

Solution Using

$$A^3 - B^3 = (A - B)(A^2 + AB + B^2)$$

$$8x^3 - 125y^3 = (2x)^3 - (5y)^3 = (2x - 5y)(4x^2 + 10xy + 25y^2)$$

EXAMPLE 0.5.8 Factor completely over the integers

$$5x^6 + 15x^4 - 20x^2$$

Solution

$$5x^6 + 15x^4 - 20x^2 = 5x^2(x^4 + 3x^2 - 4)$$
$$= 5x^2(x^2 - 1)(x^2 + 4)$$
$$= 5x^2(x - 1)(x + 1)(x^2 + 4)$$

EXERCISES 0.5

In Exercises 1–40, simplify, expressing each result as a single polynomial.

1. $(7x^2 - 5x + 2) + (x^2 + x - 6)$ **2.** $(3a + 5b - 7) + (2a - 5b + 7)$

3. $(-2x - 3y - 4) + (2x - 3y - 4)$

4. $3(2a - 5) + 4(7 - 3a)$

5. $(5y^2 - 6y + 9) + (5y - 8 - 4y^2)$

6. $(2a - b - c) - (a + b + c)$ **7.** $(-r - s + t) - (r + s - t)$

8. $(8 - 2x + y) - (9 - 2x + y)$ **9.** $(1 + x - x^2) - (2 - x^2 + x^4)$

10. $3(2a - b) - 4(b - 2a)$

11. $(2x - 3y + 4z) + (5x + 7y - 9z)$

12. $(x^2 - xy + y^2) + (x^2 + xy - y^2)$

13. $7(x - 2) - 3(2 - x)$ **14.** $-5(2y + 1) + 6(2y - 3)$

15. $(a^2 - a^2b^2 + 2b^2) - (-a^2 + a^2b^2 - 2b^2)$

16. $(3t^2 + 4t - 7) + (-5t^2 + t - 1)$

17. $(x + y - 1) - (x - y + 1)$ **18.** $(3x - 1) - (1 - 3x)$

19. $(x^2 - x + 1) - (x^2 - x + 1)$ **20.** $-5(t + 5) + 2(5t + 1)$

21. $x^3(2x^2 + x - 2)$ **22.** $-y^4(3y^3 - 2y - 4)$

23. $2xy^2(x^2 - 3xy + y^2)$ **24.** $x^n(x^2 + x + 1)$

25. $(x + 4)(x - 7)$ **26.** $(2x + y)(3x - 5y)$

27. $x(x + 6)(2x - 3)$ **28.** $(x - 4)(x^2 + 3x - 2)$

29. $2y(3y + 4)^2$ **30.** $(x + 6y)(x^2 - 6xy + 36y^2)$

31. $(7x + 1)(7x - 1)$ **32.** $(t - 5)(2t - 9)(2t + 9)$

33. $(n - 5)(n^2 + 5n + 25)$ **34.** $(3r + 2s)^3$

35. $(2x^2 - x + 3)(3x + 1)$ **36.** $(x^2 + 5x + 6)(x^2 - 3x + 1)$

37. $(x + 2)(x - 2)(x^2 + 4)$ **38.** $(x^2 + xy - y^2)(x^2 - xy + y^2)$

39. $(x - 1)(x + 1)(x^2 - x + 1)(x^2 + x + 1)$

40. $(5x^2 - 6)^2$

In Exercises 41–60, factor completely over the integers.

41. $3x^2 - 6x + 24xy$

42. $t^2 + 11t + 24$

43. $x^2 + 10x + 25$

44. $49y^4 - 1$

45. $x^3 - 1$

46. $a^3 + 1$

47. $-3x^3 + 27x$

48. $6a^2 - 7a - 3$

49. $36x + 3x^3 - 3x^5$

50. $100p^2 - 25q^2$

51. $64y^3 - 36y$

52. $x^6 - y^6$

53. $12x^2y - 22xy - 20y$

54. $8a^2 + 5ab - 3b^2$

55. $2y^5 - 162y$

56. $15ax^6 + 42ax^5 - 9ax^4$

57. $25r^2 + 10r + 1$

58. $36x^2 - 12x + 1$

59. $4p^2 - 20pq + 25q^2$

60. $x^4 - 21x^2 - 100$

0.6   FRACTIONS

DEFINITION OF EQUAL QUOTIENTS

If $bd \neq 0$, then $\dfrac{a}{b} = \dfrac{c}{d}$ if and only if $ad = bc$.

THEOREMS

The Fundamental Theorem of Fractions

If n, d, and k are real numbers and $d \neq 0$ and $k \neq 0$, then

$$\frac{n}{d} = \frac{nk}{dk}$$

The Product of Reciprocals Theorem

If d and k are nonzero real numbers, then

$$\frac{1}{d} \cdot \frac{1}{k} = \frac{1}{dk}$$

The Addition of Quotients Theorem

If n, m, and d are any real numbers and $d \neq 0$, then

$$\frac{n}{d} + \frac{m}{d} = \frac{n + m}{d}$$

The Subtraction of Quotients Theorem

If n, m, and d are any real numbers and $d \neq 0$, then

$$\frac{n}{d} - \frac{m}{d} = \frac{n - m}{d}$$

The Product of Quotients Theorem

If n, d, r, and s are any real numbers and $d \neq 0$ and $s \neq 0$, then

$$\frac{n}{d} \cdot \frac{r}{s} = \frac{nr}{ds}$$

The Reciprocal of a Quotient Theorem

If n and d are any real numbers and $n \neq 0$ and $d \neq 0$, then

$$\frac{1}{\dfrac{n}{d}} = \frac{d}{n}$$

The Quotient of Quotients Theorem

If n, d, r, and s are any real numbers and $n \neq 0$, $d \neq 0$, and $s \neq 0$, then

$$\frac{r}{s} \div \frac{n}{d} = \frac{rd}{sn}$$

EXAMPLE 0.6.1   Express as a single fraction

$$\frac{5}{6x^2} + \frac{1}{9x}$$

Solution

1. Find the L.C.D. (least common denominator) by factoring $6x^2$ and $9x$:

$$6x^2 = 2 \cdot 3 \cdot x \cdot x \quad \text{and} \quad 9x = 3 \cdot 3 \cdot x$$

Thus the L.C.D. is $2 \cdot 3 \cdot 3 \cdot x \cdot x = 18x^2$.

2. Rename the fractions by using the fundamental theorem of fractions:

$$\frac{5}{6x^2} = \frac{5 \cdot 3}{6x^2 \cdot 3} = \frac{15}{18x^2} \quad \text{and} \quad \frac{1}{9x} = \frac{1 \cdot 2x}{9x \cdot 2x} = \frac{2x}{18x^2}$$

3. Apply the theorem for the addition of quotients:

$$\frac{5}{6x^2} + \frac{1}{9x} = \frac{15}{18x^2} + \frac{2x}{18x^2} = \frac{15 + 2x}{18x^2}$$

EXAMPLE 0.6.2 Simplify

$$\frac{5}{x^2 - 6x + 9} - \frac{4}{x^2 - 9}$$

Solution

1. $x^2 - 6x + 9 = (x - 3)(x - 3)$
   $x^2 - 9 = (x - 3)(x + 3)$

   Thus the L.C.D. $= (x - 3)^2(x + 3)$

2. $\dfrac{5}{(x - 3)^2} = \dfrac{5(x + 3)}{(x - 3)^2(x + 3)} = \dfrac{5x + 15}{(x - 3)^2(x + 3)}$

   $\dfrac{4}{x^2 - 9} = \dfrac{4(x - 3)}{(x - 3)(x + 3)(x - 3)} = \dfrac{4x - 12}{(x - 3)^2(x + 3)}$

3. $\dfrac{5}{(x - 3)^2} - \dfrac{4}{x^2 - 9} = \dfrac{(5x + 15)}{(x - 3)^2(x + 3)} - \dfrac{(4x - 12)}{(x - 3)^2(x + 3)}$

   $= \dfrac{(5x + 15) - (4x - 12)}{(x - 3)^2(x + 3)}$

   $= \dfrac{x + 27}{(x - 3)^2(x + 3)}$

EXAMPLE 0.6.3 Simplify

$$\frac{3x^2 - 75y^2}{3x^2 - 21xy + 30y^2}$$

Solution $= \dfrac{3(x^2 - 25y^2)}{3(x^2 - 7xy + 10y^2)}$

$= \dfrac{3(x - 5y)(x + 5y)}{3(x - 5y)(x - 2y)}$

$= \dfrac{x + 5y}{x - 2y}$

EXAMPLE 0.6.4 Simplify

$$\frac{5x^2 - 9x - 2}{30x^3 + 6x^2} \div \frac{x^4 - 3x^2 - 4}{2x^8 + 6x^7 + 4x^6}$$

Solution

$$\frac{5x^2 - 9x - 2}{30x^3 + 6x^2} \cdot \frac{2x^8 + 6x^7 + 4x^6}{x^4 - 3x^2 - 4}$$

$$= \frac{(x - 2)(5x + 1)}{6x^2(5x + 1)} \cdot \frac{2x^6(x^2 + 3x + 2)}{(x^2 - 4)(x^2 + 1)}$$

$$= \frac{(x - 2)(5x + 1)(2x^6)(x + 2)(x + 1)}{6x^2(5x + 1)(x - 2)(x + 2)(x^2 + 1)}$$

$$= \frac{2x^6(x - 2)(x + 2)(5x + 1)(x + 1)}{6x^2(x - 2)(x + 2)(5x + 1)(x^2 + 1)}$$

$$= \frac{x^4(x + 1)}{3(x^2 + 1)}$$

EXAMPLE 0.6.5 Divide $3x^4 - 2x^3 + 5x - 6$ by $x - 3$ by using the long-division algorithm.

Solution

$$
\begin{array}{r}
3x^3 + 7x^2 + 21x + 68 \\
x - 3)\overline{3x^4 - 2x^3 \qquad\quad + 5x - 6} \\
3x^4 - 9x^3 \\
\hline
+ 7x^3 \qquad\quad + 5x - 6 \\
7x^3 - 21x^2 \\
\hline
21x^2 + 5x - 6 \\
21x^2 - 63x \\
\hline
68x - 6 \\
68x - 204 \\
\hline
198
\end{array}
$$

Thus $\dfrac{3x^4 - 2x^3 + 5x - 6}{x - 3} = 3x^3 + 7x^2 + 21x + 68 + \dfrac{198}{x - 3}.$

EXAMPLE 0.6.6   Simplify

$$\frac{\dfrac{1}{x}+\dfrac{1}{2}}{\dfrac{1}{4}}$$

Solution

$$\frac{\dfrac{1}{x}+\dfrac{1}{2}}{\dfrac{1}{4}}=\frac{4x\left(\dfrac{1}{x}+\dfrac{1}{2}\right)}{4x\left(\dfrac{1}{4}\right)}=\frac{4+2x}{x}=\frac{2x+4}{x}$$

Alternate Solution

$$\frac{\dfrac{1}{x}+\dfrac{1}{2}}{\dfrac{1}{4}}=\left(\dfrac{1}{x}+\dfrac{1}{2}\right)\div\dfrac{1}{4}$$

$$=\frac{x+2}{2x}\cdot\frac{4}{1}$$

$$=\frac{2(x+2)}{x}=\frac{2x+4}{x}$$

EXAMPLE 0.6.7   State the restricted values of y and solve

$$\frac{2}{y-5}+\frac{1}{y+5}=\frac{11}{y^2-25}$$

Solution $y\neq5$ and $y\neq-5$ since $y-5=0$ for $y=5$ and $y+5=0$ for $y=-5$ and division by zero is undefined.

Multiplying each side by the least common denominator, $(y-5)(y+5)$,

$$2(y+5)+(y-5)=11$$
$$3y+5=11$$
$$3y=6$$
$$y=2$$

Check: For $y=2$,

$$\frac{2}{y-5}+\frac{1}{y+5}=\frac{2}{2-5}+\frac{1}{2+5}=-\frac{2}{3}+\frac{1}{7}=-\frac{14}{21}+\frac{3}{21}=-\frac{11}{21}$$

$$\frac{11}{y^2-25}=\frac{11}{4-25}=\frac{11}{-21}=-\frac{11}{21}$$

EXAMPLE 0.6.8 State the restricted values of the variable and solve

$$\frac{x+1}{x+2} - \frac{x+1}{x-3} = \frac{5}{x^2 - x - 6}$$

Solution   Restricted values: $x \neq -2$ and $x \neq 3$.

$$\left(\frac{x+1}{x+2}\right)(x+2)(x-3) - \left(\frac{x+1}{x-3}\right)(x+2)(x-3)$$

$$= \frac{5}{x^2 - x - 6}(x+2)(x-3)$$

$$(x+1)(x-3) - (x+1)(x+2) = 5$$
$$(x^2 - 2x - 3) - (x^2 + 3x + 2) = 5$$
$$-5x - 5 = 5$$
$$-5x = 10$$
$$x = -2$$

Thus $x \neq -2$ and $x \neq 3$ and $x = -2$. Since there is no common solution, the solution set is empty.

EXAMPLE 0.6.9 State the restricted values of the variable and solve

$$\frac{x}{(x-4)^2} + \frac{2}{x-4} = \frac{3x-8}{(x-4)^2}$$

Solution   Restricted value: $x \neq 4$.

$$(x-4)^2 \frac{x}{(x-4)^2} + (x-4)^2 \frac{2}{x-4} = \frac{3x-8}{(x-4)^2}(x-4)^2$$

$$x + 2(x-4) = 3x - 8$$
$$3x - 8 = 3x - 8$$
$$3x = 3x$$
$$x = x$$

Since $x = x$ is true for all real numbers, the solution set is the set of all real numbers except 4.

EXAMPLE 0.6.10 Work problem. A man working alone can complete a certain job in 12 hours. His helper, working alone, requires 20 hours to do the job. How long would it take them to complete the job if they worked together?

Solution Let x = time to complete job, working together.

Formulas: Rate of A + rate of B = rate together; $r_A + r_B = r_T$
Rate $\times$ time = fraction of work done; $rt = w$
If the job is completed, $w = 1$ and $r = 1/t$.

	Rate	Time	Work Done
A (man)	$\dfrac{1}{12}$	12	1
B (helper)	$\dfrac{1}{20}$	20	1
Together	$\dfrac{1}{x}$	x	1

Equation: $r_A + r_B$ = rate together

$$\frac{1}{12} + \frac{1}{20} = \frac{1}{x}$$

$$60x\left(\frac{1}{12} + \frac{1}{20}\right) = 60x\left(\frac{1}{x}\right)$$

$$5x + 3x = 60$$

$$8x = 60$$

$$x = \tfrac{15}{2} = 7\tfrac{1}{2} \text{ hours or 7 hours 30 minutes}$$

EXERCISES 0.6

Simplify Exercises 1–10.

1. $\dfrac{156}{390}$

2. $\dfrac{15x^2y^2}{27xy^3z}$

3. $\dfrac{x - 7}{7 - x}$

4. $\dfrac{9x^2 - 36y^2}{x^2 - 4xy + 4y^2}$

5. $\dfrac{2x^2 + 5x - 150}{2x^2 + 17x - 30}$

6. $\dfrac{x^2 - 2x + 1}{15x^2} \cdot \dfrac{20x^5}{1 - x^2}$

7. $\dfrac{x^3 + 5x^2 + 6x}{x^2 - x} \cdot \dfrac{x^2 - 3x + 2}{(x^2 - 4)(x + 3)}$

8. $\dfrac{5xy + 15y}{15y} - \dfrac{x^2 + 4x + 3}{x^2 - 1}$

9. $\dfrac{\dfrac{1}{x} - \dfrac{1}{3}}{x - \dfrac{9}{x}}$

10. $\left(x + \dfrac{3}{y}\right) - \left(x - \dfrac{3}{y}\right)$

Write each of Exercises 11–20 as a single fraction and reduce to lowest terms.

11. $\dfrac{5}{x + 3} - \dfrac{x}{x + 3}$

12. $\dfrac{x}{x + 3} + \dfrac{5x^2}{x^2 - 9}$

13. $\dfrac{x}{x^2 - 6x + 5} + \dfrac{3}{x - 1}$

14. $\dfrac{2}{x - 1} - \dfrac{4}{x} + \dfrac{2}{x + 1}$

15. $\dfrac{1}{x} + \dfrac{1}{2} + \dfrac{5}{x + 2}$

16. $\dfrac{3}{x^2 + 5x + 4} - \dfrac{2}{x^2 + 4x + 3}$

17. $\dfrac{1}{(x + 2)^2} - \dfrac{2}{x^2 - 4} + \dfrac{1}{(x - 2)^2}$

18. $\left(\dfrac{3x - 1}{x} - \dfrac{3x}{3x + 1}\right) \div \left(\dfrac{5x - 2}{5x} - \dfrac{5x}{5x + 2}\right)$

19. $3x - \dfrac{x}{4x - \dfrac{5x}{2}}$

20. $4x + \dfrac{6}{2x - \dfrac{6x}{3}}$

In Exercises 21–25, divide by using the long-division algorithm. Check by multiplication.

21. $\dfrac{x^3 - 4x^2 + 3x - 1}{x - 3}$

22. $\dfrac{x^4 + x^2 - 2}{x^2 + 3}$

23. $\dfrac{39 + 20x^2 + 15x^3}{5x + 10}$

24. $\dfrac{a^3 - 1}{a - 1}$

25. $\dfrac{7x^3 - 3x + x^2 + 3x^3 - 5x - 10}{1 + 2x}$

26. Given the algebraic expression

$$\dfrac{x - 2}{x^2 + 1} \div \dfrac{k - 3}{x + 3}$$

a. For what value(s) of x will the expression be undefined?

b. For what value(s) of k will the expression be undefined?

c. For what value(s) of x will the expression equal 0? Why?

Solve and check Exercises 27–35. State the restricted values of the variable.

27. $\dfrac{5}{x-6} + \dfrac{3}{x+6} = \dfrac{8}{x^2-36}$

28. $\dfrac{x+3}{x+2} = \dfrac{3}{2}$

29. $\dfrac{2}{x+3} - \dfrac{2}{x-3} = \dfrac{1}{3-x}$

30. $\dfrac{2x+3}{x-1} = \dfrac{2x-5}{x+3}$

31. $\dfrac{5}{3x+1} - \dfrac{x+2}{4x} + \dfrac{1}{4} = 0$

32. $\dfrac{x}{x+4} - \dfrac{4}{x-4} = \dfrac{x^2+16}{x^2-16}$

33. $\dfrac{1}{1-x} + \dfrac{x}{x-1} = 1$

34. $\dfrac{x-a}{x+b} = \dfrac{x+b}{x-a};$ solve for x

35. $\dfrac{p-x}{x} = c + \dfrac{1}{x};$ solve for x

36. A mechanic can do a certain repair job in $3\frac{1}{2}$ hours. He and his assistant, working together, can complete the job in 2 hours. How long would it take the assistant working alone?

37. A plane travels 1140 miles with a wind of 35 mph in the same time it travels 860 miles against the wind. Find the speed of the plane in still air.

38. One machine requires 80 minutes to process a certain amount of data, while another machine requires 120 minutes to process the same amount of data. How long would it take them working together to process this data?

39. The numerator of a fraction is 3 less than the denominator. If 1 is added to the denominator and 2 is subtracted from the numerator, the value of the new fraction formed is $\frac{1}{4}$. Find the original fraction.

40. It takes a boat 2 hours longer to travel 120 miles upstream than it takes to travel the same distance downstream. If the rate of the boat in still water is 32 mph, find the rate of the current.

41. A swimming pool can be filled by a pipe in 10 hours and by a hose in 15 hours. How long does it take to fill the pool using both pipe and hose, assuming there is no loss in water pressure?

42. It takes one train 5 hours longer to travel 500 miles than another traveling at the same rate to go 400 miles. Find the time of each train.

43. For what value of x is $\dfrac{x+2}{x+16} = \dfrac{5}{12}$ true?

44. If a car uses 16 gallons of gasoline on a 720-mile journey, how much gas will be used on a 450-mile trip? (Assume that the mileage per gallon will be the same for both trips.)

0.7 GRAPHING AND LINEAR SYSTEMS

DEFINITIONS

An **ordered pair** is an expression of the form (a, b), where a is called the **first component** and b is called the **second component**.

A **solution of an open equation in two variables** x and y is an ordered pair (a, b) such that the equation becomes true when x is replaced by a and y is replaced by b.

Two perpendicular number lines in a plane intersecting at their origins form a **rectangular** (or **Cartesian**) **coordinate system**.

Every point in the plane is named by an ordered pair, called the **coordinates** of the point.

If a point P has coordinates (x, y), then x, the first coordinate, is called the **abscissa** of P, and y, the second coordinate, is called the **ordinate** of P.

The number lines are called the **vertical axis** and the **horizontal axis**, and the intersection of these axes is called the **origin**.

An equation of the form $Ax + By + C = 0$, where A and B are not both zero, is called a **linear equation in two variables**.

The **graph of a linear equation in two variables is a straight line**, the graph of its solution set. Every point on the line is a solution of the equation, and conversely.

The **solution set of a system of equations in two variables** is the set of ordered pairs that are common solutions to all the equations in the system.

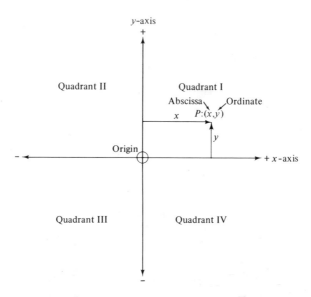

FIGURE 0.7.1 Rectangular coordinate system.

37

The coordinates (a, b) of the point of intersection of two intersecting lines is the solution of the system of two linear equations in two variables having these lines for their graphs.

A system of two linear equations in two variables may be solved algebraically by the **addition method** or the **substitution method**.

EXAMPLE 0.7.1   Graph $2x + 3y = 6$.

Solution

1. Make a table of values:

x	y	
0	2	$2(0) + 3y = 6$, $3y = 6, y = 2$
3	0	$2x + 3(0) = 6$, $2x = 6, x = 3$
-3	4	$2(-3) + 3y = 6$, $3y = 12, y = 4$

2. Plot the ordered pairs and join with a straight line.

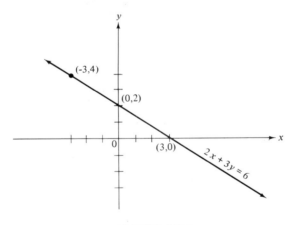

FIGURE 0.7.2

EXAMPLE 0.7.2   Solve the following system of equations:

a. By the graphical method
b. By the addition method
c. By the substitution method

$$3x + 2y = 7$$
$$2x + y = 6$$

Solution

a. Graphical:

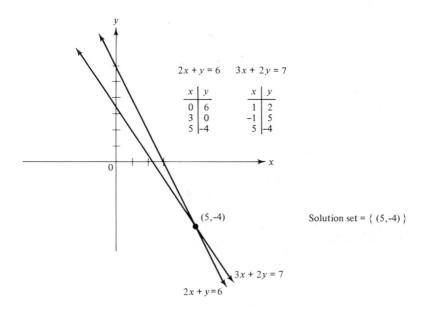

FIGURE 0.7.3

b. Addition:

$$3x + 2y = 7$$
$$2x + y \ = 6$$

Multiply each side of the second equation by -2:

$$3x + 2y = 7$$
$$-4x - 2y = -12$$

Adding the left-hand members and the right-hand members of the equations yields the equivalent equation

$$-x = -5$$

and

$$x = 5$$

If $x = 5$, then $2(5) + y = 6$, and $y = -4$. Therefore, the solution set is $\{(5, -4)\}$.

c. Substitution:

$$(1) \quad 3x + 2y = 7$$
$$(2) \quad 2x + y = 6$$

Solve equation (2) for y:

$$y = 6 - 2x$$

Substitute $6 - 2x$ for y in equation (1)

$$3x + 2(6 - 2x) = 7$$

Solve the equation for x:

$$3x + 12 - 4x = 7$$
$$-x = -5$$
$$x = 5$$

If $x = 5$, $y = -4$; therefore, the solution set is $\{(5, -4)\}$.

EXERCISES 0.7

In Exercises 1–10, solve by the addition method and check by the graphical method.

1. $x + y = 3$
   $x - y = 7$
2. $2x - y = 3$
   $3x + 2y = 8$
3. $x - 2y = 11$
   $x - 3y = 18$
4. $3a + 8b = 14$
   $a + 7b = 22$
5. $4c - 3d = 11$
   $6c - 3d = 12$
6. $4x - 5y = 48$
   $5x - 4y = 51$
7. $2p + 3q = 4$
   $5p + 6q = 7$
8. $6r - 2t = 9$
   $9r - 8t = 1$
9. $7x - 5y = 36$
   $12x = 15y + 108$
10. $9a = 5b + 30$
    $3b = 7a + 10$

In Exercises 11–20, solve by the substitution method and check by the graphical method.

11. $2x + 3y = 16$
    $y = 2x$
12. $y = 3x - 4$
    $5x + 2y = 25$
13. $4a - 7b = 9$
    $b = 12 + 5a$
14. $2a - 3b = 1$
    $4a + b = 23$
15. $3x - 5y = 49$
    $4x - y = 3$
16. $5x - 4y = 22$
    $x - 2y = 5$

17. $2r - 3s = 17$
$\quad r + 4s = 3$

18. $3p + 4q = 10$
$\quad p + 6q = 1$

19. $3x - 3y = 1$
$\quad\quad y = x + 5$

20. $3x - 9y = 3$
$\quad\quad 3y = x - 1$

REVIEW EXERCISES

Each of Exercises 1–25 contains a common error. Find the error and replace the right-hand side of the equation by a simplified algebraic expression so that the resulting statement is correct.

1. $5\left(\dfrac{a}{b}\right) = \dfrac{5a}{5b}$

2. $(x - 2)(x + 5) = x^2 - 10$

3. $(x + 3)^2 = x^2 + 9$

4. $1 - (x - y) = 1 - x - y$

5. $\dfrac{-4}{x - 4} = \dfrac{4}{x + 4}$

6. $x^2 - 3x - 28 = (x - 4)(x + 7)$

7. If $\dfrac{x}{4} - \dfrac{x}{5} = 3$, then $5x - 4x = 3$

8. $(3x - 2)^2 = 9x^2 - 6x + 4$

9. $\dfrac{(x + 1)(x - 2)}{(x - 2)(x + 1)} = 0$

10. $x^2 + 7xy + 12y^2 = (x + 3)(x + 4)$

11. If $\dfrac{n}{5} - \dfrac{n - 1}{2} = \dfrac{3}{10}$, then $2n - 5n - 5 = 3$

12. $\dfrac{x + 3}{x + 2} = \dfrac{3}{2}$

13. If $x = 2$ and $y = 5$, then $xy^2 = (2 \cdot 5)^2 = 100$

14. $2y^2 - 50 = 2(y^2 - 25)$
$\quad\quad\quad\quad\quad = (y + 5)(y - 5)$

15. If $(x - 2)(x - 3) = 6$, then $x - 2 = 2$ and $x - 3 = 3$

16. $(-5x)^2 = -5x^2$

17. $\dfrac{(2x)(7x)}{5x} = \dfrac{14}{5}$

18. If $3x = 7$, then $x = 7 - 3 = 4$

19. $x - 3(x - 2) = x - 3 + x - 2$

20. $x - 3(x - 2) = x^2 - 5x + 6$

21. $\dfrac{x^2 - 3x}{x^2 - 9} = \dfrac{x}{3}$

22. If $\dfrac{x}{3} = \dfrac{5}{6}$, then $\dfrac{6x}{(6)(3)} = 5$

23. If $-3x = 5$, then $x = 5 + 3$

24. $x^2 + 25 = (x + 5)(x + 5)$

25. If $\dfrac{x}{x - 3} = \dfrac{3}{x - 3}$, then $x = 3$

Complete Exercises 26–30.

26. If $(0, k)$ is a solution of $2x - 3y = 12$, then $k = $ _____.

27. The point with coordinates $(0, 0)$ is called the _____.

28. The line $y = 4$ intersects the line $3x + 2y + 1 = 0$ at the point whose abscissa is _____.

29. The line $y = 3x$ intersects the line $3x + y = 6$ at the point whose coordinates are _____.

30. The line $5x + 4y = 8$ intersects the line $5x + 3y = 2$ at the point whose ordinate is _____.

Simplify Exercises 31–35, expressing the result as a single polynomial.

31. $(x^2 + 2 - [x - 5])(x^2 + 2 + [x - 5])$

32. $3(2x^2 - 4x - 5) - 2(3x^2 - 5x + 8)$

33. $5x^2(2x - 3)(2x + 3)$ **34.** $\dfrac{3x^3 + 9x^2 - 30x}{x^2 + 5x}$

35. $4(x + y) - [x + 2(x + 2y)]$

Write an algebraic expression for each of Exercises 36–45.

36. The larger of two numbers if their sum is 20 and the smaller is x

37. The sum of the reciprocals of two consecutive integers if the smaller is represented by x

38. The quotient obtained when 3 less than x is divided by 3 more than x

39. The time it takes to fly 900 miles against the wind with a speed in still air of 200 mph and a wind speed of x mph

40. The width of a rectangle whose perimeter is 30 feet and whose length is x feet

41. The rate at which A and B work together if A can do the job in x hours and B can do the job in $2\frac{1}{2}$ hours

42. The total value of x pounds of seed worth 80 cents per pound and y pounds of seed worth $1.20 per pound

43. The sum of the ages of A and B ten years ago if A is x years old now and B is seven years older than A

44. The cost of x pounds of beans if 3 pounds cost 35 cents

45. The volume V of the frustrum of a pyramid with square bases whose sides are a and b, respectively, is equal to one third the product of the height h and the sum of the three terms—the square of a, the product of a and b, the square of b.

MULTIPLE-CHOICE EXAMINATION

1. If $a = 3$ and $b = -2$, then $a^2 - 4ab =$

     a. 30   b. 33   c. -15 d. 105   e. None of these

2. $4(x - 5) - 3x =$

     a. $-12x - 20$ b. $-12x - 5$ c. $x - 20$ d. $x - 5$

     e. None of these

3. $12x - [10 - (3x - 4)] =$

     a. $9x - 6$ b. $9x - 14$ c. $15x - 6$ d. $15x - 14$

     e. None of these

4. $\dfrac{12x^8}{-2x^2} =$

     a. $6x^4$ b. $-6x^6$ c. $10x^4$ d. $-10x^6$ e. None of these

5. $(-2y^2)^3 =$

     a. $-2y^5$ b. $-2y^6$ c. $-6y^5$ d. $-8y^6$ e. None of these

6. From $2x - 3y - 8z$ subtract $5x - 6y + 2z$:

     a. $-3x + 3y - 10z$ b. $3x - 9y - 10z$ c. $-3x - 9y - 6z$

     d. $7x - 9y - 6z$ e. None of these

7. $(2x + 5)(x - 2) =$

     a. $3x + 3$ b. $2x^2 - 10$ c. $2x^2 + x - 10$ d. $2x^2 + 7x - 10$

     e. None of these

8. $(3 + y)(5 - y) =$

     a. 8   b. 15   c. $15 - y^2$ d. $3 + 5y - y^2$ e. $15 + 2y - y^2$

9. $\dfrac{-35n^3 + 14n^2}{-7n^2} =$

     a. $5n - 2$ b. $5n + 2$ c. $-5n - 2$ d. $-5n + 2$

     e. None of these

10. Which of the following statements is true?

     a. $x^2x^3 = x^6$ b. $(-y)^2 = -y^2$ c. $(-x)^3 = -x^3$

     d. $x^2 + y^2 = (x + y)^2$ e. None of these

11. The factors of $x^2 - 4x - 12$ are

     a. $x - 4$ and $x - 12$ b. $x - 2$ and $x + 6$ c. x and 4

     d. $x - 4$ and $x + 3$ e. None of these

12. If $3t - 1 = 2(t + 3)$, then $t =$

     a. 1   b. 7   c. -7 d. 4   e. None of these

13. If $5x - (7x + 2) = 8$, then $x =$

     a. -3 b. -5 c. 8   d. 12   e. None of these

14. If $3y - 5 = 8y + 10$, then $y =$

   a. 0   b. 2   c. 3   d. -2 e. -3

15. If the equation $9x - 4 = 7x + 6$ is solved and then checked by substitution, the number to which each side of the equation will reduce in the check will be

   a. 0   b. 5   c. -49 d. 41   e. None of these

16. After simplification $\dfrac{x^2 - 25}{x^2 - 4x - 5}$ will be

   a. $\dfrac{5}{-4x}$ b. $\dfrac{25}{4x + 5}$ c. $\dfrac{x + 5}{x + 1}$ d. $\dfrac{x - 5}{x - 1}$ e. None of these

17. The difference $\dfrac{a + b}{a} - \dfrac{b + 4}{b} =$

   a. $\dfrac{a + 4}{a - b}$ b. $\dfrac{a - 4}{ab}$ c. $\dfrac{b - 4}{a + b}$ d. $\dfrac{b^2 - 4a}{ab}$ e. None of these

18. The product of $\dfrac{t^2 - 36}{9}$ and $\dfrac{15}{5t - 30} =$

   a. $\dfrac{t + 6}{3}$ b. $\dfrac{t - 6}{3}$ c. $t + 6$ d. $\dfrac{-2t}{5}$ e. None of these

19. If $\dfrac{x}{4} - \dfrac{x}{6} = 4$, then $x =$

   a. 48   b. 2   c. 3   d. 24   e. None of these

20. The value of $\dfrac{x - y}{x + y}$ for $x = 3$ and $y = \frac{1}{2}$ is

   a. -1 b. $\frac{5}{7}$ c. $\frac{35}{4}$ d. $\frac{5}{28}$ e. None of these

21. One number is 3 times another. The sum of the two numbers is 52. If $n =$ the smaller number, then an equation that can be used to find these numbers is

   a. $n + 3 = 52$ b. $3n = 52$ c. $2n + 3 = 52$ d. $n + 3n = 52$
   e. None of these

22. A child's age is one third the age of her mother. The difference between their ages is 24 years. If $M =$ the age of the mother, then an equation that can be used to find M is

   a. $M - \dfrac{M}{3} = 24$ b. $3M = 24$ c. $\dfrac{M}{3} = 24$ d. $M - \frac{1}{3} = 24$

   e. None of these

23. Two scouts 60 miles apart hike toward each other until they meet. If one averages 3 mph and the other 4 mph, and if $t =$ the number of hours until they meet, then an equation that can be used to find t is

a. $4t - 3t = 60$ b. $4t + 3t = 60$ c. $\dfrac{t}{4} - \dfrac{t}{3} = 60$

d. $t = \frac{60}{3} + \frac{60}{4}$ e. None of these

24. A grocer has x pounds of peanuts that sell for 40 cents a pound and y pounds of walnuts that sell for 60 cents a pound. If a 100-pound mixture of these sells for 50 cents a pound, then

a. $40x + 60y = 50$ b. $x + y = 50$ c. $40x + 60y = 5000$

d. $40x + 60y = 100$ e. None of these

25. A man can do a job in 10 days, and his son can do the same job in 15 days. How long would it take them to do this job if they worked together?

a. 6 days   b. $7\frac{1}{2}$ days   c. 9 days   d. $12\frac{1}{2}$ days.   e. 25 days

26. Which of the following is the graph of $y = x + 3$? .

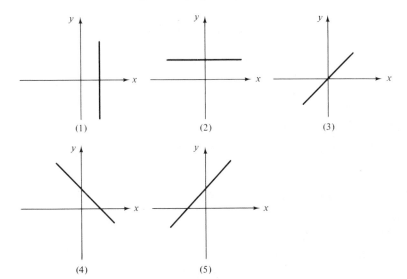

(1)   (2)   (3)

(4)   (5)

FIGURE 0.8.1

27. Which of the following is a solution of $2x - y = 5$?

a. $(0, 5)$ b. $(-5, 0)$ c. $(3, 1)$ d. $(1, 3)$ e. $(1, 4)$

28. The value of x at the point of intersection of the graphs of $3x + 2y = 11$ and $2x - y = 12$ is

a. 1   b. 2   c. 3   d. 4   e. 5

29. The coordinates of the point of intersection of the lines shown on the accompanying graph are

a. $(8, 2)$ b. $(-8, 6)$ c. $(2, 6)$ d. $(4, 3)$ e. None of these

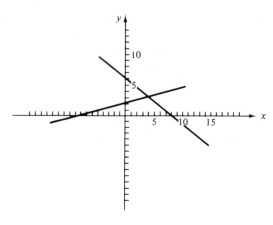

FIGURE 0.8.2

30. The symbol $>$ means "is greater than." A value of x that makes $2x + 5 > 14$ a true statement is

a. 1   b. 2   c. 3   d. 4   e. 5

CHAPTER **1**

EXPONENTS AND RADICALS

Many problems in mathematics, physics, engineering, chemistry, biology, economics, and other areas involve equations. Solving equations is a major concern of algebra.

A polynomial equation can readily be solved if the linear factors of the polynomial can be found easily. However, not all polynomials with integral coefficients can be factored over the set of integers; $x^2 - 2$ and $x^2 + x - 1$ are examples. Finding the factors in these cases is more difficult. It is desirable, then, to develop other techniques for solving equations.

Earlier it was noted that if $x^2 = 2$ and if x is positive, then $x = \sqrt{2}$. Similarly, if $x^3 = 2$ and x is a real number, then $x = \sqrt[3]{2}$. In general, if x and p are positive real numbers and if $x^n = p$, then $x = \sqrt[n]{p}$, where the natural number n used in x^n is called an exponent and where $\sqrt[n]{p}$ is called a radical.

Since a polynomial involves terms of the form cx^n, it is reasonable to conclude that the solutions of polynomial equations will involve exponents and radicals. This indeed is the case. Thus to develop methods for solving equations, it is first necessary to study exponents and radicals.

This chapter is concerned with exponents and radicals, their meaning, and how to use them in algebraic calculations.

I.I POSITIVE INTEGRAL EXPONENTS

The definition of x^n, introduced earlier, is restated below.

DEFINITION OF x^n

If x is a real number and if n is a natural number, then

$$x^n = \underbrace{x \cdot x \cdots x}_{n \text{ factors}} \quad \text{and} \quad x^1 = x$$

The natural number n is called the **exponent**, the real number x is called the **base**, and the number x^n is called the **nth power of x**.

How to simplify and calculate with expressions involving nth powers is explained by the exponent theorems. These five theorems are introduced, one at a time, and then summarized at the end of this section. In the discussion that follows, x and y are real numbers and m and n are positive integers (that is, natural numbers).

THEOREM I

$$x^m x^n = x^{m+n}$$

The product of two powers having the same base is a power having the same base and an exponent equal to the sum of the exponents.

$$x^m x^n = \underbrace{(x \cdot x \cdots x)}_{m \text{ factors}}\underbrace{(x \cdot x \cdots x)}_{n \text{ factors}} = \underbrace{(x \cdot x \cdots x)}_{m+n \text{ factors}} = x^{m+n}$$

EXAMPLE I.I.I

$$x^4 x^3 = (xxxx)(xxx) = x^7 = x^{4+3}$$

$$2^5 \cdot 2^2 = 2^{5+2} = 2^7 = 128$$

$$yy^7 = y^1 y^7 = y^{1+7} = y^8$$

THEOREM 2

$$(x^m)^n = x^{mn}$$

A power of a power is a power having the same base and an exponent equal to the product of the exponents.

$$(x^m)^n = \underbrace{x^m \cdot x^m \cdots x^m}_{n \text{ factors}}$$

$$= \underbrace{(x \cdot x \cdots x)}_{m \text{ factors}} \underbrace{(x \cdot x \cdots x)}_{m \text{ factors}} \cdots \underbrace{(x \cdot x \cdots x)}_{m \text{ factors}}$$

$$= \underbrace{x \cdot x \cdots x}_{mn \text{ factors}} = x^{mn}$$

EXAMPLE 1.1.2

$$(x^2)^3 = (x^2)(x^2)(x^2)$$
$$= (x \cdot x)(x \cdot x)(x \cdot x)$$
$$= x^6 = x^{2 \cdot 3}$$
$$(x^3)^4 = x^{3 \cdot 4} = x^{12}$$

$$(2^3)^2 = 2^{3 \cdot 2} = 2^6$$

THEOREM 3

1. $\dfrac{x^m}{x^n} = x^{m-n}$ if $m > n$ and $x \neq 0$

2. $\dfrac{x^m}{x^n} = \dfrac{1}{x^{n-m}}$ if $n > m$ and $x \neq 0$

3. $\dfrac{x^m}{x^n} = \dfrac{x^n}{x^n} = 1$ if $m = n$

The quotient of two powers having the same base is:

1. A power having the same base and an exponent equal to the difference of the exponents (larger minus smaller), or
2. A reciprocal of a power having the same base and an exponent equal to the difference of the exponents (larger minus smaller), or
3. The number 1 if the exponents are equal.

EXAMPLE 1.1.3

$$\frac{x^5}{x^2} = x^{5-2} = x^3$$

$$\left(\frac{xxxxx}{xx} = xxx = x^3\right)$$

$$\frac{2^7}{2^3} = 2^{7-3} = 2^4 = 16$$

$$\frac{2^3}{2^7} = \frac{1}{2^{7-3}} = \frac{1}{2^4} = \frac{1}{16}$$

$$\frac{x^2}{x^5} = \frac{1}{x^{5-2}} = \frac{1}{x^3}$$

$$\frac{x^5}{x^5} = 1$$

THEOREM 4

$$(xy)^n = x^n y^n$$

A power of a product is a product of two powers whose bases are factors of the product and each of whose exponents is the same as the exponent on the product.

EXAMPLE 1.1.4

$$(xy)^3 = x^3 y^3$$
$$(xy)^3 = (xy)(xy)(xy) = (xxx)(yyy) = x^3 y^3$$
$$(-2x)^5 = (-2)^5 x^5 = -32x^5$$
$$(2^6)(5^6) = (2 \cdot 5)^6 = 10^6 = 1,000,000$$

THEOREM 5

$$\left(\frac{x}{y}\right)^n = \frac{x^n}{y^n} \quad \text{if} \quad y \neq 0$$

A power of a quotient is a quotient of two powers whose bases are the numerator and denominator of the quotient and each of whose exponents is the same as the exponent on the quotient.

EXAMPLE 1.1.5

$$\left(\frac{x}{y}\right)^4 = \frac{x^4}{y^4}$$

$$\left(\frac{x}{y}\right)^4 = \left(\frac{x}{y}\right)\left(\frac{x}{y}\right)\left(\frac{x}{y}\right)\left(\frac{x}{y}\right) = \frac{xxxx}{yyyy} = \frac{x^4}{y^4}$$

$$\left(\frac{-2}{x}\right)^6 = \frac{(-2)^6}{x^6} = \frac{64}{x^6}$$

$$\frac{(98)^5}{(49)^5} = \left(\frac{98}{49}\right)^5 = 2^5 = 32$$

The exponent theorems are summarized below. The examples that follow show how these theorems are used to simplify certain algebraic expressions.

THE EXPONENT THEOREMS

Theorem 1 $x^n x^m = x^{n+m}$

Theorem 2 $(x^n)^m = x^{nm}$

Theorem 3 $\dfrac{x^n}{x^m} = x^{n-m}$ if $n > m$ and $x \neq 0$

$\dfrac{x^n}{x^m} = \dfrac{1}{x^{m-n}}$ if $m > n$ and $x \neq 0$

$\dfrac{x^n}{x^m} = 1$ if $m = n$ and $x \neq 0$

Theorem 4 $(xy)^n = x^n y^n$

Theorem 5 $\left(\dfrac{x}{y}\right)^n = \dfrac{x^n}{y^n}$ if $y \neq 0$

EXAMPLE 1.1.6   Simplify $(5x^2)^3$.

Solution $(5x^2)^3 = 5^3(x^2)^3$ (Theorem 4)
                                     $= 5^3 x^6$ (Theorem 2)
                                     $= 125x^6$ (Definition)

EXAMPLE 1.1.7 Simplify $\dfrac{a}{b^2}\left(\dfrac{a^2}{b}\right)^3$.

Solution

$$\frac{a}{b^2}\left(\frac{a^2}{b}\right)^3 = \frac{a}{b^2}\cdot\frac{(a^2)^3}{b^3} \qquad \text{(Theorem 5)}$$

$$= \frac{a(a^6)}{b^2(b^3)} \qquad \text{(Theorem 2)}$$

$$= \frac{a^7}{b^5} \qquad \text{(Theorem 1)}$$

EXAMPLE 1.1.8 Simplify $\dfrac{(abc)^5}{a^2b^5c^8}$.

Solution

$$\frac{(abc)^5}{a^2b^5c^8} = \frac{a^5b^5c^5}{a^2b^5c^8} \qquad \text{(Theorem 4)}$$

$$= \frac{a^5}{a^2}\cdot\frac{b^5}{b^5}\cdot\frac{c^5}{c^8}$$

$$= a^{5-2}\cdot 1\cdot\frac{1}{c^{8-5}} \qquad \text{(Theorem 3)}$$

$$= a^3\cdot\frac{1}{c^3}$$

$$= \frac{a^3}{c^3}$$

EXAMPLE 1.1.9 Simplify $\dfrac{(9^2x^2)^3}{(3^3x^3)^2}$.

Solution

$$\frac{(9^2x^2)^3}{(3^3x^3)^2} = \frac{9^6x^6}{3^6x^6} \qquad \text{(Theorems 4 and 2)}$$

$$= \left(\frac{9}{3}\right)^6\frac{x^6}{x^6} \qquad \text{(Theorem 5)}$$

$$= 3^6\cdot 1$$
$$= 729$$

Note that $(x^2)^3 = (x^3)^2$ since $2\cdot 3 = 3\cdot 2 = 6$.
In general, $(x^m)^n = (x^n)^m$ since $mn = nm$.

EXERCISES I.I A

Simplify Exercises 1–35 by using one or more exponent theorems. In Exercises 1–10, state which exponent theorem was used.

1. $(x^2)^4$

2. $x^3 x^4$

3. $\dfrac{x^8}{x^3}$

4. $(-2y)^3$

5. $\left(\dfrac{-3}{y}\right)^4$

6. $\dfrac{x^4}{x^8}$

7. $2^4 5^4$

8. $-5(2^3)^2$

9. $\dfrac{(100)^6}{(50)^6}$

10. $(-a)^4(-a)^6$

11. $(-5x^2)^3$

12. $-(5x^3)^2$

13. $(a^2b^3)(ab^5)$

14. $\dfrac{3^8 x^7}{3^5 x^{10}}$

15. $\dfrac{12y^{12}}{4y^4}$

16. $\dfrac{(8^2)^3}{(4^3)^2}$

17. $(-x^2)(-x)^2$

18. $\dfrac{(10a^2)^3}{(10a^3)^2}$

19. $(5xy^2)^3$

20. $(x^4y^2)(x^3y^4)$

21. $\left(\dfrac{12a^2bc}{10ab^3}\right)^2$

22. $(-2a^2b^5)^5$

23. $\dfrac{(-y^4)^4}{-(y^4)^4}$

24. $\dfrac{10^5 10^7}{10^4}$

25. $\dfrac{(10^5)^2}{10^2 10^5}$

26. $\dfrac{2^6 \cdot 5^4}{2^3 \cdot 5^6}$

27. $(2^2)^3(5^2)^2$

28. $\dfrac{3^5 \cdot 3^{12}}{3^7 \cdot 3^{10}}$

29. $\dfrac{3^5 \cdot 4^5}{6^5}$

30. $\dfrac{x^n}{x}$

31. $(y^n)^4$

32. $\left(\dfrac{x^2}{5^2}\right)^n$

33. $\dfrac{x^n x}{x^{n+1}}$

34. $\dfrac{(x^{n+1})^2}{x^2 x^n}$

35. $x^{n+1} x^n$

EXERCISES I.I B

Simplify Exercises 1–35 by using one or more exponent theorems. In Exercises 1–10, state which exponent theorem was used.

1. $\dfrac{x^7}{x^3}$ **2.** $\dfrac{c^3}{c^6}$

3. $(t^4)^3$ **4.** $\dfrac{x^2}{x^5}$

5. $(-5t)^4$ **6.** $a^5 a^7$

7. $\left(\dfrac{-n}{2}\right)^5$ **8.** $5^6 2^6$

9. $-7(x^2)^5$ **10.** $\dfrac{(24)^7}{(48)^7}$

11. $(-2y^3)^4$ **12.** $-2(y^4)^3$

13. $(ab^2)^3(a^2b)^4$ **14.** $\dfrac{6y^6}{2y^2}$

15. $\dfrac{a^3b^5}{a^4b^2}$ **16.** $\left(\dfrac{a^3b^5}{ab^3}\right)^4$

17. $(-x^3)(-x)^3$ **18.** $\dfrac{(10^4)^3}{(10^2)^5}$

19. $(-3x^2y^3z^4)^4$ **20.** $-3(a^2b^3c^4)^2$

21. $(c^3d^5)(cd^3)$ **22.** $\dfrac{5x^5}{2y^4}\cdot\dfrac{3y^6}{10x^3}$

23. $\dfrac{-(x^3)^2}{(-x^3)^2}$ **24.** $\dfrac{10^6}{10^2 10^8}$

25. $\dfrac{(10^6)^3}{10^3 10^6}$ **26.** $\dfrac{(2^3\cdot3^4)^3}{(2^2\cdot3^2)^5}$

27. $\dfrac{(2^4)^3(5^6)^2}{(10^3)^2}$ **28.** $\dfrac{9^6\cdot4^6}{6^6}$

29. $\dfrac{x}{x^n}$ **30.** $\dfrac{3^{2n}}{3^n}$

31. $\left(\dfrac{x^3}{y^2}\right)^n$ **32.** $(x^2y)^n$

33. $\dfrac{(x^n)^2}{(x^2)^{n+1}}$ **34.** $x(x^2)^n$

35. $\left(\dfrac{x^n}{y^n}\right)^n$

1.2 INTEGRAL EXPONENTS

It is desirable to establish a meaning for the expressions x^0 and x^{-n}, where x is any nonzero real number and n is a natural number. If the first theorem of exponents is to remain valid when zero is used as an exponent, then

$$x^n x^0 = x^{n+0} = x^n$$

But
$$x^n \cdot 1 = x^n$$

and there is only one multiplicative identity. Therefore, if x^0 is to be defined, it must be defined as the number 1.

DEFINITION OF x^0

$x^0 = 1$ if x is a real number and $x \neq 0$.

As examples, $5^0 = 1$, $(-3)^0 = 1$, and $(x + 2)^0 = 1$ if $x \neq -2$.

It can be shown that all the exponent theorems remain valid when x^0 is defined as 1.

Now if the first theorem of exponents is to remain valid when an exponent is a negative integer, then

$$x^n x^{-n} = x^{n+(-n)} = x^0 = 1$$

However,
$$x^n \cdot \frac{1}{x^n} = 1$$

Since each nonzero real number has exactly one reciprocal, if x^{-n} is to be defined, x^{-n} must be defined as $1/x^n$, the reciprocal of x^n.

DEFINITION OF x^{-n}

$$x^{-n} = \frac{1}{x^n}$$

if x is any real number and $x \neq 0$.

Noting that

$$\frac{1}{x^n} = \frac{1}{\underbrace{x \cdot x \cdots x}_{n \text{ factors}}} = \underbrace{\frac{1}{x}\frac{1}{x} \cdots \frac{1}{x}}_{n \text{ factors}} = \left(\frac{1}{x}\right)^n$$

the following theorem can be stated:

$$\frac{1}{x^n} = \left(\frac{1}{x}\right)^n \qquad \text{if} \quad x \neq 0$$

As examples,

$$5^{-1} = \tfrac{1}{5}$$

$$2^{-3} = \frac{1}{2^3} = \frac{1}{8}$$

$$\left(\frac{1}{3}\right)^{-2} = \left(\frac{1}{1/3}\right)^2 = 3^2 = 9$$

$$x^{-4} = \frac{1}{x^4}$$

$$\left(-\frac{2}{5}\right)^{-3} = \left(-\frac{5}{2}\right)^3 = -\frac{125}{8} \quad \text{since} \quad \frac{1}{-2/5} = -\frac{5}{2}$$

$$\frac{1}{3^{-4}} = \left(\frac{1}{3}\right)^{-4} = 3^4 = 81$$

It can be shown that the five theorems of exponents remain valid when the exponent is a negative integer. These theorems are illustrated in the following examples.

EXAMPLE 1.2.1

(Theorem 1) $2^{-3}2^{-2} = 2^{(-3)+(-2)} = 2^{-5} = \dfrac{1}{2^5} = \dfrac{1}{32}$

$$2^4 2^{-7} = 2^{4+(-7)} = 2^{-3} = \frac{1}{2^3} = \frac{1}{8}$$

EXAMPLE 1.2.2

(Theorem 2) $(2^{-3})^2 = 2^{-6} = \dfrac{1}{2^6} = \dfrac{1}{64}$

$$(2^{-2})^{-4} = 2^{(-2)(-4)} = 2^8 = 256$$

EXAMPLE 1.2.3

(Theorem 3) $\dfrac{3^4}{3^7} = 3^{4-7} = 3^{-3} = \dfrac{1}{3^3} = \dfrac{1}{27}$

$$\frac{x^4}{x^6} = x^{4-6} = x^{-2} = \frac{1}{x^2}$$

$$\frac{x^{-2}}{x^{-5}} = x^{-2-(-5)} = x^{-2+5} = x^3$$

EXAMPLE 1.2.4

(Theorem 4) $(2x^{-2})^{-3} = 2^{-3}(x^{-2})^{-3} = 2^{-3}x^6 = \dfrac{x^6}{8}$

EXAMPLE 1.2.5

(Theorem 5) $\left(\dfrac{x^{-1}}{5}\right)^{-4} = \dfrac{(x^{-1})^{-4}}{5^{-4}} = \dfrac{x^4}{5^{-4}} = 5^4 x^4 = 625x^4$

$\left(\dfrac{2^{-3}}{y^{-3}}\right)^{-2} = \dfrac{(2^{-3})^{-2}}{(y^{-3})^{-2}} = \dfrac{2^6}{y^6}$

EXAMPLE 1.2.6 Simplify $(x^{-2}y^{-2})^{-2}$.

Solution

$\begin{aligned}(x^{-2}y^{-2})^{-2} &= (x^{-2})^{-2}(y^{-2})^{-2} \qquad &\text{(Theorem 4)}\\ &= x^4 y^4 &\text{(Theorem 2)}\end{aligned}$

EXAMPLE 1.2.7 Simplify $(2^{-3} \cdot 2^{-5})^{-1}$.

Solution $\begin{aligned}(2^{-3} \cdot 2^{-5})^{-1} &= (2^{-3-5})^{-1} = (2^{-8})^{-1} \qquad &\text{(Theorem 1)}\\ &= 2^8 &\text{(Theorem 2)}\\ &= 256 &\text{(Definition)}\end{aligned}$

EXAMPLE 1.2.8 Simplify $(2^{-3} + 2^{-5})^{-1}$.

Solution There is no exponent theorem that applies to a power of a sum (or difference), so the definition must be used.

$$(2^{-3} + 2^{-5})^{-1} = \left(\frac{1}{2^3} + \frac{1}{2^5}\right)^{-1}$$

$$= \left(\frac{2^2}{2^5} + \frac{1}{2^5}\right)^{-1}$$

$\qquad\qquad$ (2^5 is the L.C.D. for the two fractions)

$$= \left(\frac{4}{32} + \frac{1}{32}\right)^{-1}$$

$$= \left(\frac{5}{32}\right)^{-1}$$

$$= \frac{32}{5} \qquad \left(\text{Definition } x^{-1} = \frac{1}{x}\right)$$

EXAMPLE 1.2.9   Simplify $\dfrac{2^{-3} + 2^{-5}}{2^{-7}}$.

Solution $\dfrac{2^{-3} + 2^{-5}}{2^{-7}} = \left(\dfrac{2^{-3} + 2^{-5}}{2^{-7}}\right)\left(\dfrac{2^7}{2^7}\right)$

$$= \dfrac{2^7(2^{-3} + 2^{-5})}{(2^7)(2^{-7})}$$

$$= \dfrac{(2^7 2^{-3}) + (2^7 2^{-5})}{2^7 2^{-7}} \qquad \text{(Distributive axiom)}$$

$$= \dfrac{2^4 + 2^2}{2^0} \qquad \text{(Theorem 1)}$$

$$= \dfrac{16 + 4}{1} = 20 \qquad \text{(Definition)}$$

The preceding example illustrates the use of the distributive axiom to distribute the operation of multiplication over a sum. Theorem 1 for exponents is also used in the simplification. It is important to note the differences illustrated by the last four examples. It is especially important to remember that the exponent theorems apply to products and quotients and *not* to sums and differences.

EXAMPLE 1.2.10   Does $(x + 2)^2 = x^2 + 4$? Why?

Solution   By the definition of squaring,

$$(x + 2)^2 = (x + 2)(x + 2) = x^2 + 4x + 4$$

If $(x + 2)^2 = x^2 + 4$

then $x^2 + 4x + 4 = x^2 + 4$

and $4x = 0$ and $x = 0$

Thus $(x + 2)^2 = x^2 + 4$ only if $x = 0$.

EXERCISES 1.2 A

Use the definitions and theorems to express Exercises 1–55 in simplest form without zero or negative exponents. Assume all variables to be nonzero.

1. 2^{-1}

2. 5^{-3}

3. $(\frac{1}{2})^{-4}$

4. $(\frac{1}{10})^{-5}$

5. $\dfrac{1}{3^{-2}}$

6. $\dfrac{1}{5^{-4}}$

7. 10^{-3}

8. $(\frac{3}{5})^{-1}$

9. $\left(-\frac{10}{3}\right)^{-4}$

10. $3^5 3^{-2}$

11. $2^5 2^{-9}$

12. $10^6 10^{-8}$

13. $10^{-2} 10^{-3}$

14. $5^4 5^{-4}$

15. $\dfrac{10^2}{10^{-4}}$

16. $\dfrac{10^{-3}}{10^2}$

17. $\dfrac{10^{-4}}{10^{-4}}$

18. $\dfrac{10^{-5}}{10^{-2}}$

19. $(2^{-3})^2$

20. $(3^{-2}5)^{-2}$

21. $(-2^{-2})^{-1}$

22. $\left(\dfrac{5^{-2}}{2^{-2}}\right)^{-1}$

23. $\dfrac{2^{-5}3^4}{2^{-4}3^{-1}}$

24. $\left(\dfrac{2^{-3}}{5}\right)^{-2}$

25. $10^{-3}(10^4 10^{-1})$

26. $5^{-2}(5^3 + 5^2)$

27. $4^7(4^{-7} - 4^{-5})$

28. $\left(\dfrac{2^{-2}}{5^0}\right)^{-3}$

29. $(2^{-1}5^{-1})^{-1}$

30. $(2^{-1} + 5^{-1})^{-1}$

31. $10^{-2}(2^{-1}5^{-1})$

32. $10^{-2}(2^{-1} + 5^{-1})$

33. $(2x)^{-3}$

34. $2x^{-3}$

35. $(4^{-1}x)^{-2}$

36. $\left(\dfrac{5}{x}\right)^{-3}$

37. $(2^{-1}x^{-1})^{-2}$

38. $(2^{-1} + x^{-1})^{-2}$

39. $\dfrac{y}{y^{-1}}$

40. $\dfrac{1}{2x^{-2}}$

41. $(xx^{-4}x^3)^{10}$

42. $(x^{-2}y)^{-1}$

43. $(x^{-3} + x^{-5})^0$

44. $(x^{-1} + y^{-1})^{-1}$

45. $x^0 + y^0$

46. $\left(\dfrac{2x^3}{y^{-4}}\right)\left(\dfrac{x^{-2}y^5}{8}\right)$

47. $\left(\dfrac{-1}{x^{-1}}\right)^{-1}$

48. $\dfrac{(a^{-3}b)(a^4b^2)}{a^{-4}b^4}$

49. $(3y^{-3})^{-2}$

50. $(3y^{-3})(y^{-2})$

★ **51.** $\dfrac{xx^{-n}}{(x^{-1})^n}$

★ **52.** $x^{-n}(x^n - x^{n-1})$

★ **53.** $(x^{-n} + y^{-n})^0(x^{-n}y^n)^{-1}$

★ **54.** $\dfrac{(x^{n+1}x^{2n-1})^{-2}}{x^{4n}}$

★ **55.** $\dfrac{x^{-n}y^{2n}}{x^{2n}y^{-n}}$

EXERCISES I.2 B

Use the definitions and theorems to express Exercises 1–55 in simplest form without zero or negative exponents. Assume all variables to be nonzero.

1. 10^{-1}

2. 6^{-2}

3. $(\frac{1}{5})^{-3}$

4. $(\frac{1}{10})^{-6}$

5. $\dfrac{1}{2^{-3}}$

6. $\dfrac{1}{10^{-5}}$

7. 10^{-4}

8. $(\frac{4}{3})^{-2}$

9. $(-\frac{3}{16})^{-1}$

10. $5^4 5^{-7}$

11. $10^{-4} 10^2$

12. $10^{-5} 10^{-4}$

13. $10^6 10^{-6}$

14. $6^{-5} 6^5$

15. $\dfrac{10^6}{10^{-3}}$

16. $\dfrac{10^{-5}}{10^{-7}}$

17. $(10^{-2})^{-1}$

18. $\dfrac{10^{-6}}{10^{-6}}$

19. $(2^{-1} 3^{-2})^2$

20. $(-3^{-1})^2$

21. $(-5^{-1})^{-1}$

22. $\left(\dfrac{5}{8^{-1}}\right)^{-2}$

23. $\dfrac{3^{-2} 5^{-1}}{3^2 5^{-2}}$

24. $\left(\dfrac{5^{-3}}{3^{-4}}\right)^{-1}$

25. $5^4(5^{-3} 5^{-1})$

26. $10^8(10^{-6} + 10^{-8})$

27. $3^{-5}(6^5 - 3^6)$

28. $\left(\dfrac{4^0}{4^{-3}}\right)^{-1}$

29. $(2^{-2} 5^{-2})^{-2}$

30. $(2^{-2} + 5^{-2})^{-2}$

31. $\dfrac{2^{-2} 5^{-2}}{10^{-2}}$

32. $\dfrac{2^{-2} + 5^{-2}}{10^{-2}}$

33. $-3(x^2)^{-4}$

34. $(-3x^2)^{-4}$

35. $(10^{-3} x^3)^{-2}$

36. $\left(\dfrac{x}{4}\right)^{-2}$

37. $(x^{-2} + 5^{-2})^{-1}$

38. $(x^{-2} \cdot 5^{-2})^{-1}$

39. $\dfrac{x^{-3}}{y^{-2}}$

40. $\dfrac{1}{4x^{-3}}$

41. $\left(\dfrac{x^{-3}}{y^{-2}}\right)\left(\dfrac{x^{-5}}{x^7}\right)^0$

42. $(y^{-5} - y^2)^{-3}(y^{-5} - y^2)^3$

43. $\left(\dfrac{x}{y}\right)^{-4}\left(\dfrac{y^{-4}}{x^{-2}}\right)$

44. $\dfrac{x^{-1} + y^{-1}}{x^{-1} - y^{-1}}$

45. $(x^{-1} - y^{-1})(x - y)^{-1}$

46. $\dfrac{a + b^{-1}}{a^{-1} + b}$

47. $\dfrac{a^4 b - a^{-1} b^{-4}}{a^5 b^{-5}}$

48. $(-2x^{-2}y^{-2})^{-3}$

49. $(-2x^{-2}y^{-2})y^{-3}$

50. $\dfrac{2x^3 y^{-2}}{3x^{-2} y^3}$

★ 51. $(x^{-2} + y^{-2})^n (x^{-2} + y^{-2})^{-n}$

★ 52. $x^{-n-1}(x^{2n+1} + x^{n+2})$

★ 53. $(2^n + 2^{-n})(2^n - 2^{-n})$

★ 54. $(2^n + 2^n)^{-n}$

★ 55. $\dfrac{2^{-3n}3^{-2n}}{2^{2n}3^{-n}}$

1.3   SQUARE ROOTS

1.3.1   Symbolic Representation of Square Roots

DEFINITION

The number a is a square root of b if and only if $a^2 = b$.

This definition implies that the operations of squaring and extracting square roots are inverses of each other, just as addition and subtraction are inverse operations and multiplication and division are inverse operations.

For example,

$$3 \text{ is a square root of } 9 \text{ since } 3^2 = 9$$
$$-3 \text{ is a square root of } 9 \text{ since } (-3)^2 = 9$$
$$\tfrac{5}{8} \text{ is a square root of } \tfrac{25}{64} \text{ since } (\tfrac{5}{8})^2 = \tfrac{25}{64}$$
$$-\tfrac{5}{8} \text{ is a square root of } \tfrac{25}{64} \text{ since } (-\tfrac{5}{8})^2 = \tfrac{25}{64}$$

and 0 is a square root of 0 since $0^2 = 0$

Examination of these examples and similar ones reveals the following properties:

1. If x is a real number, then $x^2 \geq 0$. (In other words, the square of a real number is never negative.)
2. Every positive real number has two square roots, a positive real number and its negative. Thus if a is a positive real number ($a > 0$), then the square roots of a^2 are a and $-a$.

It is useful to denote a square root by the symbol $\sqrt{x}$. However, the symbolic expression $\sqrt{x}$ must represent exactly one number. Suppose, for example, that $\sqrt{9} = 3$ and $\sqrt{9} = -3$. Then, by the transitive axiom of the equal relation, it would follow that $3 = -3$, and this is impossible. Therefore, $\sqrt{9}$ and, in general, $\sqrt{x}$ must represent exactly one number in order to avoid contradictions. Mathematicians agree that $\sqrt{9}$ shall be used to designate $+3$, the positive square root of 9. The negative square root, -3, is indicated by $-\sqrt{9}$.

In general, the following definition is made:

DEFINITION OF $\sqrt{x}$

If $x > 0$, then $\sqrt{x}$ is the unique positive real number such that $(\sqrt{x})^2 = x$.

Note: If $x < 0$, then $\sqrt{x}$ is not a real number.

The following useful theorem aids in the solution of problems involving square roots.

THEOREM

$\sqrt{x^2} = x$ if and only if $x \geq 0$.

By introducing the concept of absolute value, a theorem can be stated that holds for all real numbers x.

DEFINITION OF ABSOLUTE VALUE

Let x be a real number. Then $|x| = x$ if $x \geq 0$ and $|x| = -x$ if $x < 0$.

As examples, $|5| = 5$, $|0| = 0$, and $|-5| = -(-5) = 5$. In other words, if x is not zero, $|x|$ is always positive.

THEOREM

For all real numbers x, $\sqrt{x^2} = |x|$.

Note that since $|x|$ is always positive or zero, this theorem states that $\sqrt{x^2}$ is always positive or zero.

For example, $\sqrt{5^2} = 5$ but $\sqrt{(-5)^2} = \sqrt{25} = \sqrt{5^2} = 5$—that is, $\sqrt{(-5)^2} = |-5| = 5$.

The positive real number $\sqrt{x}$ is called the principal square root of x. Its negative is also a square root and is designated by the symbol $-\sqrt{x}$.

Thus $\sqrt{25} = 5$, and 5 is the principal square root of 25.

Also, $-\sqrt{25} = -5$, and -5 is the other square root of 25.

Thus there is exactly one symbolic representation for each of the square roots of a positive real number.

I.3.2   Irrational Square Roots

Although every positive real number x has two real square roots— namely, $\sqrt{x}$ and $-\sqrt{x}$—it does not follow that every positive rational number has two rational roots.

Rational numbers, such as 49, 81, and $\frac{9}{16}$, are called perfect squares because their square roots are also rational.

On the other hand, the number 2 is not a perfect square, and its square roots, $\sqrt{2}$ and $-\sqrt{2}$, are not rational but irrational.

Q, the set of rational numbers, was defined as the set of quotients of integers, p/q, where $q \neq 0$. It may be shown that $\sqrt{2}$ cannot be expressed as the quotient of two integers, and thus $\sqrt{2}$ is irrational.

The terminating decimals and the nonterminating repeating decimals represent the rational numbers. Therefore, it is convenient to think of the irrational numbers as the set of nonterminating, nonrepeating decimals and the real numbers as the union of these two sets—that is, the set of all decimals.

The table of squares and square roots on the inside cover of this book is useful in providing a first approximation to the square root of a number. The square root values in the table are not exact but are approximations to the nearest thousandth.

I.3.3   Geometric Interpretation of Square Roots

The set of real numbers has a property called the axiom of completeness.

THE AXIOM OF COMPLETENESS

Each point on the number line corresponds to exactly one real number, and each real number corresponds to exactly one point on the number line.

A point on the number line that corresponds to an irrational square root can be found by using the theorem of Pythagoras.

THE THEOREM OF PYTHAGORAS

The square of the length of the hypotenuse of a right triangle is equal to the sum of the squares of the lengths of the legs of the right triangle.

The hypotenuse of a right triangle, the longest side, is opposite the right angle. The legs of a right triangle are the other two sides—that is, the sides that form the right angle.

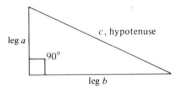

FIGURE 1.3.1

Thus if c represents the length of the hypotenuse in Figure 1.3.1 and if a and b represent the lengths of the legs, then

$$c^2 = a^2 + b^2$$

Therefore, a length of $\sqrt{2}$ can be represented by the diagonal of a square (see Figure 1.3.2).

Since
$$c^2 = 1^2 + 1^2$$
$$c^2 = 2$$

Thus
$$c = \sqrt{2}$$

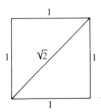

FIGURE 1.3.2

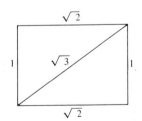

FIGURE 1.3.3

A length of $\sqrt{3}$ can be represented as the diagonal of a rectangle whose sides are 1 and $\sqrt{2}$ (see Figure 1.3.3).

Since
$$c^2 = (\sqrt{2})^2 + 1^2$$
$$c^2 = 2 + 1 = 3$$

Thus
$$c = \sqrt{3}$$

In a similar way, the numbers $\sqrt{4} = 2$, $\sqrt{5}$, $\sqrt{6}$, and so on can be represented as the lengths of the diagonals of rectangles. Now using circles with centers at the origin and with radii equal successively to these diagonals, the square roots can be located on the number line. This is illustrated in Figure 1.3.4.

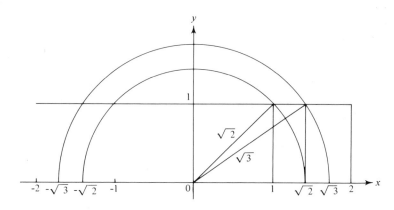

FIGURE 1.3.4

EXAMPLE 1.3.1   Simplify $\sqrt{64} - \sqrt{(-4)^2}$.

Solution $\sqrt{64} - \sqrt{(-4)^2} = \sqrt{64} - \sqrt{16}$
$$= 8 - 4 = 4$$

EXAMPLE 1.3.2   Approximate $\sqrt{12} - \sqrt{20}$ to the nearest hundredth by using a table.

Solution   From the table,
$$\sqrt{12} = 3.464$$
and
$$\sqrt{20} = 4.472$$
$$\sqrt{12} - \sqrt{20} = 3.464 - 4.472$$
$$= -1.008$$
$$= -1.01 \text{ to the nearest hundredth}$$

EXAMPLE 1.3.3   Find the hypotenuse c of a right triangle if the lengths of its legs a and b are such that $a = 12$ and $b = 35$.

Solution   Using the theorem of Pythagoras,

$$c^2 = a^2 + b^2$$
$$c^2 = (12)^2 + (35)^2$$
$$c^2 = 144 + 1225 = 1369$$
$$c = \sqrt{1369} = 37 \qquad \text{(Refer to the table on the inside cover)}$$

EXAMPLE 1.3.4   If a, b, and c are the sides of a right triangle whose hypotenuse is c, find b if $a = 16$ and $c = 18$.

Solution   Using

$$a^2 + b^2 = c^2$$
$$(16)^2 + b^2 = (18)^2$$
$$b^2 = (18)^2 - (16)^2 = (18 + 16)(18 - 16)$$
$$b^2 = 68$$
$$b = 8.25 \text{ to the nearest hundredth}$$

EXAMPLE 1.3.5   Solve:

$$\text{a. } x = \sqrt{9} \qquad \text{b. } x^2 = 9$$

Solution

a. Since $\sqrt{9} = 3$, the positive square root of 9,

$$x = \sqrt{9} \text{ means } x = 3$$

The solution set is $\{3\}$.

b. Since $(+3)^2 = 9$ and $(-3)^2 = 9$, the solution set of $x^2 = 9$ is $\{3, -3\}$.

An alternate solution to (b) is as follows:

$$x^2 = 9$$
$$x^2 - 9 = 0$$
$$(x + 3)(x - 3) = 0$$
$$x + 3 = 0 \quad \text{or} \quad x - 3 = 0$$
$$x = -3 \quad \text{or} \quad x = 3$$

and the solution set is $\{3, -3\}$.

HISTORICAL NOTE

Although there are many stories about the life of Pythagoras (ca. 580–501 B.C.) little is known for certain. He was probably born on the island of Samos, probably traveled to Egypt and Babylonia, and is known to have settled in Crotona on the Italian coast. In Crotona he founded a brotherhood composed of some 300 wealthy young aristocrats. This group, known as the Pythagoreans, became the prototype of all the secret societies of Europe and America. Their motto, "Number rules the universe," expressed the combination of mathematics and mysticism in which they believed. Shakespeare refers to the Pythagorean belief in immortality and transmigration of the soul in *The Merchant of Venice:*

> Thou almost mak'st me waver in
>   my faith,
> To hold opinion with Pythagoras,
> That souls of animals infuse
>   themselves
> Into the trunks of men.

The name of Pythagoras is most famous in connection with the relationship of the squares of the sides of a right triangle. While Pythagoras did not discover this property (it was already known to the Babylonians), he may have offered the first proof of this statement.

EXERCISES I.3 A

Simplify Exercises 1–10. (If necessary, use the table of squares and square roots.)

1. $\sqrt{16}$

2. $-\sqrt{36}$

3. $\sqrt{(-3)^2}$

4. $\sqrt{\frac{4}{25}}$

5. $\sqrt{400}$

6. $\sqrt{5^2 - 4^2}$

7. $\sqrt{5^2} - \sqrt{4^2}$

8. $\sqrt{(-5)^2 + (12)^2}$

9. $\sqrt{(-5)^2} + \sqrt{(12)^2}$

10. $\dfrac{3\sqrt{64} + 36}{3(\sqrt{64} + \sqrt{36})}$

Use the table of squares and square roots to approximate Exercises 11–20 to the nearest hundredth.

11. $\sqrt{45}$

12. $\sqrt{10}$

13. $\sqrt{6500}$

14. $\sqrt{0.05}$

15. $2\sqrt{5}$

16. $5 - \sqrt{3}$

17. $\sqrt{7} + \sqrt{2}$

18. $\dfrac{1 + \sqrt{6}}{2}$

19. $\dfrac{-5 + \sqrt{12}}{6}$

20. $\sqrt{2}\sqrt{6}$

If a and b represent the lengths of the legs of a right triangle and c represents the length of the hypotenuse of that triangle, evaluate the length of the missing side in Exercises 21–30.

21. $a = 12, b = 5$

22. $b = 5, c = 13$

23. $a = 1, b = 1$

24. $a = 15, c = 17$

25. $a = \sqrt{3}, b = \sqrt{5}$

26. $c = \sqrt{3}, a = 1$

27. $a = 2, c = 6$

★ **28.** $a = 2t, c = t^2 + 1$

★ **29.** $a = 4u^2 - v^2, b = 4uv$

★ **30.** $a = x^2, b = \sqrt{6x^2 + 9}$

Solve Exercises 31–35 over the set of real numbers.

31. $x = \sqrt{25}$

32. $x^2 = 25$

33. $x^2 = 36$

34. $x = -\sqrt{36}$

35. $x^2 = -36$

In Exercises 36–40, express each without using the radical symbol.

36. $\sqrt{x^2}$

37. $\sqrt{4x^2}$

38. $\sqrt{(-x)^2}$

39. $\sqrt{(-3x)^2}$

40. $\sqrt{x^2 - 2xy + y^2}$

EXERCISES 1.3 B

Simplify Exercises 1–10. (If necessary, use the table of squares and square roots.)

1. $\sqrt{169}$

2. $-\sqrt{1225}$

3. $\sqrt{\frac{25}{64}}$

4. $\sqrt{(-7)^2}$

5. $\sqrt{0.01}$

6. $\sqrt{(25)^2 - 7^2}$

7. $\sqrt{(25)^2} - \sqrt{7^2}$

8. $\sqrt{(-8)^2 + (-15)^2}$

9. $\sqrt{(-8)^2} + \sqrt{(-15)^2}$

10. $\dfrac{4\sqrt{225} - 144}{4\sqrt{225} - \sqrt{144}}$

Use the table of squares and square roots to approximate Exercises 11–20 to the nearest hundredth.

11. $\sqrt{86}$ **12.** $\sqrt{2300}$

13. $\sqrt{18}$ **14.** $\sqrt{0.57}$

15. $2 + \sqrt{2}$ **16.** $\sqrt{5} - \sqrt{3}$

17. $5\sqrt{8}$ **18.** $\dfrac{4 - \sqrt{10}}{2}$

19. $\dfrac{-6 - \sqrt{15}}{3}$ **20.** $\sqrt{5}\sqrt{10}$

If a and b represent the lengths of the legs of a right triangle and c represents the length of the hypotenuse of that triangle, evaluate the length of the missing side in Exercises 21–30.

21. $a = 14, b = 51$ **22.** $b = 5, c = 15$

23. $c = 10, a = b$ **24.** $a = 2, c = \sqrt{3}$

25. $b = \sqrt{10}, c = \sqrt{30}$ **26.** $a = 1, b = 2$

★ **27.** $a = p^2 - q^2, b = 2pq$ ★ **28.** $b = t^2 - 9, c = t^2 + 9$

★ **29.** $a = x^2, b = \sqrt{2x^2 + 1}$ ★ **30.** $a = 4u^2, b = \sqrt{25 + 40u^2}$

Solve Exercises 31–35 over the set of real numbers.

31. $x = \sqrt{16}$ **32.** $x^2 = 16$

33. $x^2 = 49$ **34.** $x = -\sqrt{49}$

35. $x^2 = -49$

In Exercises 36–40, express each of the following without using the radical symbol.

36. $\sqrt{x^4}$ **37.** $\sqrt{x^6}$

38. $\sqrt{(x + y)^2}$ **39.** $\sqrt{(-x - 2)^2}$

40. $\sqrt{y^2 - 14y + 49}$

1.4 CUBE ROOTS AND nth ROOTS

1.4.1 Cube Roots

DEFINITION

The number a is a **cube root** of b if and only if $a^3 = b$.

For example, 2 is a cube root of 8 because $2^3 = 8$, and -3 is a cube root of -27 because $(-3)^3 = -27$.

Since the product of three positive numbers is positive, and since the product of three negative numbers is negative, then a positive real number has exactly one real cube root and it is positive, and a negative real number has exactly one real cube root and it is negative. Therefore, the radical sign can be used to designate this unique real cube root of a real number.

DEFINITION OF $\sqrt[3]{x}$

$\sqrt[3]{x}$ is the unique real number such that $(\sqrt[3]{x})^3 = x$.

Again, a useful theorem follows the definition.

THEOREM

$\sqrt[3]{x^3} = x$ for all real numbers x.

For example, $\sqrt[3]{8} = 2$ and $\sqrt[3]{-27} = -3$.

Also, $\sqrt[3]{\frac{8}{27}} = \frac{2}{3}$ because $(\frac{2}{3})^3 = \frac{2}{3} \cdot \frac{2}{3} \cdot \frac{2}{3} = \frac{8}{27}$

and $\sqrt[3]{-\frac{64}{125}} = -\frac{4}{5}$ because $(-\frac{4}{5})^3 = (-\frac{4}{5})(-\frac{4}{5})(-\frac{4}{5}) = -\frac{64}{125}$

$\sqrt[3]{0} = 0$ because $0^3 = 0 \cdot 0 \cdot 0 = 0$

Rational numbers such as 8, -27, and $-\frac{64}{125}$ are called perfect cubes because their cube roots are rational numbers.

As is the case with square roots, some cube roots are irrational and can be named only by using the radical sign.

For example, $\sqrt[3]{2}$, $\sqrt[3]{-3}$, and $\sqrt[3]{\frac{2}{7}}$ are irrational real numbers. Their nonterminating, nonrepeating decimal representation can be approximated by arithmetical calculations. A table of cubes and cube roots appears inside the front cover of this book. The values of the cube roots in the table are approximations to the nearest thousandth.

1.4.2 nth Roots

> **DEFINITION**
>
> The number a is an nth **root** of b if and only if $a^n = b$.

For example, 5 is a 4th root of 625 because $5^4 = 625$; -2 is a 5th root of -32 because $(-2)^5 = -32$.

Roots such as the square roots, the 4th roots, and the 6th roots have the following common properties:

1. Every positive real number has exactly two real nth roots for n even: one positive and one negative. For example, 5 and -5 are 4th roots of 625.
2. Negative real numbers do *not* have real nth roots for n even. For example, there is no number x such that $x^4 = -625$.

Roots such as the cube root, the 5th root, and the 7th root have the following common properties:

1. Every real number has exactly one real nth root for n odd.
2. The real nth root of a positive number is positive for n odd.
3. The real nth root of a negative number is negative for n odd. For example, 2 is the real 5th root of 32, and -2 is the real 5th root of -32.

The symbol $\sqrt[n]{x}$ (read "the principal nth root of x") is used to indicate exactly one nth root of the real number x.

> **DEFINITION OF $\sqrt[n]{x}$, x POSITIVE**
>
> $\sqrt[n]{x}$ is the unique positive real number such that $(\sqrt[n]{x})^n = x$.

> **DEFINITION OF $\sqrt[n]{-x}$, x POSITIVE, n ODD**
>
> $$\sqrt[n]{-x} = -\sqrt[n]{x}$$

Note that $\sqrt[n]{-x}$ is not defined for n even and $-x$ negative.

The symbol $\sqrt[n]{x}$ is called a radical, the natural number n is called the **index**, and the real number x is called the **radicand**.

THEOREM

If x is a positive real number and n is a natural number, then

$$\sqrt[n]{x^n} = x$$

EXAMPLE 1.4.1   Simplify, if possible:

a. $\sqrt[3]{8}$ b. $\sqrt[3]{-8}$ c. $\sqrt[4]{81}$ d. $\sqrt[4]{-81}$ e. $\sqrt[5]{243}$

f. $\sqrt[5]{-243}$

Solution

a. $\sqrt[3]{8} = \sqrt[3]{2^3} = 2$

b. $\sqrt[3]{-8} = \sqrt[3]{(-2)^3} = -2$

c. $\sqrt[4]{81} = \sqrt[4]{3^4} = 3$ since 3 is positive

d. $\sqrt[4]{-81}$ is not a real number

e. $\sqrt[5]{243} = \sqrt[5]{3^5} = 3$

f. $\sqrt[5]{-243} = \sqrt[5]{(-3)^5} = -3$

EXAMPLE 1.4.2   Simplify $\dfrac{(\sqrt[3]{25})^3}{\sqrt[3]{64}}$.

Solution $\dfrac{(\sqrt[3]{25})^3}{\sqrt[3]{64}} = \dfrac{25}{\sqrt[3]{4^3}} = \dfrac{25}{4}$

EXAMPLE 1.4.3   Approximate $2\sqrt[3]{15} - \sqrt[3]{6}$, correct to the nearest hundredth.

Solution   Using a table,

$$\sqrt[3]{15} = 2.466$$
$$2\sqrt[3]{15} = 4.932$$
$$\sqrt[3]{6} = 1.817$$
$$\overline{2\sqrt[3]{15} - \sqrt[3]{6} = 3.115}$$
$$= 3.12 \text{ to the nearest hundredth}$$

EXERCISES I.4 A

Simplify Exercises 1–15. (If necessary, use a table.)

1. $\sqrt[3]{64}$ 2. $\sqrt[3]{-27}$
3. $\sqrt[4]{16}$ 4. $\sqrt[3]{(-8)^3}$
5. $\sqrt[4]{(-7)^4}$ 6. $\sqrt[5]{-1}$
7. $\sqrt[6]{10^6}$ 8. $(\sqrt[3]{15})^3$
9. $\sqrt[4]{5}\sqrt[4]{5}\sqrt[4]{5}\sqrt[4]{5}$ 10. $\sqrt[5]{\frac{1}{32}}$
11. $\sqrt[3]{\frac{27}{125}}$ 12. $\sqrt[3]{1000}$
13. $\sqrt[3]{1,000,000}$ 14. $\sqrt[4]{0.0001}$
15. $(\sqrt[4]{25x^2})^4$

In Exercises 16–25, use the table inside the cover of this book to approximate each, correct to the nearest hundredth.

16. $\sqrt[3]{90}$ 17. $\sqrt[3]{31}$
18. $\sqrt[3]{57}$ 19. $\sqrt[3]{12}$
20. $2\sqrt[3]{5}$ 21. $\dfrac{\sqrt[3]{4}}{2}$

22. $\sqrt[3]{20} - \sqrt[3]{10}$ 23. $\dfrac{\sqrt[3]{316} - 100}{8}$

24. $\sqrt[3]{\dfrac{316 - 100}{8}}$ 25. $\sqrt[3]{8}\sqrt[3]{2}$

List the elements in the sets in Exercises 26–32, where x is a real number.

26. $\{x \mid x^3 = 64\}$ 27. $\{x \mid x^3 = -64\}$
28. $\{x \mid x = \sqrt[3]{64}\}$ 29. $\{x \mid x = \sqrt[3]{-64}\}$
30. $\{x \mid x = \sqrt[4]{16}\}$ 31. $\{x \mid x^4 = 16\}$
32. $\{x \mid x^4 = -16\}$

EXERCISES I.4 B

Simplify Exercises 1–15. (If necessary, use the table.)

1. $\sqrt[3]{-125}$ 2. $\sqrt[3]{729}$
3. $\sqrt[4]{625}$ 4. $\sqrt[4]{(-6)^4}$

5. $\sqrt[3]{(-12)^3}$

6. $\sqrt[5]{-32}$

7. $\sqrt[8]{10^8}$

8. $(\sqrt[4]{24})^4$

9. $\sqrt[3]{7}\sqrt[3]{7}\sqrt[3]{7}$

10. $\sqrt[5]{\dfrac{1}{5^3 \cdot 5^2}}$

11. $\sqrt[3]{\dfrac{343}{8}}$

12. $\sqrt[5]{100,000}$

13. $\sqrt[3]{0.001}$

14. $\sqrt[4]{0.0016}$

15. $(\sqrt[3]{-3x^3y^6})^3$

In Exercises 16–25, use the table to approximate each, correct to the nearest hundredth.

16. $\sqrt[3]{80}$

17. $\sqrt[3]{96}$

18. $\sqrt[3]{65}$

19. $\sqrt[3]{29}$

20. $5\sqrt[3]{10}$

21. $\dfrac{\sqrt[3]{100}}{10}$

22. $\sqrt[3]{6} + \sqrt[3]{2}$

23. $\sqrt[3]{\dfrac{100 - 612}{8}}$

24. $\dfrac{\sqrt[3]{100} - 612}{8}$

25. $\sqrt[3]{125}\sqrt[3]{4}$

List the elements in the sets in Exercises 26–32, where x is a real number.

26. $\{x \mid x^5 = 10^5\}$

27. $\{x \mid x^5 = -10^5\}$

28. $\{x \mid x = \sqrt[5]{10^5}\}$

29. $\{x \mid x = \sqrt[5]{-10^5}\}$

30. $\{x \mid x^6 = 64\}$

31. $\{x \mid x = \sqrt[6]{64}\}$

32. $\{x \mid x^6 = -64\}$

I.5 FRACTIONAL EXPONENTS

It is useful to assign a meaning to expressions such as $9^{1/2}$, $8^{1/3}$, $x^{3/4}$, and $x^{-2/3}$. To see if it is possible to define a power having a fractional exponent and still have the five exponent theorems remain valid, the second theorem of exponents, $(x^m)^n = x^{mn}$, is examined first.

If the second theorem is to be valid, then

$$(9^{1/2})^2 = 9^{1/2 \cdot 2} = 9^1 = 9$$

But $$(9^{1/2})^2 = 9^{1/2} \cdot 9^{1/2} = 9$$

and $$\sqrt{9}\sqrt{9} = 9$$

Thus a possibility is $9^{1/2} = \sqrt{9} = 3$.

Similarly, $$(8^{1/3})^3 = 8^{1/3 \cdot 3} = 8^1 = 8$$

$$(8^{1/3})(8^{1/3})(8^{1/3}) = 8$$

$$\sqrt[3]{8}\sqrt[3]{8}\sqrt[3]{8} = 8$$

This suggests the possibility $8^{1/3} = \sqrt[3]{8} = 2$.

Accordingly, if x is a nonnegative real number and n is a natural number, then $x^{1/n}$ is defined as $\sqrt[n]{x}$, the principal real root of x.

DEFINITIONS OF $x^{1/n}$ AND $(-x)^{1/n}$

x is a positive real number, n is a natural number.

$$x^{1/n} = \sqrt[n]{x}$$
$$(-x)^{1/n} = -\sqrt[n]{x} \text{ if } n \text{ is odd } (1, 3, 5, 7, \ldots)$$
$$(-x)^{1/n} \text{ is undefined if } n \text{ is even } (2, 4, 6, 8, \ldots)$$
$$0^{1/n} = 0$$

For example, $$(64)^{1/3} = \sqrt[3]{64} = 4$$
$$(-64)^{1/3} = \sqrt[3]{-64} = -4$$
$$(64)^{1/2} = \sqrt{64} = 8$$
$$(-64)^{1/2} = \sqrt{-64}$$

The last equation is undefined since the index, 2, is even and the radicand, -64, is negative.

Similarly,

$$x^{m/n} \text{ is defined as } (x^m)^{1/n} = \sqrt[n]{x^m}$$

and

$$x^{-m/n} \text{ is defined as } \frac{1}{x^{m/n}}$$

Note that $x^{m/n} = (x^{1/n})^m = (\sqrt[n]{x})^m$.

It can be shown that the five exponent theorems remain valid with these definitions. The results on exponents developed so far are summarized in the following table.

TABLE 1.1   EXPONENT DEFINITIONS AND THEOREMS

Assumptions: x and y are real numbers, $x \neq 0$ and $y \neq 0$.

n and m are natural numbers.

a and b are rational numbers.

Definitions	*Theorems*
1. $x^n = xx \cdots x$ (n factors)	1. $x^a x^b = x^{a+b}$
2. $x^1 = x$	2. $(x^a)^b = x^{ab}$
3. $x^0 = 1$	3. $\dfrac{x^a}{x^b} = x^{a-b}$
4. $x^{-n} = \dfrac{1}{x^n}$	4. $(xy)^a = x^a y^a$
5. $x^{1/n} = \sqrt[n]{x}$ ($x > 0$ if n even)	5. $\left(\dfrac{x}{y}\right)^a = \dfrac{x^a}{y^a}$
6. $x^{m/n} = \sqrt[n]{x^m}$ ($x > 0$ if n even)	
7. $x^{-m/n} = \dfrac{1}{\sqrt[n]{x^m}}$ ($x > 0$ if n even)	
8. $0^a = 0$ if $a \neq 0$	

EXAMPLE 1.5.1   Simplify $(16)^{3/4}$.

Solution   Factoring the base, $16 = 2^4$, then
$$(16)^{3/4} = (2^4)^{3/4} = 2^{4 \cdot 3/4} = 2^3 = 8$$

EXAMPLE 1.5.2   Simplify $(64)^{-2/3}$.

Solution   Factoring the base, $64 = 2^6$, then
$$(64)^{-2/3} = (2^6)^{-2/3} = 2^{6(-2/3)} = 2^{-4} = \frac{1}{2^4} = \frac{1}{16}$$

EXAMPLE 1.5.3   Simplify $(\frac{1}{16})^{-1/2}$.

Solution $(\frac{1}{16})^{-1/2} = (16)^{1/2} = \sqrt{16} = 4$

EXAMPLE 1.5.4   Simplify $[(-6)^2]^{1/2}$.

Solution $[(-6)^2]^{1/2} = \sqrt{(-6)^2} = \sqrt{36} = 6$

It is important to note that the theorems do not apply whenever the base is negative and an exponent indicates a root whose index is even. Thus
$$[(-6)^2]^{1/2} \neq (-6)^1$$

EXAMPLE I.5.5 Simplify $(10^{-2/3} \cdot 10^{5/6})^3$.

Solution $(10^{-2/3} \cdot 10^{5/6})^3 = (10^{-4/6} \cdot 10^{5/6})^3$

$$= (10^{1/6})^3 = 10^{3/6} = 10^{1/2} = \sqrt{10}$$

EXAMPLE I.5.6 $\left(\dfrac{2^{3/4}x^{-3/4}}{2^{1/2}}\right)^{-4}$

Solution $\left(\dfrac{2^{3/4}x^{-3/4}}{2^{1/2}}\right)^{-4} = (2^{3/4-2/4}x^{-3/4})^{-4}$

$$= (2^{1/4})^{-4}(x^{-3/4})^{-4}$$

$$= 2^{-1}x^3 = \frac{x^3}{2}$$

EXERCISES I.5 A

In Exercises 1–5, verify each example of an exponent theorem by simplifying each side. State which theorem is illustrated.

1. $16^{1/2}16^{1/4} = 16^{1/2 + 1/4}$ **2.** $(64^{1/2})^{1/3} = 64^{1/2 \cdot 1/3}$

3. $\dfrac{16^{3/4}}{16^{1/4}} = 16^{3/4 - 1/4}$ **4.** $(16 \cdot 625)^{1/4} = (16)^{1/4}(625)^{1/4}$

5. $\left(\dfrac{10,000}{16}\right)^{1/4} = \dfrac{10,000^{1/4}}{16^{1/4}}$

In Exercises 6–50, express each as a simplified rational number or as a simplified quotient. Assume all variables to be positive. No exponents should appear in the answer.

6. $4^{1/2}$ **7.** $16^{1/2}$

8. $16^{1/4}$ **9.** $27^{1/3}$

10. $32^{1/5}$ **11.** $\left(\frac{27}{125}\right)^{1/3}$

12. $\left(\frac{81}{256}\right)^{1/4}$ **13.** $64^{2/3}$

14. $64^{3/2}$ **15.** $8^{-2/3}$

16. $25^{-3/2}$ **17.** $100,000^{-3/5}$

18. $\left(\frac{1}{27}\right)^{-2/3}$ **19.** $\left(-\frac{1}{125}\right)^{-2/3}$

20. $\left(\frac{216}{343}\right)^{2/3}$ **21.** $\left(\frac{1}{16}\right)^{3/4}$

22. $\left(\frac{1}{16}\right)^{-3/4}$ **23.** $(-8)^{2/3}$

24. $-8^{2/3}$ **25.** $(-8)^{-2/3}$

26. $-8^{-2/3}$

27. $(5^{3/2})^4$

28. $(8^{3/2})^{-2/3}$

29. $(10^{-3/5})^{-5/6}$

30. $(4^2)^{1/6}$

31. $(\frac{1}{14})^{-1/4}$

32. $8^{1/2} \cdot 8^{1/6}$

33. $5^{1/2} \cdot 5^{-1/6}$

34. $10^{1.2} \cdot 10^{3.8}$

35. $10^{-1.4} \cdot 10^{1.9}$

36. $10^{-0.4} \cdot 10^{-0.6}$

37. $(10^{-2} \cdot 10^{-1/2})^{-1/2}$

38. $(5 \cdot 5^{1/2})^{1/3}$

39. $\dfrac{16^{1/3}}{4^{1/6}}$

40. $\dfrac{10^{2/3}}{10^{1/6}}$

★ **41.** $2^{1/2} \cdot 4^{1/3} \cdot 32^{1/6}$

42. $(x^2 y^4)^{1/4}$ $x > 0, y > 0$

43. $(4x^{2/3})^{-3}$

44. $4(x^{2/3})^{-3}$

45. $(9x^{-4}y^2)^{1/2}$

46. $(10x^{1/6}y^{5/6})^{-6}$

★ **47.** $(x^{1/2} + y^{1/2})^2$

★ **48.** $(x^{1/2} + 3^{1/2})(x^{1/2} - 3^{1/2})$

★ **49.** $x^{1/2}(2x^{1/2} + 5x^{-1/2})$

50. $(3^2 + 4^2)^{1/2}$

EXERCISES I.5 B

In Exercises 1–5, verify each example of an exponent theorem by simplifying each side. State which theorem is illustrated.

1. $(64)^{1/2}(64)^{1/6} = (64)^{1/2 + 1/6}$

2. $(81^{1/2})^{1/2} = (81)^{1/2 \cdot 1/2}$

3. $\dfrac{64^{1/2}}{64^{1/6}} = 64^{1/2 - 1/6}$

4. $(8 \cdot 125)^{1/3} = 8^{1/3} \cdot 125^{1/3}$

5. $\left(\dfrac{1000}{125}\right)^{1/3} = \dfrac{1000^{1/3}}{125^{1/3}}$

In Exercises 6–50, express each as a simplified rational number or as a simplified quotient. Assume all variables to be positive. No exponents should appear in the answer.

6. $25^{1/2}$

7. $81^{1/4}$

8. $81^{1/2}$

9. $216^{1/3}$

10. $243^{1/5}$

11. $(\frac{64}{81})^{1/2}$

12. $(\frac{1000}{27})^{1/3}$

13. $32^{3/5}$

14. $100^{3/2}$

15. $16^{-3/2}$

16. $125^{-2/3}$

17. $32^{-2/5}$

18. $(\frac{1}{125})^{-2/3}$

19. $(-\frac{729}{512})^{-2/3}$

20. $\left(\frac{1}{16}\right)^{3/2}$

21. $\left(\frac{1}{16}\right)^{-3/2}$

22. $\left(\frac{25}{16}\right)^{-3/2}$

23. $-125^{2/3}$

24. $(-125)^{2/3}$

25. $-125^{-2/3}$

26. $(-125)^{-2/3}$

27. $(6^{3/4})^{-4/3}$

28. $(7^{-2/5})^{-5/4}$

29. $(5^6)^{-1/2}$

30. $(25^3)^{1/6}$

31. $(25^3)^{-1/6}$

32. $2^{-2/3} \cdot 2^{-1/3}$

33. $4^{3/4} \cdot 4^{-1/4}$

34. $10^{3.5} \cdot 10^{-1.5}$

35. $10^{-1.47} \cdot 10^{1.72}$

36. $10^{-0.08} \cdot 10^{-0.17}$

37. $(2 \cdot 2^{1/3})^{1/2}$

38. $(2^3 \cdot 2^{1/2})^{1/2}$

39. $\dfrac{7^{1/2}}{7^{1/4}}$

40. $\dfrac{(216)^{1/4}}{6^{1/12}}$

41. $7^{1/2} \cdot 7^{1/3} \cdot 7^{1/6}$

42. $8x^{-2/3}$

43. $(8x)^{-2/3}$

44. $(x^{-3/4})^{-4/3}$

★ 45. $\{[(81)(81)^{1/2}]^{1/3}\}^{1/2}$

★ 46. $(x^{1/2} - x^{-1/2})^2$

★ 47. $(5 - x^{1/2})(5 + x^{1/2})$

★ 48. $(x^{1/2} - y^{1/2})(x^{1/2} + y^{1/2})$

★ 49. $x^{-1/2}(x^{-1/2} + x^{1/2})$

★ 50. $x^{1/4}(x^{3/4} + x^{-3/4} + x^{-1/4})$

1.6 RADICALS: SIMPLIFICATION AND PRODUCTS

A radical of the form $\sqrt{M}$, where M is a monomial with an integer for its coefficient, is said to be **simplified if no perfect squares are factors of the radicand** M.

Similarly, a radical of the form $\sqrt[3]{M}$ is said to be **simplified if no perfect cubes are factors of the radicand**.

Renaming a number denoted by a radical so that any resulting radical is simplified is called **reducing the radicand**.

To reduce the radicand of a square root radical, the product of square roots theorem is used.

THE PRODUCT OF SQUARE ROOTS THEOREM

If r and s are nonnegative real numbers, then

$$\sqrt{r}\sqrt{s} = \sqrt{rs} \quad \text{and} \quad \sqrt{rs} = \sqrt{r}\sqrt{s}$$

Since $r \geq 0$ and $s \geq 0$, then by using exponents,

$$\sqrt{r}\sqrt{s} = r^{1/2}s^{1/2} = (rs)^{1/2} = \sqrt{rs}$$

EXAMPLE 1.6.1   Simplify $\sqrt{3}\sqrt{12}$.

Solution $\sqrt{3}\sqrt{12} = \sqrt{3\cdot12} = \sqrt{36} = 6$
The following theorem is useful for simplifying square root radicals:

THEOREM

If x and y are any nonnegative real numbers,
$$\sqrt{x^2y} = \sqrt{x^2}\sqrt{y} = x\sqrt{y}$$

EXAMPLE 1.6.2   Simplify $\sqrt{75}$.

Solution   The basic idea is to factor the radicand to find the perfect square factors:
$$75 = 3\cdot5^2$$
$$\sqrt{75} = \sqrt{5^2\cdot3} = \sqrt{5^2}\sqrt{3} = 5\sqrt{3}$$
Alternate Solution $\sqrt{75} = (5^2\cdot3)^{1/2} = (5^2)^{1/2}(3^{1/2}) = 5\sqrt{3}$

EXAMPLE 1.6.3   Simplify $\sqrt{6x}\sqrt{12xy^3}$ where $x \geq 0$ and $y \geq 0$.

Solution
$$\begin{aligned}
\sqrt{6x}\sqrt{12xy^3} &= \sqrt{6x(12xy^3)} \\
&= \sqrt{(2\cdot3)(2\cdot2\cdot3)x^2y^2y} \\
&= \sqrt{2^23^2x^2y^2}\sqrt{2y} \\
&= \sqrt{(6xy)^2}\sqrt{2y} \\
&= 6xy\sqrt{2y}
\end{aligned}$$

Alternate Solution
$$\begin{aligned}
(6x)^{1/2}(12xy^3)^{1/2} &= (72x^2y^3)^{1/2} \\
&= (2^2\cdot3^2\cdot2x^2y^2y)^{1/2} \\
&= (6^2x^2y^2)^{1/2}(2y)^{1/2} \\
&= 6xy\sqrt{2y}
\end{aligned}$$

Reducing the radicand of a cube root involves the product of cube roots theorem.

THE PRODUCT OF CUBE ROOTS THEOREM

If a and b are any real numbers, then

$$\sqrt[3]{a}\sqrt[3]{b} = \sqrt[3]{ab} \text{ and } \sqrt[3]{ab} = \sqrt[3]{a}\sqrt[3]{b}$$

$$\sqrt[3]{a}\sqrt[3]{b} = a^{1/3}b^{1/3} = (ab)^{1/3} = \sqrt[3]{ab}$$

EXAMPLE 1.6.4 Simplify $\sqrt[3]{20}\sqrt[3]{50}$.

Solution $\sqrt[3]{20}\sqrt[3]{50} = \sqrt[3]{20(50)} = \sqrt[3]{1000} = \sqrt[3]{10^3} = 10$

The following theorem is useful for simplifying cube root radicals:

THEOREM

If x and y are any real numbers, then

$$\sqrt[3]{x^3 y} = \sqrt[3]{x^3}\sqrt[3]{y} = x\sqrt[3]{y}$$

EXAMPLE 1.6.5 Simplify $\sqrt[3]{500}$.

Solution $500 = 2\cdot 2\cdot 125 = 2^2(125) = 2^2 5^3$

$$\sqrt[3]{500} = \sqrt[3]{5^3 2^2} = \sqrt[3]{5^3}\sqrt[3]{2^2}$$
$$= 5\sqrt[3]{4}$$

Alternate Solution $\sqrt[3]{500} = (5^3 \cdot 2^2)^{1/3} = (5^{3\cdot 1/3})(2^{2/3}) = 5\sqrt[3]{4}$

EXAMPLE 1.6.6 Simplify $\sqrt[3]{-81x^6}$.

Solution $\sqrt[3]{-81x^6} = -\sqrt[3]{81x^6}$
$$= -\sqrt[3]{3^4 x^6}$$
$$= -\sqrt[3]{3^3 \cdot 3 \cdot (x^2)^3}$$
$$= -\sqrt[3]{3^3 (x^2)^3}\sqrt[3]{3}$$
$$= -3x^2\sqrt[3]{3}$$

Alternate Solution $\sqrt[3]{-81x^6} = -(3^4 x^6)^{1/3}$
$$= -(3^{4/3})(x^{6/3})$$
$$= -(3^{1+1/3})(x^2)$$
$$= -(3\cdot 3^{1/3})(x^2)$$
$$= -3x^2\sqrt[3]{3}$$

EXERCISES I.6 A

Simplify Exercises 1–40. Assume all variables to be positive.

1. $\sqrt{5}\sqrt{20}$

2. $\sqrt{50}\sqrt{2}$

3. $\sqrt{3x}\sqrt{12x}$

4. $\sqrt{y}\sqrt{y^3}$

5. $\sqrt{12}$

6. $\sqrt{125}$

7. $\sqrt{405}$

8. $\sqrt{9x}$

9. $\sqrt{24x^2}$

10. $\sqrt{40y^3}$

11. $\sqrt{x}\sqrt{x^3}$

12. $\sqrt{y^3}\sqrt{y^4}$

13. $\sqrt{6x}\sqrt{6x^2}$

14. $\sqrt{40x^5}$

15. $\sqrt{32x}$

16. $\sqrt{1452}$

17. $\sqrt{343x^3}$

18. $5\sqrt{72}$

19. $2x\sqrt{54x^4}$

20. $10xy\sqrt{800x^2y}$

21. $(\sqrt{5} + \sqrt{2})(\sqrt{5} - \sqrt{2})$

22. $(4 - \sqrt{3})(4 + \sqrt{3})$

23. $(\sqrt{6} - 2)(\sqrt{6} + 2)$

24. $(1 - \sqrt{2})(1 + \sqrt{2})$

25. $(\sqrt{8} - \sqrt{2})(\sqrt{8} + \sqrt{2})$

26. $\sqrt[3]{54}$

27. $\sqrt[3]{-54}$

28. $\sqrt[3]{72}$

29. $\sqrt[3]{16}$

30. $\sqrt[3]{2}\sqrt[3]{4}$

31. $\sqrt[3]{9}\sqrt[3]{81}$

32. $\sqrt[3]{20x}\sqrt[3]{50x^2}$

33. $\sqrt[3]{32x}$

34. $\sqrt[3]{108y^2}$

35. $\sqrt[3]{40y^4}$

36. $\sqrt[3]{x}\sqrt[3]{x^2}$

37. $\sqrt[3]{4y^2}\sqrt[3]{4y^4}$

38. $2y\sqrt[3]{54y^2}$

39. $\sqrt[3]{6x^2}\sqrt[3]{36x}$

40. $-2\sqrt[3]{-81}$

EXERCISES I.6 B

Simplify Exercises 1–40. Assume all variables to be positive.

1. $\sqrt{2}\sqrt{18}$

2. $\sqrt{27}\sqrt{3}$

3. $\sqrt{5x}\sqrt{45x}$

4. $\sqrt{xy^3}\sqrt{x^3y}$

5. $\sqrt{24}$

6. $\sqrt{128}$

7. $\sqrt{500}$

8. $\sqrt{49y}$

9. $\sqrt{18y^3}$

10. $\sqrt{28y^4}$

11. $\sqrt{x^3}\sqrt{x^5}$ **12.** $\sqrt{2x}\sqrt{8x^3}$

13. $\sqrt{x^3y^5}$ **14.** $\sqrt{54y^7}$

15. $\sqrt{108y}$ **16.** $\sqrt{1849}$

17. $\sqrt{112x^3}$ **18.** $4\sqrt{363}$

19. $3x\sqrt{200x^6}$ **20.** $10xy\sqrt{81xy^2z^3}$

21. $(\sqrt{6} - \sqrt{2})(\sqrt{6} + \sqrt{2})$ **22.** $\dfrac{\sqrt{3} + 1}{2} \cdot \dfrac{\sqrt{3} - 1}{2}$

23. $(\sqrt{x} - \sqrt{3})(\sqrt{x} + \sqrt{3})$ **24.** $(2\sqrt{x} + y)(2\sqrt{x} - y)$

25. $(\sqrt{3x} - \sqrt{2x})(\sqrt{3x} + \sqrt{2x})$ **26.** $\sqrt[3]{128}$

27. $\sqrt[3]{-128}$ **28.** $\sqrt[3]{500}$

29. $\sqrt[3]{625}$ **30.** $\sqrt[3]{5}\sqrt[3]{25}$

31. $\sqrt[3]{12}\sqrt[3]{18}$ **32.** $\sqrt[3]{98x^2}\sqrt[3]{28x}$

33. $\sqrt[3]{24x^2}$ **34.** $\sqrt[3]{256y}$

35. $\sqrt[3]{56y^5}$ **36.** $\sqrt[3]{x^2}\sqrt[3]{x^4}$

37. $\sqrt[3]{25y^3}\sqrt[3]{25y^6}$ **38.** $5y^2\sqrt[3]{80y^4}$

39. $\sqrt[3]{49x}\sqrt[3]{7x^2}$ **40.** $-5\sqrt[3]{-625}$

1.7 RADICALS: RATIONALIZING

In order to rename an expression that has a radical in a denominator or a fraction in a radicand, the quotient of square roots theorem is used for expressions involving square roots. The process is called **rationalizing the denominator**, or **simplification**.

THE QUOTIENT OF SQUARE ROOTS THEOREM

If r and s are positive real numbers, then

$$\sqrt{\frac{r}{s}} = \frac{\sqrt{r}}{\sqrt{s}} \quad \text{and} \quad \frac{\sqrt{r}}{\sqrt{s}} = \sqrt{\frac{r}{s}}$$

Since r and s are positive,

$$\sqrt{\frac{r}{s}} = \left(\frac{r}{s}\right)^{1/2} = \frac{r^{1/2}}{s^{1/2}} = \frac{\sqrt{r}}{\sqrt{s}}$$

EXAMPLE 1.7.1   Simplify $\sqrt{\tfrac{2}{3}}$.

Solution $\sqrt{\dfrac{2}{3}} = \dfrac{\sqrt{2}}{\sqrt{3}}$ (By the quotient of square roots theorem)

$\qquad\qquad = \dfrac{\sqrt{2}\sqrt{3}}{\sqrt{3}\sqrt{3}}$ (By the Fundamental Theorem of Fractions—the numerator and denominator are multiplied by that number that causes the radicand of the denominator to become a perfect square)

Thus $\sqrt{\dfrac{2}{3}} = \dfrac{\sqrt{6}}{3}$

EXAMPLE 1.7.2   Rationalize $\dfrac{2}{\sqrt{27}}$.

Solution $\dfrac{2}{\sqrt{27}} = \dfrac{2}{3\sqrt{3}}$ (Reducing the radicand)

$\qquad\qquad = \dfrac{2\sqrt{3}}{3\sqrt{3}\sqrt{3}}$

$\qquad\qquad = \dfrac{2\sqrt{3}}{9}$

If the denominator is a sum or difference of terms involving a radical, then the difference of squares theorem provides a technique for rationalizing denominators involving a square root. In other words, a fraction having an irrational denominator may be renamed as a fraction with a rational denominator. The basic idea involved is:

$$X^2 - Y^2 = (X + Y)(X - Y)$$
$$a - b = (\sqrt{a} + \sqrt{b})(\sqrt{a} - \sqrt{b})$$

If the denominator has the form $\sqrt{a} + \sqrt{b}$, then multiplication by $\sqrt{a} - \sqrt{b}$ (called its *conjugate*) will produce the rational number, $a - b$. This assumes, of course, that a and b are positive rational numbers.

Similarly, multiplying $\sqrt{a} - \sqrt{b}$ by its conjugate, $\sqrt{a} + \sqrt{b}$, will produce the rational number $a - b$.

EXAMPLE I.7.3　Rationalize $\dfrac{1}{\sqrt{5} - 1}$.

Solution　$\dfrac{1}{\sqrt{5} - 1} \cdot \dfrac{\sqrt{5} + 1}{\sqrt{5} + 1} = \dfrac{\sqrt{5} + 1}{5 - 1} = \dfrac{\sqrt{5} + 1}{4}$

EXAMPLE I.7.4　Simplify $\dfrac{2}{6 + \sqrt{2}}$.

Solution　$\dfrac{2}{6 + \sqrt{2}} \cdot \dfrac{6 - \sqrt{2}}{6 - \sqrt{2}} = \dfrac{2(6 - \sqrt{2})}{36 - 2} = \dfrac{2(6 - \sqrt{2})}{34} = \dfrac{6 - \sqrt{2}}{17}$

EXAMPLE I.7.5　Express with a rational denominator $\dfrac{3}{\sqrt{5} - \sqrt{2}}$.

Solution　$\dfrac{3}{\sqrt{5} - \sqrt{2}} \cdot \dfrac{\sqrt{5} + \sqrt{2}}{\sqrt{5} + \sqrt{2}} = \dfrac{3(\sqrt{5} + \sqrt{2})}{5 - 2} = \dfrac{3(\sqrt{5} + \sqrt{2})}{3}$

$$= \sqrt{5} + \sqrt{2}$$

EXERCISES I.7 A

In Exercises 1–25, rationalize the denominator and write in simplest radical form. Assume all variables to be positive.

1. $\sqrt{\frac{1}{2}}$

2. $\sqrt{\frac{1}{6}}$

3. $\sqrt{\frac{5}{12}}$

4. $\dfrac{1}{\sqrt{5}}$

5. $\dfrac{1}{\sqrt{20}}$

6. $\sqrt{\frac{2}{25}}$

7. $\sqrt{\frac{3}{50}}$

8. $\dfrac{8}{\sqrt{12}}$

9. $\dfrac{14\sqrt{9}}{3\sqrt{7}}$

10. $\dfrac{\sqrt{64x^4y^6}}{\sqrt{128x^6y^6}}$

11. $\sqrt{\dfrac{2}{x}}$

12. $\sqrt{\dfrac{8}{y}}$

13. $\dfrac{6}{\sqrt{72x^3}}$

14. $3\sqrt{\frac{1}{3}} \cdot 5\sqrt{\frac{3}{5}}$

15. $\dfrac{\sqrt{3}}{\sqrt{6x}}$

16. $\dfrac{1}{\sqrt{2} - 1}$

17. $\dfrac{4}{\sqrt{7} + \sqrt{3}}$

18. $\dfrac{\sqrt{5}}{2 - \sqrt{5}}$

19. $\dfrac{5}{\sqrt{11} + 1}$

20. $\dfrac{2}{\sqrt{x} - 1}$

21. $\dfrac{\sqrt{6}}{\sqrt{3} + \sqrt{27}}$

22. $\dfrac{\sqrt{25}}{\sqrt{16} + \sqrt{9}}$

23. $\dfrac{\sqrt{25}}{\sqrt{16} + 9}$

★ 24. $\dfrac{x - y}{\sqrt{x} - \sqrt{y}}$

★ 25. $\dfrac{x - 4y}{\sqrt{x} + \sqrt{2y}}$

EXERCISES I.7 B

In Exercise 1–25, rationalize the denominator and write in simplest radical form. Assume all variables to be positive.

1. $\sqrt{\frac{1}{3}}$

2. $\sqrt{\frac{3}{5}}$

3. $\sqrt{\frac{3}{8}}$

4. $\dfrac{1}{\sqrt{10}}$

5. $\dfrac{1}{\sqrt{18}}$

6. $\sqrt{\frac{3}{12}}$

7. $\sqrt{\frac{3}{32}}$

8. $\dfrac{9}{\sqrt{45}}$

9. $\dfrac{4}{2\sqrt{200}}$

10. $\dfrac{21\sqrt{8y^3}}{\sqrt{49y}}$

11. $\sqrt{\dfrac{1}{3x}}$

12. $\sqrt{\dfrac{125}{y^3}}$

13. $\dfrac{4}{\sqrt{32x}}$

14. $\sqrt{\dfrac{7}{8x}}\sqrt{\dfrac{2x^3}{49}}$

15. $\dfrac{\sqrt{2x}}{\sqrt{10x}}$

16. $\dfrac{1}{1 + \sqrt{3}}$ $(1 - \sqrt{3})$

17. $\dfrac{\sqrt{3}}{\sqrt{7} - \sqrt{3}}$

18. $\dfrac{4}{\sqrt{7} + \sqrt{5}}$

19. $\dfrac{4}{1 - \sqrt{x}}$

20. $\dfrac{12}{\sqrt{10} - 4}$

21. $\dfrac{x - 4y}{\sqrt{x} + 2\sqrt{y}}$

22. $\dfrac{\sqrt{10}}{\sqrt{5} + \sqrt{45}}$

23. $\dfrac{\sqrt{4} - \sqrt{9}}{\sqrt{2} - \sqrt{3}}$

★ 24. $\dfrac{1}{\sqrt{4x} - \sqrt{x}}$

★ 25. $\dfrac{9x^2 - 25y^2}{\sqrt{3x} - \sqrt{5y}}$

1.8 SUMS AND DIFFERENCES OF RADICALS

It is often possible to express a sum or difference of two radical terms as a single term. For example, since $\sqrt{5}$ is a real number, so are $3\sqrt{5}$ and $4\sqrt{5}$ and $3\sqrt{5} + 4\sqrt{5}$ by the closure axioms for real numbers. Now using the distributive axiom,

$$3\sqrt{5} + 4\sqrt{5} = (3 + 4)\sqrt{5} = 7\sqrt{5}$$

Radicals that have the same radicands *and* the same indices are called **like radicals**. Radicals that are not like are called unlike. For example,

$$3\sqrt{5x} \text{ and } 4\sqrt{5x} \text{ contain like radicals}$$
$$\sqrt{5x} \text{ and } \sqrt[3]{5x} \text{ are unlike radicals}$$
$$\sqrt{5x} \text{ and } \sqrt{3x} \text{ are unlike radicals}$$

Sometimes it is necessary to reduce each radical term to simplified radical form in order to identify like radicals.

EXAMPLE 1.8.1 Simplify $\sqrt{50x} + \sqrt{18x}$ if possible.

Solution $\sqrt{50x} + \sqrt{18x} = 5\sqrt{2x} + 3\sqrt{2x}$
(Simplifying each radical)
$$= (5 + 3)\sqrt{2x}$$
(Using the distributive axiom)
$$= 8\sqrt{2x}$$

EXAMPLE 1.8.2   Simplify $\sqrt{48} + \sqrt{18} - \sqrt{12}$.

Solution $\sqrt{48} + \sqrt{18} - \sqrt{12} = 4\sqrt{3} + 3\sqrt{2} - 2\sqrt{3}$
$$\text{(Simplifying)}$$
$$= (4\sqrt{3} - 2\sqrt{3}) + 3\sqrt{2}$$
$$\text{(Collecting like radicals)}$$
$$= (4 - 2)\sqrt{3} + 3\sqrt{2}$$
$$\text{(Using distributive axiom)}$$
$$= 2\sqrt{3} + 3\sqrt{2}$$
$$\text{(Simplified form)}$$

EXAMPLE 1.8.3   Simplify $\sqrt{40} - \sqrt{\frac{2}{5}}$.

Solution $\sqrt{40} - \sqrt{\frac{2}{5}} = \sqrt{4 \cdot 10} - \sqrt{\frac{2}{5} \cdot \frac{5}{5}}$
$$= 2\sqrt{10} - \frac{\sqrt{10}}{5}$$
$$= (2 - \tfrac{1}{5})\sqrt{10}$$
$$= \frac{9\sqrt{10}}{5}$$

EXERCISES 1.8 A

Write Exercises 1–28 in simplest radical form. Assume all variables to be positive.

1. $6\sqrt{5} + 2\sqrt{5}$ **2.** $7\sqrt{2} - 4\sqrt{2}$

3. $5\sqrt{6} + \sqrt{6}$ **4.** $9\sqrt{10} - 8\sqrt{10}$

5. $3\sqrt{5} + 5\sqrt{3}$ **6.** $5\sqrt{3} - 4\sqrt{3} + 2\sqrt{3}$

7. $6\sqrt{14x} - 4\sqrt{14x}$ **8.** $2\sqrt{2x} + 3\sqrt{3x}$

9. $\sqrt{27} + \sqrt{3}$ **10.** $\sqrt{40} - \sqrt{10}$

11. $3\sqrt{24} + 2\sqrt{54}$ **12.** $\sqrt{32x} + \sqrt{98x}$

13. $\sqrt{52x} - \sqrt{13x}$ **14.** $\sqrt{63x^2} + \sqrt{28x^2}$

15. $4\sqrt{9x^2} - 3\sqrt{4x^2}$ **16.** $\sqrt{4y^2} + 4\sqrt{y^2}$

17. $\sqrt{\frac{2}{5}} + \sqrt{\frac{1}{10}}$ **18.** $\sqrt{18} - \sqrt{\frac{1}{18}}$

19. $\sqrt{20} + \sqrt{12} + \sqrt{45}$

20. $\sqrt{16x} - \sqrt{4y} - \sqrt{9y}$

21. $\sqrt{\frac{42}{25}} + \sqrt{2\frac{5}{8}}$

22. $\sqrt{3x} - \sqrt{18x} + \sqrt{12x}$

23. $\sqrt{224} + 4\sqrt{\frac{1}{2}} - \frac{1}{2}\sqrt{50}$

24. $(\sqrt{3} + \sqrt{2})^2$

25. $(2\sqrt{3} - 3\sqrt{2})^2$

26. $(\sqrt{6x} + \sqrt{3x})^2$

27. $\sqrt{3}(\sqrt{6} + \sqrt{12} + \sqrt{24})$

28. $\sqrt{5}(\sqrt{10} - \sqrt{15} - \sqrt{40})$

29. Find the value of $x^2 - 10x + 23$ for $x = 5 + \sqrt{2}$.

30. Find the value of $x^2 + x - 1$ for $x = \sqrt{5} - 1$.

EXERCISES 1.8 B

Write Exercises 1–28 in simplest radical form. Assume all variables to be positive.

1. $7\sqrt{3} - 5\sqrt{3}$

2. $5\sqrt{7} + 4\sqrt{7}$

3. $8\sqrt{6} - 7\sqrt{6}$

4. $2\sqrt{15} + \sqrt{15}$

5. $2\sqrt{30} + 5\sqrt{30} - 4\sqrt{30}$

6. $7\sqrt{5} - 5\sqrt{7}$

7. $8\sqrt{3x} - 5\sqrt{3y}$

8. $6\sqrt{7x} + 2\sqrt{7x}$

9. $\sqrt{20} + \sqrt{5}$

10. $\sqrt{72} - \sqrt{50}$

11. $5\sqrt{18} - 4\sqrt{8}$

12. $\sqrt{75x} + \sqrt{108x}$

13. $\sqrt{20x} - \sqrt{45x}$

14. $\sqrt{150x^2} - \sqrt{24x^2}$

15. $3\sqrt{25x^2} - 5\sqrt{9x^2}$

16. $\sqrt{9y^2} + 9\sqrt{y^2}$

17. $\sqrt{24} + \sqrt{\frac{3}{8}}$

18. $\sqrt{\frac{2}{5}} - \sqrt{\frac{5}{2}}$

19. $\sqrt{54} + \sqrt{32} - \sqrt{24}$

20. $\sqrt{25x} - \sqrt{36y} - \sqrt{x}$

21. $\sqrt{\frac{7}{8}} + \sqrt{\frac{5}{6}} - \sqrt{\frac{7}{2}}$

22. $\sqrt{\frac{2}{x}} + \sqrt{\frac{1}{2x}} + \sqrt{2x}$

23. $\dfrac{2}{\sqrt{12}} + 2\sqrt{81} + \dfrac{\sqrt{27}}{\sqrt{3}}$

24. $(\sqrt{5} - \sqrt{3})^2$

25. $(2\sqrt{5} + 1)^2$

26. $(\sqrt{10x} - \sqrt{5x})^2$

27. $\sqrt{2}(\sqrt{8} - \sqrt{18} - \sqrt{12})$

28. $\sqrt{6}(\sqrt{12} + \sqrt{3} - \sqrt{2})$

29. Find the value of $x^2 - x + 2$ for $x = \sqrt{2} + \sqrt{3}$.

30. Find the value of $x^2 + 2x - 2$ for $x = \sqrt{3} - 1$.

1.9 FURTHER SIMPLIFICATION OF RADICALS (Optional)

If a radical has the form $\sqrt[mn]{x^m}$, where m and n are natural numbers, then by changing to exponential notation,

$$\sqrt[mn]{x^m} = (x^m)^{1/mn} = x^{1/n} = \sqrt[n]{x}$$

This process is called **lowering the index**, and $\sqrt[n]{x}$ is the simplified form of $\sqrt[mn]{x^m}$.

EXAMPLE 1.9.1 Simplify $\sqrt[6]{8x^3}$ where $x \geq 0$.

Solution $\sqrt[6]{8x^3} = (2x)^{3(1/6)} = (2x)^{1/2} = \sqrt{2x}$

EXAMPLE 1.9.2 Simplify $\sqrt[6]{25y^4}$.

Solution $\sqrt[6]{25y^4} = \sqrt[6]{(5y^2)^2} = (5y^2)^{2(1/6)} = (5y^2)^{1/3} = \sqrt[3]{5y^2}$

EXAMPLE 1.9.3 Simplify $\sqrt[4]{36}$.

Solution $\sqrt[4]{36} = (6^2)^{1/4} = 6^{1/2} = \sqrt{6}$

A product or quotient of radicals is considered simplified when it is written as an expression contining at most one radical in the numerator. Certain products and quotients of radicals are most easily simplified by rewriting the expression in exponential form and by applying the theorems on exponents.

EXAMPLE 1.9.4 Simplify $\sqrt{2} \cdot \sqrt[3]{2}$.

Solution $\sqrt{2} \cdot \sqrt[3]{2} = 2^{1/2}2^{1/3} = 2^{1/2 + 1/3} = 2^{5/6} = \sqrt[6]{2^5} = \sqrt[6]{32}$

EXAMPLE 1.9.5 Simplify $\dfrac{\sqrt[3]{16}}{\sqrt[6]{4}}$.

Solution $\dfrac{\sqrt[3]{16}}{\sqrt[6]{4}} = \dfrac{(2^4)^{1/3}}{(2^2)^{1/6}} = \dfrac{2^{4/3}}{2^{1/3}} = 2^{4/3 - 1/3} = 2^1 = 2$

EXAMPLE 1.9.6 Simplify $\dfrac{\sqrt[12]{3}}{\sqrt[4]{27}}$.

Solution $\dfrac{\sqrt[12]{3}}{\sqrt[4]{27}} = \dfrac{3^{1/12}}{(3^3)^{1/4}} = \dfrac{3^{1/12}}{3^{3/4}} \cdot \dfrac{3^{1/4}}{3^{1/4}}$

$$= \dfrac{3^{1/12 + 3/12}}{3}$$

$$= \dfrac{3^{4/12}}{3} = \dfrac{\sqrt[3]{3}}{3}$$

EXAMPLE 1.9.7 Simplify $\sqrt[3]{7\sqrt{7}}$.

Solution $\sqrt[3]{7\sqrt{7}} = (7 \cdot 7^{1/2})^{1/3}$

$$= (7^{3/2})^{1/3}$$

$$= 7^{1/2} = \sqrt{7}$$

EXERCISES 1.9 A

Simplify Exercises 1–30. Assume all variables to be nonnegative.

1. $\sqrt[6]{125}$

2. $\sqrt[6]{100}$

3. $\sqrt[4]{49}$

4. $\sqrt[4]{121x^2}$

5. $\sqrt[8]{81t^4}$

6. $\sqrt[6]{16t^4}$

7. $\sqrt[4]{400}$

8. $\sqrt[4]{324a^2}$

9. $\sqrt{\sqrt[3]{4c^2}}$

10. $\sqrt[3]{25\sqrt{5}}$

11. $\sqrt{6}\sqrt[4]{6}$

12. $\sqrt[3]{4y^2}\sqrt[6]{4y^2}$

13. $\sqrt{5}\sqrt[6]{5}$

14. $\sqrt[12]{10}\sqrt[4]{10}$

15. $\dfrac{\sqrt[3]{81}}{\sqrt[6]{9}}$

16. $\dfrac{\sqrt{7}}{\sqrt[4]{7}}$

17. $\dfrac{\sqrt[4]{8}}{\sqrt{8}}$

18. $\dfrac{\sqrt[6]{81x^4}}{\sqrt[3]{3x}}$

19. $\dfrac{\sqrt{10}}{\sqrt[3]{5}}$

20. $\sqrt[3]{5\sqrt{5}}$

21. $\sqrt{2\sqrt[3]{2}}$

22. $\sqrt{6\sqrt[3]{6}\sqrt[6]{6}}$

23. $\sqrt[3]{9}\sqrt[4]{3}\sqrt[12]{3}$

24. $\sqrt{\sqrt[3]{\sqrt[4]{2^8}}}$

25. $\sqrt{81\sqrt[3]{81\sqrt{81}}}$

26. $6\sqrt{\frac{1}{3}} + \sqrt{18} - \sqrt[4]{9}$

27. $\sqrt[6]{8x^3} + \sqrt[4]{400x^2} + \sqrt{50x}$

28. $\dfrac{\sqrt[4]{7}\sqrt[12]{7}}{\sqrt[3]{49}}$

29. $\sqrt[6]{10y}(\sqrt{10y} + \sqrt[3]{10y})$

30. $\sqrt[a]{\sqrt[b]{\sqrt[c]{x^{abc}}}}$

EXERCISES 1.9 B

Simplify Exercises 1–30. Assume all variables to be nonnegative.

1. $\sqrt[6]{49}$

2. $\sqrt[6]{1000}$

3. $\sqrt[4]{9x^2}$

4. $\sqrt[6]{27y^3}$

5. $\sqrt[9]{216x^3}$

6. $\sqrt[4]{900y^6}$

7. $\sqrt[4]{256t^2}$

8. $\sqrt[6]{512y^2}$

9. $\sqrt[8]{64r^6}$

10. $\sqrt[3]{\sqrt[4]{27n^3}}$

11. $\sqrt[4]{3}\,\sqrt[12]{3}$

12. $\sqrt{8x^3}\sqrt[3]{4x^2}$

13. $\sqrt{7\sqrt[3]{49}}$

14. $\sqrt[6]{25}\sqrt[3]{25}$

15. $\dfrac{\sqrt[3]{5}}{\sqrt[6]{5}}$

16. $\dfrac{\sqrt[6]{100x^2}}{\sqrt[6]{10x}}$

17. $\dfrac{\sqrt[12]{6}}{\sqrt[4]{216}}$

18. $\dfrac{\sqrt{15}}{\sqrt[3]{15}}$

19. $\dfrac{\sqrt{10}}{\sqrt[3]{2}\sqrt[6]{2}}$

20. $\sqrt[3]{6x\sqrt{6x}}$

21. $\sqrt{2y\sqrt[3]{2y}}$

22. $\sqrt{2}\sqrt[3]{4}\sqrt[6]{32}$

23. $\sqrt[6]{x}\,\sqrt[10]{x}\,\sqrt[15]{x}$

24. $\sqrt{\sqrt[3]{\sqrt[4]{5^{12}}}}$

25. $\sqrt[3]{25\sqrt{25\sqrt[3]{25}}}$

26. $\sqrt[6]{x^3} + \sqrt[4]{16x^2} - \sqrt{9x}$

27. $\dfrac{6}{\sqrt{12}} + \sqrt[4]{729} + \sqrt[6]{27}$

28. $\dfrac{\sqrt[3]{4}\sqrt[6]{4}}{\sqrt[4]{4}}$

29. $\sqrt[3]{4}(\sqrt[6]{4} - \sqrt[12]{16})$

30. $\sqrt{\sqrt[3]{\sqrt[4]{x^{60}}}}$

1.10 SCIENTIFIC NOTATION

Many scientific measurements require the use of extremely large or extremely small numbers. For example, the speed of light is about 30,000,000,000 centimers per second; the number of molecules in 1 cubic centimer of gas at 0 degrees centigrade is about 30,000,000,000,000,000,000,000; the time for an electronic computer to do a certain arithmetical operation is 0.000 000 0024 second; and the mass of the hydrogen atom is about 0.000 000 000 000 000 000 000 001 672 gram.

A very convenient and effective system for expressing such numbers is called scientific notation.

A number is expressed in scientific notation when it is written as a product of a decimal fraction between 1 and 10 and an integral power of 10. In symbols, a number written in scientific notation has the form

$$N \times 10^k$$

where N is a number between 1 and 10 in decimal form and k is an integer.

EXAMPLE 1.10.1

Ordinary Notation	Scientific Notation
3.14	3.14×10^0
20.5	2.05×10^1
608	6.08×10^2
5,000,000	5×10^6
0.14	1.4×10^{-1}
0.025	2.5×10^{-2}
0.000 000 167	1.67×10^{-7}

It may be noted that the exponent on 10 for the scientific notation indicates the number of places to move the decimal point to obtain the ordinary notation (to the right if the exponent is positive, to the left if the exponent is negative and no change if the exponent is 0).

EXAMPLE 1.10.2 Express 30,000,000,000 centimers per second, the speed of light, in scientific notation.

Solution 3.00×10^{10}

EXAMPLE 1.10.3 Express 3.1×10^{-5} inches, the diameter of an average red blood corpuscle, in ordinary notation.

Solution 0.000 031

EXAMPLE 1.10.4 Calculate and express the answer in ordinary notation:

$$\frac{(2.5 \times 10^{-3})(4.2 \times 10^5)}{1.4 \times 10^{-1}}$$

Solution First multiply each number in scientific notation by $10^k \times 10^{-k}$ to eliminate the decimal point:

$$\frac{(2.5 \times 10^{-3})(4.2 \times 10^5)}{1.4 \times 10^{-1}} = \frac{(25 \times 10^{-4})(42 \times 10^4)}{14 \times 10^{-2}}$$

$$= \frac{(25)(42)}{14} \cdot \frac{(10^{-4})(10^4)}{10^{-2}}$$

$$= (25)(3) \times 10^{-4+4-(-2)}$$

$$= 75 \times 10^2$$

$$= 7500$$

EXAMPLE 1.10.5 Evaluate

$$\sqrt{(3.0 \times 10^{-3})^2 - 4(4.8 \times 10^{-4})(0.3 \times 10^{-2})}$$

Solution $\sqrt{(3.0 \times 10^{-3})^2 - 4(4.8 \times 10^{-4})(0.3 \times 10^{-2})}$

$$= \sqrt{(9 \times 10^{-6}) - (5.76)(10^{-6})}$$

$$= \sqrt{(9.00 - 5.76)10^{-6}}$$

$$= \sqrt{3.24 \times 10^{-6}}$$

$$= \sqrt{(324)(10^{-8})}$$

$$= 18 \times 10^{-4}$$

$$= 0.0018$$

EXERCISES 1.10 A

In Exercises 1–10, write each number in scientific notation.

1. 92,900,000 miles; distance from the earth to the sun
2. 6,600,000,000,000,000,000,000 tons; weight of the earth
3. 11,400,000; population of Tokyo
4. 0.000 000 095 centimeter; wave length of certain X rays
5. 0.000 000 0024 second; time for an electronic computer to do an addition

6. 2,210,000,000; heartbeats per normal lifetime
7. 0.00061 atmosphere; a gas pressure
8. 120,000; seating capacity of a football stadium
9. 0.002 205 pound; weight of one gram
10. 0.000 000 0667; constant of gravitation

In Exercises 11–20, write each number in ordinary notation.

11. 2.3×10^3; pounds of pollution per car per year
12. -4.60×10^2 degrees Fahrenheit; absolute zero
13. 1.80×10^{-5}; ionization constant of acetic acid
14. 8.64×10^5 miles; diameter of the sun
15. 3.03×10^{-8} centimeters; grating space in calcite crystals
16. 1.745×10^{-2}; number of radians in 1 degree
17. 1.6667×10^{-1} inches; width of an em space (printing industry)
18. 4.80×10^{-10} absolute electrostatic units; electronic charge
19. 1.87×10^9 dollars; a congressional appropriation
20. 6.3×10^{18} electrons per second; for one ampere of current

Calculate Exercises 21–25, using scientific notation. Express the answer in ordinary notation.

21. $(5.4 \times 10^{-3})(2.0 \times 10^5)$

22. $\sqrt{(1.25 \times 10^{-2})(8.0 \times 10^{-3})}$

23. $\dfrac{6.9 \times 10^{-8}}{2.3 \times 10^{-6}}$

24. $\dfrac{(3.75 \times 10^{-6})(2.00 \times 10^9)}{2.5 \times 10^{-2}}$

25. $\dfrac{(1.2 \times 10^{-2})^3}{3.6 \times 10^{-5}}$

26. Find the number of radians in 60 degrees if 1 degree $= 1.745 \times 10^{-2}$ radians.
27. Find the number of em spaces across the width of a page 7 inches wide if 1 em space $= 1.6667 \times 10^{-1}$ inches.
28. How long does it take a spaceship traveling 2.8×10^4 mph to travel the 2.48×10^5 miles from the earth to the moon?
29. (Nuclear physics) Find the force F with which a helium nucleus and a neon nucleus repel each other when separated by a distance of 4×10^{-9} meters using

$$F = \frac{khn}{d^2}$$

where
$$k = 9 \times 10^9$$
$$h = 3.2 \times 10^{-19}$$
$$n = 1.6 \times 10^{-18}$$
$$d = 4 \times 10^{-9}$$

30. (Chemistry) Use $[H^+] = \dfrac{K}{ac}$ to calculate the hydrogen ion concentration

$[H^+]$ of a certain solution, given that

$$K = 1.0 \times 10^{-14}$$
$$a = 7.5 \times 10^{-5}$$
$$c = 0.1$$

EXERCISES 1.10 B

In Exercises 1–10, write each number in scientific notation.

1. 2,000,000,000 light-years; probable diameter of the universe
2. 5,870,000,000,000 miles; the distance light travels in a year, called a light-year
3. 300,000,000,000 dollars; national debt
4. 0.000 000 015 centimeter; radius of an atom
5. 0.000 011 feet per 0 degrees centigrade; expansion of steel pipe
6. 0.000 005 centimeter; size of a certain virus
7. 0.03 millimeter per second; rate of certain plant growth
8. 0.001 5625 square mile; area of one acre
9. 603,000,000,000,000,000,000,000; Avogadro's number
10. 3,500,000,000; approximate world population

In Exercises 11–20, write each number in ordinary notation.

11. 4.9×10^{10} dollars; federal investment in public water supplies
12. 9.11×10^{-28} grams; mass of an electron
13. 8.31×10^7 ergs per degree-mole; molar gas constant
14. 3.4×10^4 centimeters per second; velocity of sound
15. 1×10^{-8} centimeters; equals one angstrom, unit used to measure wavelengths
16. 3.937×10^{-1}; number of inches in 1 centimeter
17. 2.5×10^{-2} seconds; shutter speed of a motion picture camera

18. 7.1×10^8 years; half-life of uranium 235

19. 5×10^{-7} centimeters; thickness of an oil film

20. 1.256×10^8 cubic yards; of earth in Fort Peck Dam, largest dam in the world

Calculate Exercises 21–25, using scientific notation. Express the answer in ordinary notation.

21. $(3.6 \times 10^{-5})(1.1 \times 10^4)$ **22.** $(4.3 \times 10^{-1})(4.7 \times 10^{-3})$

23. $\sqrt[3]{\dfrac{1.5 \times 10^{-12}}{1.2 \times 10^{-16}}}$ **24.** $\dfrac{(2.4 \times 10^{-5})(1.5 \times 10^4)}{1.8 \times 10^3}$

25. $\dfrac{(1.4 \times 10^{-3})^2}{2.8 \times 10^{-4}}$

26. Find the number of inches in 100 centimeters if 1 centimeter $= 3.937 \times 10^{-1}$ inches.

27. Find how many seconds it takes a sound to travel 1.70×10^6 centimeters if sound travels 3.4×10^4 centimeters per second.

28. Find I (electric current flowing in diode) if

$$I = KE^{3/2} \text{ amperes}$$

where

$$K = 16 \times 10^{-6}$$
$$E = 225 \text{ volts}$$

29. (Chemistry) Find the pH of a solution if the electric potential E is measured as 4.57×10^{-1} volts where

$$\text{pH} = \frac{E - (2.80 \times 10^{-1})}{5.9 \times 10^{-2}}$$

30. (Television: vacuum tubes) Find f if

$$f = \frac{1}{RC + rc}$$

where

$$R = 10^4$$
$$C = 0.05 \times 10^{-6}$$
$$r = 5 \times 10^4$$
$$c = 0.03 \times 10^{-6}$$

SUMMARY

☐ **The axiom of completeness.** Each point on the number line corresponds to exactly one real number, and each real number corresponds to exactly one point on the number line.

☐ **The theorem of Pythagoras.** The square of the length of the hypotenuse of a right triangle is equal to the sum of the squares of the lengths of the legs of the right triangle.

Radicals

☐ **Definition of radicals.** If x is a real number and n is an odd natural number, then $\sqrt[n]{x^n} = x$. If x is a nonnegative real number and n is an even natural number, then $\sqrt[n]{x^n} = x$.

☐ **The product of square roots theorem.** If r and s are nonnegative real numbers, then $\sqrt{rs} = \sqrt{r}\sqrt{s}$ and $\sqrt{r}\sqrt{s} = \sqrt{rs}$.

☐ **The product of cube roots theorem.** If r and s are any real numbers, then $\sqrt[3]{rs} = \sqrt[3]{r}\sqrt[3]{s}$ and $\sqrt[3]{r}\sqrt[3]{s} = \sqrt[3]{rs}$.

☐ **The quotient of square roots theorem.** If r and s are positive real numbers, then

$$\frac{\sqrt{r}}{\sqrt{s}} = \sqrt{\frac{r}{s}} \quad \text{and} \quad \sqrt{\frac{r}{s}} = \frac{\sqrt{r}}{\sqrt{s}}$$

Exponents

Note: x and y are real numbers; a and b are rational numbers; n is a natural number.)

☐ **Definition.** $x^n = \underbrace{x \cdot x \cdots x}_{n \text{ factors}}$ and $x^1 = x$

☐ **Definition.** If $x \neq 0$, then $x^0 = 1$.

☐ **Definition.** If $x \neq 0$, then $x^{-a} = 1/x^a$.

☐ **Definition.** If $x \geq 0$, then $x^{1/n} = \sqrt[n]{x}$; if $x < 0$ and n is odd, $x^{1/n} = \sqrt[n]{x}$.

☐ **Definition.** If $x \geq 0$, then $x^{m/n} = \sqrt[n]{x^m} = (\sqrt[n]{x})^m$.

☐ **Definition.** If $x > 0$, then $x^{-m/n} = 1/\sqrt[n]{x^m}$.

Theorems

☐ **Theorem 1.** $x^a x^b = x^{a+b}$

☐ **Theorem 2.** $(x^a)^b = x^{ab}$

☐ **Theorem 3.** $\dfrac{x^a}{x^b} = x^{a-b}$ if $x \neq 0$.

☐ **Theorem 4.** $(xy)^a = x^a y^a$

☐ **Theorem 5.** $(x/y)^a = x^a/y^a$ if $y \neq 0$.

REVIEW EXERCISES

1. If a and b are the legs of a right triangle and c is the hypotenuse, find the missing side:

    a. $a = \sqrt{7}, b = \sqrt{5}$ b. $a = 24, c = 26$

2. Approximate the following to the nearest hundredth:

    a. $\sqrt{73}$ b. $\sqrt{5 \times 10^{-4}}$ c. $\sqrt[3]{25}$ d. $\sqrt[3]{4 \times 10^{-5}}$

3. Simplify:

    a. $\sqrt[3]{-81}$ b. $\sqrt{(-7)^2}$ c. $\sqrt[3]{\frac{8}{27}}$ d. $\sqrt[7]{(0.0013)^7}$

4. Write in simplest radical form:

    a. $\sqrt{192}$ b. $2\sqrt{45}$ c. $\sqrt[3]{144}$ d. $\sqrt{60} + \sqrt{1500}$

5. Rationalize the denominator and write in simplest radical form:

    a. $\dfrac{3}{\sqrt{5}}$ b. $\dfrac{\sqrt{3}}{\sqrt{5}}$ c. $\dfrac{24\sqrt{60}}{8\sqrt{5}}$ d. $\dfrac{\sqrt{24} - \sqrt{75}}{\sqrt{3}}$ e. $\dfrac{\sqrt{3}}{1 - \sqrt{2}}$

6. Simplify:

    a. $\sqrt{10}\sqrt{15}$ b. $\sqrt{5}(\sqrt{10} - \sqrt{5} - \sqrt{45})$

    c. $\sqrt{252} + 14\sqrt{\frac{1}{7}} - \frac{1}{2}\sqrt{72}$ d. $(2\sqrt{8} + 8\sqrt{2})^2$

    e. $\sqrt{6} - \frac{1}{2}\sqrt{54}$ f. $\dfrac{2}{\sqrt{120}} + \sqrt{27} + \sqrt{30}$

7. Simplify:

    a. $(a^2 b^3)(a^3 b^5)^2$ b. $\dfrac{a^3 b^4}{5m^2 n^2} \cdot \dfrac{-36a^5 b^3}{10m^6 n}$

    c. $(\frac{1}{3}a^2 b)(\frac{1}{4}a^2 b)^3$ d. $-(y)^2(-y)^2$

8. Simplify by writing equivalent statements containing only positive exponents (assume all variables to be $\neq 0$):

   a. m^{-4} b. $\dfrac{1}{p^{-2}}$ c. $\dfrac{x^0 y^{-1} z^2}{x^{-2} y^0 z}$ d. $\dfrac{a^{-2} - b^{-2}}{a^{-1} - b^{-1}}$

9. Write each of the following in radical form:

   a. $x^{1/4}$ b. $2a^{1/2}$ c. $x^{3/4}$ d. $(xy)^{-2/3}$

10. Write each of the following in exponential form:

   a. $\sqrt[3]{y}$ b. $\sqrt[6]{x^3 y^5}$ c. $x\sqrt{y}$ d. $\sqrt[3]{(xy)^2}$

11. Simplify:

   a. $2(4)^{-1/2}$ b. $2^3 \cdot 2^{-1} + 2^0$

   c. $16^{-3/4}$ d. $5^0 - 36^{-1/2} + 32^{2/5}$

12. Find the value of $x^2 - 8x + 10$ for:

   a. $x = 4 + \sqrt{6}$ b. $x = 4 - \sqrt{6}$ c. $x = 4 + 2\sqrt{6}$

13. Find the value of $\sqrt{b^2 - 4ac}$ for each of the following:

   a. $a = 3, b = -5, c = -2$ b. $a = 1, b = -7, c = 1$

   c. $a = 4, b = 4, c = -5$

14. Find the value of $100(1 - R^{-2/5})$ for

   a. $R = \frac{3125}{243}$ b. $R = 32 \times 10^5$

15. Evaluate, using scientific notation; express the answer in ordinary notation:

   a. $\sqrt{(2.4 \times 10^9)(1.5 \times 10^{-4})^3}$ b. $\dfrac{(1.60 \times 10^{-5})(2.50 \times 10^{-3})^2}{1.25 \times 10^{-15}}$

16. Simplify (assume $x \geq 0$ and $y > 0$):

   a. $\sqrt[4]{64x^6} + 3x\sqrt[6]{8x^3}$ b. $\sqrt[6]{125} + \sqrt[3]{5\sqrt{5}}$

   c. $\dfrac{\sqrt[6]{15xy}\sqrt[3]{15xy}}{\sqrt[4]{25y^2}}$

CHAPTER **2**

QUADRATIC EQUATIONS

The solution of linear equations in one variable has been treated previously. If a polynomial can be expressed as a product of linear factors, then the corresponding polynomial equation can be reduced to the solution of linear equations. So far factoring has been restricted to factors with integers for coefficients. Not all polynomials can be factored in this way.

For example, $x^2 - 2$ and $x^2 + 1$ cannot be factored over the integers. On the other hand, using the factoring theorem,

$$x^2 - a^2 = (x - a)(x + a)$$

one could write

$$x^2 - 2 = x^2 - (\sqrt{2})^2 = (x - \sqrt{2})(x + \sqrt{2})$$

and

$$x^2 + 1 = x^2 - (-1) = x^2 - (\sqrt{-1})^2 = (x - \sqrt{-1})(x + \sqrt{-1})$$

The radical $\sqrt{2}$ has been defined, and it designates a positive real irrational number. The expression $\sqrt{-1}$ has not been defined so far. If $\sqrt{-1}$ has the property $(\sqrt{-1})^2 = -1$, then $\sqrt{-1}$ cannot be a real number since the square of a real number is either positive or zero. By defining $\sqrt{-1}$ to be a new kind of number, called an *imaginary number*, factorization can be extended, and solutions can be provided for the equation $x^2 + 1 = 0$:

$$x^2 + 1 = (x - \sqrt{-1})(x + \sqrt{-1}) = 0$$
$$x - \sqrt{-1} = 0 \quad \text{or} \quad x + \sqrt{-1} = 0$$
$$x = \sqrt{-1} \quad \text{or} \quad x = -\sqrt{-1}$$

DEFINITION

Numbers having the form $a + b\sqrt{-1}$, where a and b are real numbers with $b \neq 0$, are called **imaginary numbers.**

These numbers are needed if every polynomial equation is to have a solution, and they will be discussed in this chapter.

It is now possible to provide a more general solution method for quadratic polynomial equations having the form

$$ax^2 + bx + c = 0$$

where $a \neq 0$. This equation is called a quadratic equation. Its solution and some of its many applications are the major theme of this chapter.

2.1   COMPLEX NUMBERS: DEFINITION, ADDITION, AND SUBTRACTION

Square Roots of Negative Numbers

Up to this point a meaning has not been assigned to an even root of a negative real number, for example, $\sqrt{-1}$, $\sqrt{-9}$, and $\sqrt[4]{-16}$. In the discussions of the properties of the set of real numbers, it was observed that some real numbers are not rational but irrational, such as $\sqrt{2}$, $\sqrt{3}$, $\sqrt{5}$, $\sqrt[3]{5}$, and $\sqrt[3]{-5}$. The set of rational numbers was established as being closed under the operations of addition, subtraction, multiplication, division, and raising to a power. However, it is not closed with respect to the root-extraction operation. For example, $\sqrt{2}$ is not a rational number. To have a solution for $x^2 = 2$, it is necessary to include the irrational numbers $\sqrt{2}$ and $-\sqrt{2}$ in the number system, since $(\sqrt{2})^2 = \sqrt{2}\cdot\sqrt{2} = 2$ and $(-\sqrt{2})^2 = (-\sqrt{2})(-\sqrt{2}) = 2$.

Thus the set of rationals was extended to the set of real numbers for two basic reasons:

1. To help close the number system with respect to the root-extraction operation.
2. To establish a one-to-one correspondence between the real numbers and the points of the number line (the completeness axiom).

Although the real number system has the property that exactly one real number can be assigned to each point on the number line, still the set of real numbers is not closed with respect to the operation of root extraction. There is no real number whose square is a negative real number.

In particular, there is no real number x such that $x^2 = -1$. To obtain closure, a number is invented with the property that its square is -1. This number is assigned the symbolic name $\sqrt{-1}$. Thus $(\sqrt{-1})^2 = \sqrt{-1}\sqrt{-1} = -1$. It is convenient to designate this new number by the letter i.

DEFINITION

$$i = \sqrt{-1} \quad \text{and} \quad i^2 = -1$$

Now it is still necessary to include numbers whose squares are the other negative real numbers. In other words, a meaning must be established for expressions such as $\sqrt{-4}$, $\sqrt{-5}$, and $\sqrt{-4/9}$. Therefore, the following definition is stated:

DEFINITION

$\sqrt{-a} = i\sqrt{a}$ if a is a positive real number.

The number named by the symbol $\sqrt{-a}$ is called the principal square root of $-a$. There is another square root, $-\sqrt{-a}$, because

$(-\sqrt{-a})(-\sqrt{-a})$ equals $+(\sqrt{-a})^2 = -a$. The definition states that the principal square root of a negative real number can be expressed as the product of the new number i and a positive real number.

For example,

$$\sqrt{-4} = i\sqrt{4} = i \cdot 2 = 2i$$
$$\sqrt{-5} = i\sqrt{5} = \sqrt{5}i$$

Numbers of the form bi, where b is a real number, are called **pure imaginary numbers**. Numbers of the form $a + bi$, where a and b are real numbers and $b \neq 0$, are called **imaginary numbers**.

For example, $5i$, $-\frac{2}{3}i$, $\frac{1}{2} + 6i$, and $-7 - i\sqrt{2}$ are imaginary numbers, but of these, only $5i$ and $-\frac{2}{3}i$ are pure imaginary numbers.

Note that $i\sqrt{5} = \sqrt{5}i$ and that $i\sqrt{2} = \sqrt{2}i$. When the multiplier of i is a radical, it is conventional to write $i\sqrt{5}$ rather than $\sqrt{5}i$ to avoid the accidental error of writing $\sqrt{5i}$, since $\sqrt{5i} \neq \sqrt{5}i$.

The set of **complex numbers, C** (Figure 2.1.1), is the union of the set of real numbers, R, with the set of imaginary numbers, Im.

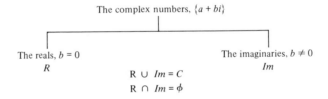

The complex numbers, $\{a + bi\}$

The reals, $b = 0$ The imaginaries, $b \neq 0$

R Im

$R \cup Im = C$
$R \cap Im = \phi$

FIGURE 2.1.1

Every complex number can be expressed in the **standard form** $a + bi$, where a and b are real numbers and i is the imaginary unit such that $i^2 = -1$.

The real number a is called the **real part** of $a + bi$.

The real number b is called the **imaginary part** of $a + bi$.

Two complex numbers are said to be equal if and only if their real parts are equal and their imaginary parts are equal.

DEFINITION

$a + bi = c + di$ if and only if $a = c$ and $b = d$

Sums and Differences of Complex Numbers

The sum and difference of two complex numbers are defined as follows:

DEFINITION OF SUM

$$(a + bi) + (c + di) = (a + c) + (b + d)i$$

DEFINITION OF DIFFERENCE

$$(a + bi) - (c + di) = (a - c) + (b - d)i$$

The above definitions state that two complex numbers are added or subtracted in the usual way where the imaginary unit i is treated as a literal constant. The addition properties for the set of complex numbers are similar to those for the set of real numbers, as can be seen in the following summary.

ADDITION PROPERTIES OF COMPLEX NUMBERS

1. Closure $(a + bi) + (c + di)$ is a complex number.

   $(a + bi) - (c + di)$ is a complex number.

2. Commutativity for Addition

   $$(a + bi) + (c + di) = (c + di) + (a + bi)$$

3. Associativity (Addition)

   $$[(a + bi) + (c + di)] + (e + fi) =$$
   $$(a + bi) + [(c + di) + (e + fi)]$$

4. Identity (Addition). There is exactly one complex number, $0 = 0 + 0 \cdot i$, so that $(a + bi) + 0 = a + bi$ for any $a + bi$.

5. Inverse (Addition). For each $a + bi$, there is exactly one complex number, $-(a + bi)$, called the additive inverse of $a + bi$, so that

   $$(a + bi) + [-(a + bi)] = 0$$

   Moreover, $-(a + bi) = -a + (-bi)$

EXAMPLE 2.1.1 Express each of the following in terms of the imaginary unit i:

a. $\sqrt{-36}$ b. $\sqrt{-5}$ c. $\sqrt{-4x^2}$ $(x > 0)$

Solution

a. $\sqrt{-36} = \sqrt{+36}\sqrt{-1} = 6i$

b. $\sqrt{-5} = \sqrt{5}\sqrt{-1} = \sqrt{5}i = i\sqrt{5}$, preferred form

c. $\sqrt{-4x^2} = \sqrt{4x^2}\sqrt{-1} = 2xi$

EXAMPLE 2.1.2   Write each of the following in standard form, $a + bi$:

a. $3 + \sqrt{-4}$ b. $5 - \sqrt{-9}$ c. $\sqrt{-12}$ d. 6

Solution

a. $3 + \sqrt{-4} = 3 + 2i$
b. $5 - \sqrt{-9} = 5 - 3i$
c. $\sqrt{-12} = \sqrt{4}\sqrt{3}\sqrt{-1} = 2\sqrt{3}i = 0 + 2i\sqrt{3}$, preferred form
d. $6 = 6 + 0i$

EXAMPLE 2.1.3   Determine the real numbers x and y for which $x + yi = 3 - 5i$ is true.

Solution   Since

$$a + bi = c + di \text{ if and only if } a = c \text{ and } b = d$$
$$x + yi = 3 - 5i \text{ if and only if } x = 3 \text{ and } y = -5$$

EXAMPLE 2.1.4   Determine the real numbers x and y for which $(x - 2y) + 2xi = 8i$.

Solution

$$a + bi = c + di \text{ if and only if } a = c \text{ and } b = d$$
$$(x - 2y) + 2xi = 0 + 8i \text{ if and only if}$$
$$x - 2y = 0 \quad \text{and} \quad 2x = 8$$
$$x = 2y \quad \text{and} \quad x = 4$$
$$2y = 4 \quad \text{and} \quad y = 2$$

EXAMPLE 2.1.5   Add $(3 - 4i) + (5 + 2i)$.

Solution $(3 - 4i) + (5 + 2i) = (3 + 5) + (-4 + 2)i = 8 - 2i$

EXAMPLE 2.1.6   Subtract $(-2 - i) - (7 + 6i)$.

Solution $(-2 - i) - (7 + 6i) = (-2 - 7) + (-1 - 6)i = -9 - 7i$

EXAMPLE 2.1.7   Simplify $(5 + 2\sqrt{-9}) - (4 - \sqrt{-25})$.

Solution   First, express each complex number in the standard form, $a + bi$:

$$
\begin{aligned}
(5 + 2\sqrt{-9}) - (4 - \sqrt{-25}) &= (5 + 2 \cdot 3i) - (4 - 5i) \\
&= (5 + 6i) - (4 - 5i) \\
&= (5 + 6i) + (-4 + 5i) \\
&= (5 - 4) + (6 + 5)i \\
&= 1 + 11i
\end{aligned}
$$

EXERCISES 2.1 A

Express Exercises 1–10 in terms of the imaginary unit i.

1. $\sqrt{-4}$ **2.** $\sqrt{-81}$

3. $\sqrt{-x^2}$ $(x \geq 0)$ **4.** $\sqrt{-\frac{1}{4}}$

5. $\sqrt{-2}$ **6.** $\sqrt{-8}$

7. $5\sqrt{-64}$ **8.** $-5\sqrt{-64}$

9. $2\sqrt{-18}$ **10.** $-2\sqrt{-18}$

Write numbers in Exercises 11–18 in standard form $(a + bi)$.

11. $4 + \sqrt{-25}$ **12.** $6 - \sqrt{-49}$

13. $7 + 2\sqrt{-9}$ **14.** $8 - \sqrt{-12}$

15. 5 **16.** 0

17. $\sqrt{-4}$ **18.** $\sqrt{-49} + \sqrt{-36}$

Determine the real numbers x and y for which each of the equations in Exercises 19–24 is true.

19. $x + yi = 4 + 2i$ **20.** $3x + yi = 12 - 5i$

21. $x - 2yi = 10i$ **22.** $4x + 3yi = 16$

23. $(x + y) + (x - y)i = 4 + 6i$

24. $(3x - y - 1) + (2x + y - 4)i = 0$

In Exercises 25–35, perform the indicated operations and express the results in standard form.

25. $(2 + 3i) + (5 - i)$ **26.** $(3 + 4i) + (2 + 3i)$

27. $(4 - 5i) - (6 - 2i)$ **28.** $(5 - 2i) - (4 + 3i)$

29. $(2 + \sqrt{-1}) + (3 + \sqrt{-4})$ **30.** $(6 - \sqrt{-9}) - (9 - 2\sqrt{-16})$

31. $(8 + \sqrt{-12}) + (1 - \sqrt{-27})$

32. $(\sqrt{2} + \sqrt{-6}) - (\sqrt{50} + \sqrt{-24})$

33. $(1 + i) + (3 - 2i) - (5 + 3i)$

34. $(-1 + \sqrt{-3}) - (1 + \sqrt{-3}) + 2$

35. $\sqrt{-4} + \sqrt{-9} + \sqrt{-16}$

★ *In Exercises 36–40, what restriction should be placed on the value of x so that the given expression denotes a real number?*

36. $\sqrt{x - 4}$ **37.** $\sqrt{9 - x}$

38. $\sqrt{x^2 - 16}$ **39.** $\sqrt{9 - x^2}$

40. $\dfrac{1}{\sqrt{x^2 - 25}}$

41. Under what conditions will $a + bi = a - bi$?

EXERCISES 2.1 B

Express Exercises 1–10 in terms of the imaginary unit i. Assume $x \geq 0$ and $y \geq 0$.

1. $\sqrt{-16}$ **2.** $\sqrt{-25y^2}$

3. $\sqrt{-\frac{36}{49}}$ **4.** $5\sqrt{-9}$

5. $-2\sqrt{-100}$ **6.** $\sqrt{-98}$

7. $4\sqrt{-27}$ **8.** $-4\sqrt{-27}$

9. $3\sqrt{-49x^2}$ **10.** $-5\sqrt{-121y^4}$

Write the numbers in Exercises 11–18 in standard form $(a + bi)$.

11. $1 - \sqrt{-36}$ **12.** $9 + \sqrt{-50}$

13. $4 - 2\sqrt{-72}$ **14.** -6

15. $-2\sqrt{-25}$ **16.** i^2

17. $\sqrt{-16} + \sqrt{-4}$ **18.** $\sqrt{-25} - \sqrt{-9}$

Determine the real numbers x and y for which each of the equations in Exercises 19–24 is true.

19. $x + yi = 3 - 4i$ **20.** $2x - 3yi = 6 + 9i$

21. $2x + 3yi = 6i$ **22.** $5x - 2yi = 10$

23. $(x - 2y) + (x + y)i = 12i$ **24.** $(3x + 2y + 1) + (x + 2y)i = 0$

In Exercises 25–35, perform the indicated operations and express the results in standard form.

25. $(6 - 4i) + (-2 + 2i)$

26. $(-6 - i) + (3 - 4i)$

27. $(2 + 5i) - (4 - i)$

28. $(3 - 2i) - (7 + 3i)$

29. $(5 + \sqrt{-9}) + (2 - \sqrt{-25})$

30. $(5 - \sqrt{-36}) - (8 + 2\sqrt{-49})$

31. $(8 - \sqrt{-8}) - (9 - \sqrt{-18})$

32. $(\sqrt{45} + \sqrt{-24}) + (\sqrt{20} - \sqrt{-54})$

33. $(4 - 2i) - (5 + 6i) - (3 - 8i)$

34. $(4 - \sqrt{-9}) + (1 + \sqrt{-16}) + (3 - \sqrt{-1})$

35. $\sqrt{-25} + \sqrt{-36} - \sqrt{-49}$

★ *In Exercises 36–40, what restriction should be placed on the values of x so that the given expression denotes a real number?*

36. $\sqrt{32 - 2x}$

37. $\sqrt{15 - 3x}$

38. $\sqrt{50 - 2x^2}$

39. $\sqrt{4x^2 - 36}$

40. $\dfrac{1}{\sqrt{100 - x^2}}$

★ **41.** If z is any complex number, under what conditions does

  a. $z + (a + bi) = 2a$?

  b. $z - (a + bi) = 2bi$?

2.2 COMPLEX NUMBERS: PRODUCTS

The product of two complex numbers is defined to be the result obtained by applying the distributive, associative, and commutative properties, treating i as a literal constant and then finally replacing i^2 by -1:

$$(a + bi)(c + di) = (a + bi)c + (a + bi)di$$

$$= (ac + bci) + (adi + bdi^2)$$

$$= ac + (bc + ad)i + bd(-1)$$

$$= (ac - bd) + (bc + ad)i$$

DEFINITION

$$(a + bi)(c + di) = (ac - bd) + (bc + ad)i$$

With this definition, the following properties can now be established for the multiplication of two complex numbers:

MULTIPLICATION PROPERTIES OF COMPLEX NUMBERS

1. Closure $(a + bi)(c + di)$ is a complex number.
2. Commutativity $(a + bi)(c + di) = (c + di)(a + bi)$
3. Associativity

$$[(a + bi)(c + di)](e + fi) = (a + bi)[(c + di)(e + fi)]$$

4. Identity. For any complex number $a + bi$ there is exactly one number, $1 = 1 + 0i$, so that $(a + bi) \cdot 1 = a + bi$.
5. Inverse. For each $a + bi \neq 0$ there is exactly one complex number, $1/(a + bi)$, called the *reciprocal* of $a + bi$, so that $(a + bi)[1/(a + bi)] = 1$.
6. Distributive Property

$$(a + bi)[(c + di) + (e + fi)] = (a + bi)(c + di) + (a + bi)(e + fi)$$

While the definition is necessary to show how these properties are logically derived, it is too difficult to remember to use directly in computations. Instead, the product is obtained by multiplying in the usual way, as for real numbers, treating i as a literal constant *but* replacing i^2 by -1.

EXAMPLE 2.2.1   Multiply $(3 + 2i)(2 - 5i)$.

Solution $(3 + 2i)(2 - 5i) = 6 - 15i + 4i - 10i^2$

(Using the distributive property twice)

$$= 6 - 11i - 10i^2$$

(Collecting like terms)

$$= 6 - 11i - 10(-1)$$

(Replacing i^2 by -1)

$$= 16 - 11i$$

EXAMPLE 2.2.2   Simplify $(4 + \sqrt{-9})(3 - \sqrt{-25})$.

Solution   Before any calculations are performed, the expression $\sqrt{-k^2}$ must be replaced by ki with $k > 0$, since all the properties are established for expressions having the form $a + bi$. Thus

$$(4 + \sqrt{-9})(3 - \sqrt{-25}) = (4 + \sqrt{9}\sqrt{-1})(3 - \sqrt{25}\sqrt{-1})$$
$$= (4 + 3i)(3 - 5i)$$
$$= 12 - 20i + 9i - 15i^2$$
$$= 12 - 11i - 15(-1)$$
$$= 27 - 11i$$

EXAMPLE 2.2.3   Multiply $\sqrt{-16}\sqrt{-25}$.

Solution $\sqrt{-16}\sqrt{-25} = (4i)(5i) = 20i^2 = -20$

Note that $\sqrt{-16}\sqrt{-25} \neq \sqrt{(-16)(-25)}$, since $\sqrt{(-16)(-25)} = \sqrt{400} = 20$. The theorem $\sqrt{a}\sqrt{b} = \sqrt{ab}$ is *not* valid when a and b are negative numbers.

EXERCISES 2.2 A

In Exercises 1–25, multiply and express the result in simplified form.

1. $\sqrt{-5}\sqrt{-20}$
2. $\sqrt{-8}\sqrt{-9}$
3. $\sqrt{-12}\sqrt{3}$
4. $\sqrt{2}\sqrt{-3}$
5. $\sqrt{-2}\sqrt{-3}\sqrt{-6}$
6. $2i(4 + 3i)$
7. $3i(5 - 2i)$
8. $\sqrt{-2}(\sqrt{18} - \sqrt{-18})$
9. $(3 + \sqrt{-4})(2 - \sqrt{-9})$
10. $(4 + 5i)(3 - 2i)$
11. $(4 + 2i)(4 - 2i)$
12. $(\sqrt{-2} + \sqrt{-3})(\sqrt{-2} - \sqrt{-3})$
13. $(3 - i)(1 + 2i)$
14. $5i(-3i)(1 - 6i)$
15. $(3 + 2i)^2$
16. $2i(3 + 4i)(5 - 6i)$
17. $(2 + i)[(3 + 2i) + (4 - 5i)]$
18. $4(3 + 2i) - 5i(3 + 2i)$
19. $i(i - 1)(i - 2)$
20. $(1 + 2i)(1 - 2i)(1 - 3i)$
21. $(1 - 3i)(1 - 2i)(1 + 2i)$
22. $(5 - 4i)^2$
23. $(2 + i)^2 - 4(2 + i) + 5$
24. $(2 - i)^2 - 4(2 - i) + 5$
25. $\left(\dfrac{-1 + \sqrt{-3}}{2}\right)^3$

26. Show that $3 + i$ and $3 - i$ are solutions of $x^2 - 6x + 10 = 0$.
27. Show that $2i$ is a solution of $x^2 + ix + 6 = 0$.

★ *In Exercises 28–30, find real values for x and y that satisfy the equations.*

28. $x + yi = i^2 + 1$ **29.** $(ix)^2 + 9 = 0$

30. $x^2 + (1 + i)x + i = 0$

Using $i^1 = i$, $i^2 = -1$, $i^3 = i^2 \cdot i = -i$, and $i^4 = i^2 \cdot i^2 = (-1)(-1) = 1$, simplify Exercises 31–38.

31. i^5 **32.** i^6

33. i^7 **34.** i^8

35. i^{15} **36.** i^{22}

37. $i^3 + i^9$ **38.** $i^3 + i^5 + i^7$

EXERCISES 2.2 B

In Exercises 1–25, multiply and express the result in simplified form.

1. $\sqrt{20}\sqrt{-5}$ **2.** $\sqrt{-20}\sqrt{-5}$

3. $\sqrt{-27}\sqrt{-4}$ **4.** $\sqrt{-27}\sqrt{4}$

5. $\sqrt{-5}\sqrt{-10}\sqrt{-20}$ **6.** $5i(6 - i)$

7. $(6 + 7i)(3 + 2i)$ **8.** $(3 - 7i)(2 - 5i)$

9. $(4 + i)(3 - 2i)$ **10.** $(3 + 2i)(2 + 3i)(5 - 4i)$

11. $(4 + \sqrt{2}i)(3 - \sqrt{5}i)$ **12.** $(4 + 2i)^2$

13. $(5 + \sqrt{-36})(5 - \sqrt{-36})$ **14.** $(3 - 2i)[(2 - 4i) + (5 + 2i)]$

15. $\sqrt{-5}(\sqrt{80} + \sqrt{-80})$ **16.** $6(5 - 4i) - 2i(5 - 4i)$

17. $(2i + 7)(2i - 7)$ **18.** $(\sqrt{-2} + \sqrt{-8})^2$

19. $(i + 1)(4 + 3i)(4 - 3i)$ **20.** $(4 - 3i)(i + 1)(4 + 3i)$

21. $(1 - \sqrt{-3})^2$

22. $(1 - \sqrt{-3})^2 - 2(1 - \sqrt{-3}) + 4$

23. $(1 - \sqrt{-3})^3$ **24.** $(3 + 4i)(1 + 2i)^2$

25. $2i[(25 - 24i)(25 + 24i)]$

26. Show that $2 + i\sqrt{5}$ and $2 - i\sqrt{5}$ are solutions of $x^2 - 4x + 9 = 0$.

27. Show that $1 + i$ is a solution of $x^2 - 2ix - 2 = 0$.

★ *In Exercises 28–30, find real values for x and y that satisfy the equations.*

28. $(x + i)^2 = y$ **29.** $(x + yi)^2 = 8i$

30. $(x + yi)^2 = i$

Using $i^1 = i,\ i^2 = -1,\ i^3 = i^2 \cdot i = -i,\ and\ i^4 = i^2 \cdot i^2 = (-1)(-1) = 1,$
simplify Exercises 31–38.

31. $i^4 + i^7$ **32.** i^{-2}

33. i^{13} **34.** i^{120}

35. i^{54} **36.** $i^2 + i^3 + i^4$

37. $(i^3)^4$ **38.** $(i^2)^{-3}$

2.3 COMPLEX NUMBERS: QUOTIENTS

The quotient of two complex numbers is defined so that division retains its meaning as the inverse operation of multiplication; thus

$$\frac{a + bi}{c + di} = (a + bi)\left(\frac{1}{c + di}\right)$$

Therefore, it is necessary to investigate the reciprocal of a complex number. To do this, the concept of the conjugate of a complex number is introduced.

DEFINITION

The **conjugate** of $a + bi$ is $a - bi$.

The **conjugate** of $a - bi$ is $a + bi$.

THE PRODUCT OF CONJUGATES THEOREM

$$(a + bi)(a - bi) = a^2 + b^2, \quad \text{a real number}$$

For example, $(3 + 4i)(3 - 4i) = 9 + 16 = 25$.

Similarly, $(-2 + 5i)(-2 - 5i) = 4 + 25 = 29$.

Now

$$(a + bi)\left(\frac{a}{a^2 + b^2} + \frac{-b}{a^2 + b^2}\, i\right) = \frac{a^2 - abi + abi - b^2i^2}{a^2 + b^2}$$

$$= \frac{a^2 - b^2(-1)}{a^2 + b^2}$$

$$= \frac{a^2 + b^2}{a^2 + b^2} = 1$$

Thus

$$\frac{1}{a + bi} = \frac{a - bi}{(a + bi)(a - bi)} = \frac{a}{a^2 + b^2} + \frac{-b}{a^2 + b^2}i$$

DEFINITION OF DIVISION

$$\frac{a + bi}{c + di} = (a + bi)\left(\frac{1}{c + di}\right)$$

EXAMPLE 2.3.1 Express $\dfrac{1}{3 + 5i}$ in the $a + bi$ form.

Solution Multiplying numerator and denominator by $3 - 5i$, the conjugate of $3 + 5i$,

$$\frac{1}{3 + 5i} = \frac{3 - 5i}{(3 + 5i)(3 - 5i)}$$

$$= \frac{3 - 5i}{9 - 25i^2} = \frac{3 - 5i}{9 + 25} = \frac{3}{34} + \frac{-5}{34}i$$

EXAMPLE 2.3.2 Divide $\dfrac{2 + \sqrt{-9}}{3 - \sqrt{-4}}$.

Solution First rewriting in terms of i,

$$\frac{2 + \sqrt{-9}}{3 - \sqrt{-4}} = \frac{2 + 3i}{3 - 2i}$$

$$= \frac{(2 + 3i)(3 + 2i)}{(3 - 2i)(3 + 2i)}$$

(Multiplying numerator and denominator by $3 + 2i$, the conjugate of $3 - 2i$)

$$= \frac{6 + 4i + 9i + 6i^2}{9 - 4i^2}$$

$$= \frac{6 + 13i + 6(-1)}{9 - 4(-1)}$$

$$= \frac{13i}{13}$$

$$= i$$

Using more advanced methods, it can be shown that the set of complex numbers is closed with respect to the operation of root extraction. Thus the set of complex numbers is closed with respect to all six operations of elementary algebra. It can also be shown that every polynomial in one variable with coefficients from the set of real numbers or complex numbers has a solution in the set of complex numbers. This statement is called the Fundamental Theorem of Algebra. The German mathematician Karl Friedrich Gauss (1777–1855) gave the first satisfactory proof of this theorem when he was only twenty-one years old.

EXERCISES 2.3 A

Find the conjugate of each of the complex numbers in Exercises 1–8.

1. $5 + 7i$

2. $6 - 4i$

3. $2 + i$

4. $4 - 3i$

5. i

6. $-i$

7. 3

8. $i - 1$

Express each of the quotients in Exercises 9–18 in standard form.

9. $\dfrac{1}{3 + 2i}$

10. $\dfrac{1}{3i - 5}$

11. $\dfrac{1 + i}{1 - i}$

12. $\dfrac{3 + 4i}{2i}$

13. $\dfrac{3i}{2 + 4i}$

14. $\dfrac{1 - \sqrt{-3}}{1 + \sqrt{-3}}$

15. $\dfrac{\sqrt{-3} + 3\sqrt{-1}}{\sqrt{-3} - 3\sqrt{-1}}$

16. $\dfrac{\sqrt{2} + 3i}{1 + i\sqrt{2}}$

17. $\dfrac{a + bi}{a - bi}$ (*a* and *b* are real numbers)

18. $\dfrac{i + 2}{i + 1}$

Simplify Exercises 19–20, expressing the results in standard form.

19. $\dfrac{2 + i}{1 + 2i} + \dfrac{2 - i}{1 - 2i}$

20. $\dfrac{2 + 3i}{3 + i} - \dfrac{1 + i}{1 + 2i}$

EXERCISES 2.3 B

Find the reciprocal of each of the complex numbers in Exercises 1–8 and write the answer in standard form.

1. $3 - 4i$ **2.** $8 + 5i$

3. $1 - i$ **4.** $1 + i$

5. 1 **6.** -1

7. i **8.** $-i$

Express each of the quotients in Exercises 9–18 in standard form.

9. $\dfrac{2}{3 + 2i}$ **10.** $\dfrac{4}{2i - 3}$

11. $\dfrac{2i}{2i + 3}$ **12.** $\dfrac{2i + 3}{2i}$

13. $\dfrac{2i + 3}{3 + 2i}$ **14.** $\dfrac{1 + 2i}{1 - 2i}$

15. $\dfrac{\sqrt{2} + i}{\sqrt{3} - 2i}$ **16.** $\dfrac{2 + 3\sqrt{-3}}{3 + 2\sqrt{-2}}$

17. $\dfrac{\sqrt{-2} + 2\sqrt{-1}}{\sqrt{-3} - 3\sqrt{-1}}$ **18.** $\dfrac{1 + 2i}{3 + 4i} \div \dfrac{2i}{2 - i}$

★ *Simplify Exercises 19–20, expressing the results in standard form.*

19. $(\sqrt{1 + \sqrt{-3}} + \sqrt{1 - \sqrt{-3}})^2$

20. $\dfrac{1}{a + bi} + \dfrac{1}{a - bi}$ (*a* and *b* are real numbers)

2.4 A GEOMETRIC MODEL FOR COMPLEX NUMBERS (Optional)

A complex number $a + bi$ may be represented graphically by interpreting the real part, a, as the distance along a horizontal axis, called the real axis, and by interpreting the imaginary part, b, as the distance along a vertical axis, called the imaginary axis.

For example, the complex numbers $3 + 2i$, $-4 + i$, $-2 - 3i$, and $3 - 3i$ are represented in Figure 2.4.1 as points A, B, C, and D, respectively.

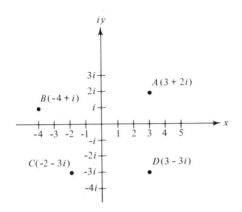

FIGURE 2.4.1

Vectors

Many physical concepts, such as force, velocity, and acceleration, can be described by vectors. A vector is a quantity that has both magnitude and direction. Since a vector can be considered as the sum of a horizontal component and a vertical component which when expressed symbolically must be identified in some way, a complex number can be interpreted as a vector. The real part of the complex number is the horizontal component of the vector, and the imaginary part is the vertical component. A geometric model of a vector is the directed line segment from the origin to the point that is the graph of the complex number representing the vector.

In Figure 2.4.2, the sum of two complex numbers (or vectors) is represented graphically:

$$(1 + 3i) + (2 + i) = (1 + 2) + (3 + 1)i = 3 + 4i$$

The methods of geometry can be used to show that the sum of two vectors is the directed diagonal of the parallelogram formed by using the two vectors as two adjacent sides.

The complex number $1 + 3i$, illustrated in Figure 2.4.2, can be interpreted as a force vector whose horizontal x-component is 1 pound and whose vertical y-component is 3 pounds.

The **magnitude of the force** is the length of the directed line segment that represents the vector. For the vector $1 + 3i$, the magnitude is $\sqrt{1^2 + 3^2} = \sqrt{10}$, using the theorem of Pythagoras.

The **direction of the force** is often given by stating the slope of the line. Thus the direction of the vector $1 + 3i$ is $\frac{3}{1}$.

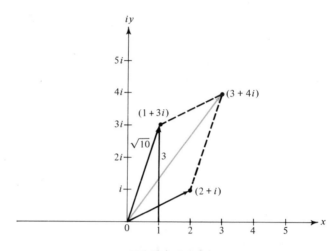

FIGURE 2.4.2

In general, if $a + bi$ represents a vector,

$$\text{the magnitude} = \sqrt{a^2 + b^2}$$

and
$$\text{the direction} = \frac{b}{a} \text{ (the slope)}$$

When two forces act on the same object, the resulting force is that represented by the sum of the complex numbers that represent the two forces. Thus the resulting force of the forces $1 + 3i$ and $2 + i$ is

$$(1 + 3i) + (2 + i) = (1 + 2) + (3 + 1)i = 3 + 4i$$

The resulting magnitude is $\sqrt{3^2 + 4^2} = \sqrt{25} = 5$, and the resulting direction is $\frac{4}{3}$ (the slope of the line from 0 to $3 + 4i$).

EXAMPLE 2.4.1   A plane directed toward the east flies with a speed in still air of 240 mph. A wind from the south blowing at 70 mph forces the plane off its course. What is the resultant speed and direction of the plane?

Solution

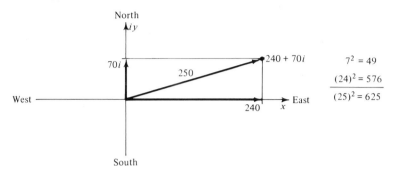

FIGURE 2.4.3

The resulting velocity is $a + bi = 240 + 70i$.

The resulting speed is $\sqrt{a^2 + b^2} = \sqrt{(240)^2 + (70)^2} = 250$ mph.

The resulting direction is $b/a = \frac{70}{240} = \frac{7}{24}$.

As a result, the plane flies 250 mph in the direction of a line from east to north with slope $\frac{7}{24}$.

HISTORICAL NOTE

The Irish mathematician William Rowan Hamilton (1805–1865) presented the modern rigorous treatment of complex numbers as number pairs in 1835. Later he extended these numbers to a space of three dimensions in his Lectures on Quaternions in 1853. Quaternions can be used to describe the three-dimensional rotations of an object in space just as multiplication of a complex number by the imaginary unit i can be interpreted as a rotation of 90 degrees about the origin (see Figure 2.4.4).

In 1752, Jean Le Rond d'Alembert (1717–1783) of France used complex numbers in his study of hydrodynamics, which led to the modern theory of aerodynamics.

In 1772, Johann Heinrich Lambert (1728–1777) of Germany used complex numbers to construct maps by a technique called "conformal conic projection."

In the twentieth century, the American electrician and mathematician Charles Proteus Steinmetz (born in Breslau, Germany, in 1865 and died in the United States in 1923) used complex numbers to develop the theory of electrical circuits.

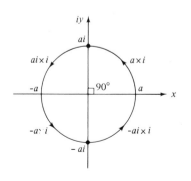

FIGURE 2.4.4

EXERCISES 2.4 A

Plot each of the points in Exercises 1–10 in a complex plane. Be sure to label the axes and to label each point you plot.

1. 2   **2.** $-4 + i$

3. $2i$ **4.** $3 + 2i$

5. $2 - 3i$ **6.** $-2 - 3i$

7. -5 **8.** $-i$

9. $2 + i^2$ **10.** 0

11. A boy who wants to row his boat across a river heads for a point directly opposite on the other shore. His rowing rate in still water is 8 mph, and the rate of the current is 6 mph.

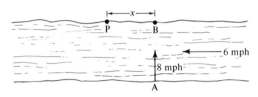

FIGURE 2.4.5

    a. Represent his resultant velocity vector as a complex number $a + bi$.
    b. Find his resultant speed (magnitude of the velocity vector).
    c. How far downstream from his intended destination does he land if it takes him $\frac{1}{2}$ hour to cross the river?

12. Two forces represented by the vectors $-4 + 20i$ and $-4 - 4i$ act on an object. Find the resultant force vector, its magnitude, and its direction. Illustrate geometrically.

EXERCISES 2.4 B

Plot each of the points in Exercises 1–10 in a complex plane.

1. $1 + 2i$ **2.** -3

3. $-i^2$ **4.** $-3 - 4i$

5. $-2i$ **6.** $1 - 3i$

7. $i^2 + 1$ **8.** $3 + 4i$

9. $3(2 - i)$ **10.** $\dfrac{2i + 1}{i}$

11. A plane flying 175 mph south suddenly encounters a 60-mph wind blowing from the east.
    a. Represent the resultant velocity vector as a complex number $a + bi$.
    b. Find the resultant speed.
    c. Find the resultant direction, as a slope.
    d. How far west from its intended destination will the plane be at the end of 45 minutes?

12. Two men pull on a heavy object at right angles to each other, each with a force of 100 pounds.
    a. Represent the resultant force vector as a complex number $a + bi$.
    b. Find the direction of the resultant force.
    c. Find the magnitude of the resultant force.

2.5 QUADRATIC EQUATIONS: SOLUTION BY FACTORING

A **quadratic equation** is an equation that can be expressed in the form

$$ax^2 + bx + c = 0 \qquad (a \neq 0)$$

Previously, the coefficients a, b, and c of the quadratic polynomial

$$ax^2 + bx + c \qquad (a \neq 0)$$

were restricted to belong to the set of integers. The factorization of such polynomials was also restricted to polynomial factors with integral coefficients. These polynomials will now be extended so that a, b, and c can be any complex numbers, and polynomial factors can also have complex coefficients. Now expressions such as $x^2 + 4$, which were not factorable over the set of integers, can be factored over the set of complex numbers:

$$x^2 + 4 = x^2 - (-4) = (x + \sqrt{-4})(x - \sqrt{-4})$$
$$= (x + 2i)(x - 2i) \qquad (i^2 = -1)$$

The solution of quadratic equations by factoring involves four basic steps:

1. Find an equivalent quadratic equation whose right side is zero.
2. Factor the quadratic polynomial.
3. Set each factor equal to zero.
4. Solve for the variable.

This method is based on the zero-product theorem, which is restated below for convenience.

THE ZERO-PRODUCT THEOREM

Let r and s be any complex numbers. If $rs = 0$, then $r = 0$ or $s = 0$.

Proof: If $r \neq 0$, then r has a reciprocal, $1/r$. Then, given $rs = 0$,

$$\frac{1}{r}(rs) = 0$$

and $$s = 0$$

Similarly, if $s \neq 0$, then $r = 0$.
Thus if $rs = 0$, then $r = 0$ or $s = 0$.

It is important to remember in solving an equation by the factoring method that the right side of the equation must be 0 before the factors on the left side can be equated to 0.

EXAMPLE 2.5.1   Solve for x: $x^2 - x = 6$.

Solution $x^2 - x = 6$

$\qquad x^2 - x - 6 = 0$ (Equivalent equation whose right side is zero)

$\qquad (x - 3)(x + 2) = 0$ (Factor the quadratic polynomial)

$x - 3 = 0$ or $x + 2 = 0$ (Zero-product theorem: If $rs = 0$, then $r = 0$ or $s = 0$)

$\qquad x = 3 \qquad\qquad x = -2$

$\quad$ *Check:* $\qquad\qquad x^2 - x = 6 \qquad\qquad x^2 - x = 6$

$\qquad\qquad\qquad\qquad (3)^2 - 3 = 6 \qquad (-2)^2 - (-2) = 6$

$\qquad\qquad\qquad\qquad\quad 9 - 3 = 6 \qquad\qquad 4 + 2 = 6$

$\qquad\qquad\qquad\qquad\qquad 6 = 6 \qquad\qquad\qquad 6 = 6$

Therefore, the solution set is $\{3, -2\}$.

EXAMPLE 2.5.2   Solve $(x - 2)(x - 5) = 40$.

Solution $(x - 2)(x - 5) = 40$

$$x^2 - 7x + 10 = 40$$

$$x^2 - 7x - 30 = 0$$

$$(x + 3)(x - 10) = 0$$

$$x + 3 = 0 \quad \text{or} \quad x - 10 = 0$$

$$x = -3 \quad \text{or} \quad x = 10$$

The solution set is $\{-3, 10\}$.

Check:

For $x = -3$, $(x - 2)(x - 5) = (-3 - 2)(-3 - 5) = (-5)(-8) = 40$.
For $x = 10$, $(x - 2)(x - 5) = (10 - 2)(10 - 5) = (8)(5) = 40$.

A special case of the quadratic equation $ax^2 + bx + c = 0$ occurs when $b = 0$. Then

$$ax^2 + c = 0$$

Since $a \neq 0$, we can divide by a:

$$x^2 + \frac{c}{a} = 0$$

$$x^2 - \left(-\frac{c}{a}\right) = 0$$

$$\left(x + \sqrt{-\frac{c}{a}}\right)\left(x - \sqrt{-\frac{c}{a}}\right) = 0$$

$$x = -\sqrt{-\frac{c}{a}} \quad \text{or} \quad x = +\sqrt{-\frac{c}{a}}$$

The term $\sqrt{-c/a}$ designates a real number if either c or a (but not both) is a negative number, and an imaginary number if a and c both are positive or both negative numbers.

A useful shortcut to this problem is:

$$ax^2 + c = 0$$

$$ax^2 = -c$$

$$x^2 = -\frac{c}{a}$$

$$x = \pm\sqrt{-\frac{c}{a}} \quad \left(\text{read } x = +\sqrt{-\frac{c}{a}} \quad \text{or} \quad x = -\sqrt{-\frac{c}{a}}\right)$$

EXAMPLE 2.5.3   Solve for x: $3x^2 - 2 = 0$.

Solution   In this problem $b = 0$.

$$3x^2 - 2 = 0$$
$$3x^2 = 2$$
$$x^2 = \tfrac{2}{3}$$
$$x = \pm\sqrt{\frac{2}{3}} = \pm\frac{\sqrt{6}}{3}$$

The solution set is $\left\{-\dfrac{\sqrt{6}}{3}, \dfrac{\sqrt{6}}{3}\right\}$.

Check: For $x = +\sqrt{\tfrac{2}{3}}$,
$$3x^2 - 2 = 3(\sqrt{\tfrac{2}{3}})^2 - 2 = 3(\tfrac{2}{3}) - 2 = 2 - 2 = 0$$
For $x = -\sqrt{\tfrac{2}{3}}$,
$$3x^2 - 2 = 3(-\sqrt{\tfrac{2}{3}})^2 - 2 = 3(\tfrac{2}{3}) - 2 = 2 - 2 = 0$$

It is important to remember that the expression $\sqrt{a^2}$ designates exactly one number, the principal square root of a^2. However, the quadratic equation $x^2 = a^2$ has two solutions, and both these solutions must be found—that is,

$$x = +\sqrt{a^2} \quad \text{and} \quad x = -\sqrt{a^2}$$

Another special case occurs when $c = 0$. The equation reads

$$ax^2 + bx = 0$$

Now there is a common factor, x, for the left member of the equation:

$$ax^2 + bx = 0$$
$$x(ax + b) = 0$$
$$x = 0 \quad \text{or} \quad ax + b = 0 \quad \text{(Zero-product theorem)}$$
$$ax = -b$$
$$x = -\frac{b}{a}$$

Solution set: $\{0, -b/a\}$.

EXAMPLE 2.5.4   Solve by factoring $3x^2 + 2x = 0$.

Solution $3x^2 + 2x = 0$
$$x(3x + 2) = 0$$
$$x = 0 \quad \text{or} \quad 3x + 2 = 0$$
$$3x = -2$$
$$x = -\tfrac{2}{3}$$

Therefore, the solution set is $\{0, -\frac{2}{3}\}$.

Check: If $x = 0$, $3(0)^2 + 2(0) = 0 + 0 = 0$.

If $x = -\frac{2}{3}$, $3x^2 + 2x = 3(-\frac{2}{3})^2 + 2(-\frac{2}{3}) = 3(\frac{4}{9}) - \frac{4}{3} = \frac{4}{3} - \frac{4}{3} = 0$.

EXAMPLE 2.5.5   Solve $4x^2 = 12x$.

Solution $4x^2 = 12x$

$\qquad 4x^2 - 12x = 0$

$\qquad 4x(x - 3) = 0$

$\qquad 4x = 0 \quad \text{or} \quad x - 3 = 0$

Since $4 \neq 0$, $x = 0$ or $x = 3$.

Check: For $x = 0$, $4x^2 = 4(0)^2 = 0$ and $12x = 12(0) = 0$.

For $x = 3$, $4(3)^2 = 4(9) = 36$ and $12x = 12(3) = 36$.

Note in the preceding examples that each quadratic equation had *two* solutions. If both sides of the equation $4x^2 = 12x$ had been divided by x, this would have yielded $4x = 12$ and $x = 3$, thus losing the solution $x = 0$. Since the equation $4x^2 = 12x$ is true for $x = 0$, dividing both sides by x is the same as dividing both sides by 0, and this is not permitted. In general, division by an expression containing a variable may lose a solution, and multiplication by an expression containing a variable may introduce an extraneous solution (a number that is not a solution of the original equation).

A quadratic equation may have only one number in its solution set. Such a solution is called a **double root.**

In solving a quadratic equation, it is always necessary to account for *two* solutions. By considering a double root as a solution counted twice, it can then be stated that a quadratic equation, an equation of degree 2, always has two solutions in the set of complex numbers.

EXAMPLE 2.5.6   Solve $x^2 - 6x + 9 = 0$.

Solution $x^2 - 6x + 9 = (x - 3)^2$

$\qquad\qquad x^2 - 6x + 9 = 0$

$\qquad\qquad\quad (x - 3)^2 = 0$

$\qquad\quad (x - 3)(x - 3) = 0$

$\qquad x - 3 = 0 \quad \text{or} \quad x - 3 = 0$

$\qquad\quad x = 3 \quad \text{or} \qquad x = 3$

The solution set is {3}, where 3 is called a double root.

EXAMPLE 2.5.7   Solve $\dfrac{t+2}{t-1} - \dfrac{t-2}{3t} = \dfrac{3}{t-1}$.

Solution   Multiplying each side by the L.C.D. $= 3t(t-1)$,

$$3t(t+2) - (t-1)(t-2) = 3(3t)$$
$$3t^2 + 6t - (t^2 - 3t + 2) = 9t$$
$$2t^2 - 2 = 0$$
$$2(t+1)(t-1) = 0$$
$$t = -1 \quad \text{or} \quad t = 1$$

Since the restricted values are $t \neq 0$ and $t \neq 1$, the only solution is $t = -1$.

EXERCISES 2.5 A

Find all the values of x that are solutions of each of Exercises 1–30. Check each solution.

1. $(x+5)(x-1) = 0$ **2.** $5x(x-4) = 0$

3. $(x+c)(x-2c) = 0$ **4.** $x^2 + 5x + 6 = 0$

5. $4x^2 = 25$ **6.** $x^2 + 3x = 0$

7. $x^2 + 3x = -2$ **8.** $3x^2 + x = 2$

9. $x^2 + 4 = 0$ **10.** $(x+3)(x-2) = 0$

11. $(x+3)(x-2) = 14$ **12.** $5x^2 = 20x$

13. $x^2 + 10x + 25 = 0$ **14.** $x^2 = kx$

15. $3x^2 - 15 = 0$ **16.** $4x^2 = a^2 \quad (a > 0)$

17. $(x-2)^2 = 1$ **18.** $(x-2)^2 = n^2 \quad (n > 0)$

19. $(x-a)^2 = n^2 \quad (n > 0)$ **20.** $(ax+b)^2 = 25 \quad (a \neq 0)$

21. $(cx-d)^2 = 0 \quad (c \neq 0)$ **22.** $x^2 - 6ax = 16a^2$

23. $x^3 - 5x^2 + 4x = 0$ **24.** $x^2 + 4bx + 4b^2 = 0$

25. $\dfrac{x-4}{3} = \dfrac{3}{x+4}$ **26.** $\dfrac{3}{x-1} - \dfrac{3}{x+1} = \dfrac{2}{5}$

27. $2 - \dfrac{1}{x^2} + \dfrac{1}{x} = 0$ **28.** $x^2 - (c+d)x + cd = 0$

29. $\dfrac{x}{x+5} + \dfrac{2}{x-5} = \dfrac{4x}{x^2-25}$

30. $\dfrac{x}{x+1} + \dfrac{3}{x+3} + \dfrac{2}{(x+1)(x+3)} = 0$

★ *Solve each of the formulas in Exercises 31–35 for the letter indicated.*

31. $s = \frac{1}{2}gt^2$ for t

32. $A = \pi(R^2 - r^2)$ for r

33. $k = \dfrac{n_1^2}{n_1^2 + n_2^2}$ for n_1

34. $cd^2 = c^2d$ for c

35. $4wd^2 - 5wdL + wL^2 = 0$ for d

EXERCISES 2.5 B

Find all the values of x that are solutions of each of Exercises 1–30. Check each solution.

1. $3x(x + 1) = 0$

2. $(ax + b)(cx + d) = 0$ $(a \neq 0, c \neq 0)$

3. $cx(dx + 1) = 0$ $(d \neq 0)$

4. $x^2 + x - 6 = 0$

5. $x^2 = 3x$

6. $x^2 + 9 = 0$

7. $2x^2 - 4x = 0$

8. $-2x^2 + 4x - 2 = 0$

9. $10x = 4x^2$

10. $(x - 2)(x - 1) = 6$

11. $2x^2 - 1 = x$

12. $2x^2 - 1 = 0$

13. $(x - 4)(x - 3) = 42$

14. $4x^2 - 4x - 8 = 0$

15. $4x^2 - 12x + 9 = 0$

16. $(ax + b)^2 = 0$ $(a \neq 0)$

17. $(ax + b)^2 = c^2$ $(a \neq 0, c > 0)$

18. $(5x - 2)^2 = 100$

19. $x^2 - 2ax - 63a^2 = 0$

20. $ax^2 - x^2 - ax = 0$ $(a \neq 1)$

21. $8mx^2 + 28mx - 60m = 0$ $(m \neq 0)$

22. $x^3 + 5x = 6x^2$

23. $x^4 - 13x^2 + 40 = 4$

24. $\dfrac{x}{a} - \dfrac{a}{x} = 0$ $(a > 0)$

25. $x + 3 = \dfrac{15}{2x - 1}$

26. $\dfrac{4}{3x + 12} + \dfrac{x}{9} = \dfrac{1}{x + 4}$

27. $x + \dfrac{12}{x} = 7$

28. $\dfrac{3}{x + 2} + \dfrac{x + 8}{x(x + 2)} = 1$

29. $1 + \dfrac{a - b}{x} = \dfrac{ab}{x^2}$

30. $\dfrac{3}{x - 2} - \dfrac{x + 1}{x} = \dfrac{2}{x(x - 2)}$

Solve each of the formulas in Exercises 31–35 for the letter indicated.

31. $K = \dfrac{mv^2}{2}$ for v **32.** $c^2 = a^2 + b^2$ for a

33. $5wd^2 - 59wd + 90w = 0$ for d **34.** $F = \dfrac{GMm}{R^2}$ for R

35. $\dfrac{c_1}{r_1^2} = \dfrac{c_2}{r_2^2}$ for r_2

2.6 QUADRATIC EQUATIONS: COMPLETING THE SQUARE

Not all quadratic polynomials can be factored easily, and the solution of quadratic equations could be very time-consuming and cumbersome if the factors had to be determined by the trial-and-error method. In order to find the solution more rapidly, other methods are available. One such method is called completing the square. The aim of this method is to make the left side of the equation a perfect square trinomial which is equal to a constant— that is, $(x + a)^2 = k$.

The equation $(x + a)^2 = k$ is equivalent to the equation

$$(x + a)^2 - k = 0$$

whose left side can now be factored by the difference of two squares theorem:

$$(x + a)^2 - (\sqrt{k})^2 = 0$$

$$(x + a - \sqrt{k})(x + a + \sqrt{k}) = 0$$

$$x + a - \sqrt{k} = 0 \quad \text{or} \quad x + a + \sqrt{k} = 0$$

$$x = -a + \sqrt{k} \qquad \text{or} \quad x = -a - \sqrt{k}$$

In practice, it is convenient to write the solution in the shorter form indicated below:

$$(x + a)^2 = k$$

$$x + a = \pm\sqrt{k}$$

$$x = -a + \sqrt{k} \quad \text{or} \quad x = -a - \sqrt{k}$$

The method of completing the square is based on the formula for a perfect square trinomial—that is,

$$x^2 + 2nx + n^2 = (x + n)^2$$

Note that the constant term n^2 is the square of one half the coefficient of x.

EXAMPLE 2.6.1   Solve $x^2 + 6x + 4 = 0$ by completing the square.

Solution

1. Subtract the constant from both sides of the equation: $x^2 + 6x = -4$
2. Add to each side the square of one half the coefficient of x: $x^2 + 6x + 9 = -4 + 9$
3. Write the left side as the square of a binomial and simplify the right side: $(x + 3)^2 = 5$
4. Take the square root of each side: $x + 3 = \pm\sqrt{5}$
5. Solve for x: $x = -3 + \sqrt{5}$ or $x = -3 - \sqrt{5}$

Check: If $x = -3 + \sqrt{5}$,

$$x^2 + 6x + 4 = 0$$

$$(-3 + \sqrt{5})^2 + 6(-3 + \sqrt{5}) + 4 = 0$$

$$9 - 6\sqrt{5} + 5 - 18 + 6\sqrt{5} + 4 = 0$$

$$0 = 0$$

If $x = -3 - \sqrt{5}$,

$$x^2 + 6x + 4 = 0$$

$$(-3 - \sqrt{5})^2 + 6(-3 - \sqrt{5}) + 4 = 0$$

$$9 + 6\sqrt{5} + 5 - 18 - 6\sqrt{5} + 4 = 0$$

$$0 = 0$$

Therefore, the solution set is $\{-3 + \sqrt{5}, \, -3 - \sqrt{5}\}$.

In the quadratic trinomial $ax^2 + bx + c$, if $a \neq 1$, then it is convenient to divide by a first.

EXAMPLE 2.6.2 Solve $3x^2 + 6x + 1 = 0$ by completing the square.

Solution

1. Divide by 3:

$$3x^2 + 6x + 1 = 0$$

$$x^2 + 2x + \tfrac{1}{3} = 0$$

2. Subtract $\tfrac{1}{3}$ from each side of the equation: $\quad x^2 + 2x = -\tfrac{1}{3}$

3. Complete the square and add this term to each side:

$$x^2 + 2x + 1 = -\tfrac{1}{3} + 1$$

4. Write the left side as the square of a binomial and simplify the right side:

$$(x + 1)^2 = \tfrac{2}{3}$$

5. Take the square root of each side of the equation:

$$x + 1 = \pm\sqrt{\tfrac{2}{3}} = \pm\frac{\sqrt{6}}{3}$$

6. Solve for x:

$$x = -1 + \frac{\sqrt{6}}{3} \quad \text{or}$$

$$x = -1 - \frac{\sqrt{6}}{3}$$

Check: For $x = -1 + \dfrac{\sqrt{6}}{3}$,

$$3x^2 + 6x + 1 = 3\left(-1 + \frac{\sqrt{6}}{3}\right)^2 + 6\left(-1 + \frac{\sqrt{6}}{3}\right) + 1$$

$$= 3\left(1 - \frac{2\sqrt{6}}{3} + \frac{6}{9}\right) - 6 + 2\sqrt{6} + 1$$

$$= 3 - 2\sqrt{6} + 2 - 5 + 2\sqrt{6}$$

$$= 5 - 5 - 2\sqrt{6} + 2\sqrt{6}$$

$$= 0$$

For $x = -1 - \dfrac{\sqrt{6}}{3}$,

$$3x^2 + 6x + 1 = 3\left(-1 - \frac{\sqrt{6}}{3}\right)^2 + 6\left(-1 - \frac{\sqrt{6}}{3}\right) + 1$$

$$= 3\left(1 + \frac{2\sqrt{6}}{3} + \frac{6}{9}\right) - 6 - 2\sqrt{6} + 1$$

$$= 3 + 2\sqrt{6} + 2 - 5 - 2\sqrt{6}$$

$$= 0$$

EXAMPLE 2.6.3 Solve the equation $x^2 + x + 1 = 0$ by completing the square.

Solution $x^2 + x + 1 = 0$

$$x^2 + x = -1$$

$$x^2 + x + \tfrac{1}{4} = -1 + \tfrac{1}{4}$$

$$(x + \tfrac{1}{2})^2 = -\tfrac{3}{4}$$

$$x + \tfrac{1}{2} = \pm \frac{i\sqrt{3}}{2}$$

$$x = \frac{-1 \pm i\sqrt{3}}{2}$$

Note that both roots are imaginary.

Check: For $x = \dfrac{-1 + i\sqrt{3}}{2}$,

$$x^2 + x + 1 = \left(\frac{-1 + i\sqrt{3}}{2}\right)^2 + \left(\frac{-1 + i\sqrt{3}}{2}\right) + 1$$

$$= \frac{1 - 2i\sqrt{3} - 3}{4} + \frac{-1 + i\sqrt{3}}{2} + 1$$

$$= \frac{-2 - 2i\sqrt{3}}{4} + \frac{-1 + i\sqrt{3}}{2} + 1$$

$$= \frac{-1 - i\sqrt{3}}{2} + \frac{-1 + i\sqrt{3}}{2} + 1 = \frac{-2}{2} + 1 = 0$$

The check is similar for $x = \dfrac{-1 - i\sqrt{3}}{2}$.

EXERCISES 2.6 A

Solve each quadratic equation in Exercises 1–22 by completing the square. Check each solution.

1. $(x - 5)^2 = 2$

2. $(x + 3)^2 = -4$

3. $x^2 - 4x + 1 = 0$

4. $x^2 - 2x + 3 = 0$

5. $x^2 - 2x - 3 = 0$

6. $y^2 - 8y + 15 = 0$

7. $y^2 - 8y + 17 = 0$

8. $x^2 = 2x + 19$

9. $x(x - 1) = 1$

10. $t + 2 = 3t^2$

11. $t^2 + 2 = 2t$

12. $4x - x^2 = 3$

13. $y^2 + 4y = 0$

14. $4x(3 - x) = 7$

15. $14z = z^2 + 53$

16. $6z^2 = 5z + 4$

17. $u(u + 6) + 2 = 0$

18. $(u + 3)^2 + 16 = 0$

19. $x - 10 + \dfrac{95}{x} = 0$

20. $\dfrac{x}{2} + \dfrac{2}{3} = \dfrac{x^2}{6}$

21. $\dfrac{y + 3}{y - 2} = \dfrac{13}{y}$

22. $\dfrac{t}{2} = \dfrac{1 - t}{2 - t}$

Solve Exercises 23–30 for the letter indicated.

23. $x^2 - 6xy + 5y^2 = 0$ for x

24. $x^2 - 4y^2 = 4$ for x

25. $4x^2 + 25y^2 = 100$ for y

26. $x^2 + xy + y^2 = 0$ for y

27. $s = vt - 16t^2$ for t

28. $P = EI + RI^2$ for I

29. $x^2 - 2bx + c = 0$ for x

30. $ax^2 + bx + c = 0$ for x

EXERCISES 2.6 B

Solve each quadratic equation in Exercises 1–22 by completing the square. Check each solution.

1. $(x - 2)^2 = 7$

2. $(x + 4)^2 = -5$

3. $x^2 - 10x + 20 = 0$

4. $x^2 - 6x - 3 = 0$

5. $x^2 - 6x + 3 = 0$

6. $x^2 + 6x - 3 = 0$

7. $x^2 + 6x + 3 = 0$

8. $y^2 = 2(5y - 4)$

9. $y(2y - 1) = 1$

10. $u^2 + 3 = u$

11. $4z = z^2 + 9$

12. $z + 10 = 2z^2$

13. $4u = 15 - 4u^2$

14. $t^2 + 8t + 20 = 0$

15. $2(6 - t^2) = 5t$

16. $3y^2 = 7y + 6$

17. $x(3 - x) = 4$

18. $x + 2 = \dfrac{3}{x}$

19. $y + \dfrac{5}{y} = 2$

20. $\dfrac{10}{x^2} = \dfrac{20}{(10 - x)^2}$

21. $\dfrac{1}{t} + \dfrac{3 - t}{1 - t} = 0$

22. $x^2 - 2\pi x + 1 = 0$

★ *Solve Exercises 23–30 for the letter indicated.*

23. $x^2 - xy - y^2 = 0$ for x

24. $4x^2 + 9y^2 = 36$ for y

25. $16x^2 - y^2 = 16$ for y

26. $x^2 - xy - 2y^2 = 0$ for x

27. $K = \dfrac{x^2}{1 - x}$ for x

28. $\dfrac{a}{x^2} = \dfrac{b}{(c - x)^2}$ for x

29. $Q = a + bt + ct^2$ for t

30. $Ax^2 + 2Bx + C = 0$ for x

2.7 THE QUADRATIC FORMULA

In the preceding section it was seen that any quadratic equation of the form

$$ax^2 + bx + c = 0 \quad (a \neq 0)$$

can be solved by completing the square. The process is sometimes lengthy, so the method is applied to the general equation and a formula is developed:

$$ax^2 + bx + c = 0$$

1. Divide each side by a:

$$x^2 + \frac{b}{a}x + \frac{c}{a} = 0$$

2. Subtract c/a from each side:

$$x^2 + \frac{b}{a}x = -\frac{c}{a}$$

3. Complete the square:

$$x^2 + \frac{b}{a}x + \left(\frac{b}{2a}\right)^2 = -\frac{c}{a} + \left(\frac{b}{2a}\right)^2$$

4. Write the left side as the square of a binomial and simplify the right side:

$$\left(x + \frac{b}{2a}\right)^2 = -\frac{c}{a} + \frac{b^2}{4a^2}$$

$$= \frac{b^2 - 4ac}{4a^2}$$

5. Take the square root of each side:

$$x + \frac{b}{2a} = \pm\frac{\sqrt{b^2 - 4ac}}{2a}$$

6. Solve for x:

$$x = \frac{-b + \sqrt{b^2 - 4ac}}{2a}$$

or

$$x = \frac{-b - \sqrt{b^2 - 4ac}}{2a}$$

If $b^2 - 4ac \geq 0$, then $\sqrt{b^2 - 4ac}$ is a real number and x designates a real number. If $b^2 - 4ac < 0$, then $\sqrt{b^2 - 4ac}$ represents an imaginary number and the solutions of the equation are imaginary.

If $b^2 - 4ac = 0$, then the equation has a double root—namely, $-b/2a$. The roots are also said to be equal in this case.

Because the expression $b^2 - 4ac$ determines whether the roots are real, imaginary, or equal, $b^2 - 4ac$ is called the **discriminant** of the quadratic equation.

Since a, b, and c were chosen completely arbitrarily except $a \neq 0$, the set of equations

$$\left\{x = \frac{-b + \sqrt{b^2 - 4ac}}{2a}, \ x = \frac{-b - \sqrt{b^2 - 4ac}}{2a}\right\}$$

can be used as a formula for the solution of a quadratic equation.

THE QUADRATIC FORMULA

The quadratic equation $ax^2 + bx + c = 0$ $(a \neq 0)$ has the solutions

$$x = \frac{-b + \sqrt{b^2 - 4ac}}{2a}, \qquad x = \frac{-b - \sqrt{b^2 - 4ac}}{2a}$$

DEFINITION

$b^2 - 4ac$ is the **discriminant** of the quadratic equation

$$ax^2 + bx + c = 0 \ (a \neq 0)$$

If $b^2 - 4ac$ **is positive, the solutions are real.**

If $b^2 - 4ac$ **is negative, the solutions are imaginary.**

If $b^2 - 4ac = 0$, **the solutions are real and equal—that is, the equation has a double root which is a real number.**

EXAMPLE 2.7.1 Solve $2x^2 + 3x + 1 = 0$ by using the quadratic formula.

Solution Comparing $2x^2 + 3x + 1 = 0$ with $ax^2 + bx + c = 0$, it is seen that $a = 2$, $b = 3$, and $c = 1$. Therefore,

$$x = \frac{-b \pm \sqrt{b^2 - 4ac}}{2a}$$

yields $$x = \frac{-3 + \sqrt{9 - 4(2)(1)}}{4} = \frac{-3 + 1}{4} = \frac{-2}{4} = -\frac{1}{2}$$

or $$x = \frac{-3 - \sqrt{9 - 4(2)(1)}}{4} = \frac{-3 - 1}{4} = \frac{-4}{4} = -1$$

Check:

If $x = -\frac{1}{2}$,   If $x = -1$,

$2x^2 + 3x + 1 = 0$ $2x^2 + 3x + 1 = 0$

$2(-\frac{1}{2})^2 + 3(-\frac{1}{2}) + 1 = 0$ $2(-1)^2 + 3(-1) + 1 = 0$

$\frac{1}{2} - \frac{3}{2} + 1 = 0$ $2 - 3 + 1 = 0$

$0 = 0$ $0 = 0$

Therefore, $\{-\frac{1}{2}, -1\}$ is the solution set.

Note in this example that the discriminant $b^2 - 4ac = 9 - 4(2)(1) = 9 - 8 = 1$, a positive number, indicating that the roots are real numbers. Moreover, the discriminant is a perfect square since $1^2 = 1$. **Whenever the discriminant is a perfect square, the roots are rational numbers.**

EXAMPLE 2.7.2 Solve $2x^2 - 3x = x^2 - 1$ by using the quadratic formula.

Solution First an equivalent equation of the form $ax^2 + bx + c = 0$ must be found:

$$2x^2 - 3x = x^2 - 1$$

$$x^2 - 3x + 1 = 0$$

$$a = 1, \quad b = -3, \quad c = 1$$

$$x = \frac{-(-3) \pm \sqrt{(-3)^2 - 4(1)(1)}}{2}$$

$$x = \frac{3 + \sqrt{5}}{2} \quad \text{or} \quad x = \frac{3 - \sqrt{5}}{2}$$

The solution set is $\left\{ \dfrac{3 + \sqrt{5}}{2}, \dfrac{3 - \sqrt{5}}{2} \right\}$.

Check: For $x = \dfrac{3 + \sqrt{5}}{2}$,

$$2x^2 - 3x = 2\left(\frac{3 + \sqrt{5}}{2}\right)^2 - 3\left(\frac{3 + \sqrt{5}}{2}\right)$$

$$= \frac{2(14 + 6\sqrt{5})}{4} - \frac{3(3 + \sqrt{5})}{2}$$

$$= \frac{14 + 6\sqrt{5} - 9 - 3\sqrt{5}}{2} = \frac{5 + 3\sqrt{5}}{2}$$

$$x^2 - 1 = \left(\frac{3 + \sqrt{5}}{2}\right)^2 - 1 = \frac{14 + 6\sqrt{5}}{4} - \frac{4}{4}$$

$$= \frac{10 + 6\sqrt{5}}{4}$$

$$= \frac{5 + 3\sqrt{5}}{2}$$

The check for $x = (3 - \sqrt{5})/2$ is similar.

Note that the discriminant $b^2 - 4ac = (-3)^2 - 4(1)(1) = 9 - 4 = 5$, a positive number, indicating that the roots are real. Also, since 5 is not a perfect square, the roots are irrational.

The solutions $(3 + \sqrt{5})/2$ and $(3 - \sqrt{5})/2$ are said to be in simplified exact form. If an approximation to a solution is desired, a table of square roots can be used, and the answer can be expressed in decimal form for any specified accuracy.

Thus, expressing $(3\sqrt{5})/2$ correct to the nearest hundredth,

$$\frac{3 + \sqrt{5}}{2} = \frac{3 + 2.236}{2} = \frac{5.236}{2} = 2.618$$

Therefore, $(3 + \sqrt{5})/2 = 2.62$ correct to the nearest hundredth.

EXAMPLE 2.7.3   Solve for x: $\dfrac{x^2}{4} - \dfrac{x}{2} + 1 = 0.$

Solution   Although this problem can be worked with $a = \frac{1}{4}$, $b = -\frac{1}{2}$, and $c = 1$, it is easier to solve by first finding an equivalent equation with integers for coefficients. Multiplying each term by 4, the desired equivalent equation is obtained:

$$x^2 - 2x + 4 = 0$$

then $a = 1, b = -2,$ and $c = 4$

$$x = \frac{-(-2) \pm \sqrt{(-2)^2 - 4(1)(4)}}{2(1)}$$

$$= \frac{2 \pm \sqrt{4 - 16}}{2}$$

$$= \frac{2 \pm \sqrt{-12}}{2}$$

$$= \frac{2 \pm 2i\sqrt{3}}{2}$$

$$= 1 \pm i\sqrt{3}$$

The solution set is $\{1 + i\sqrt{3}, 1 - i\sqrt{3}\}$.

Check: For $x = 1 + i\sqrt{3}$,

$$\frac{x^2}{4} - \frac{x}{2} + 1 = \frac{(1 + i\sqrt{3})^2}{4} - \frac{1 + i\sqrt{3}}{2} + 1$$

$$= \frac{-2 + 2i\sqrt{3}}{4} - \frac{1 + i\sqrt{3}}{2} + 1$$

$$= \frac{-1 + i\sqrt{3} - 1 - i\sqrt{3}}{2} + 1$$

$$= -1 + 1 = 0$$

The check is similar for $x = 1 - i\sqrt{3}$.

Note that the discriminant $b^2 - 4ac = (-2)^2 - 4(1)(4) = 4 - 16 = -12$, a negative number, indicating that the roots are imaginary.

EXAMPLE 2.7.4   Solve $9x^2 + 30x + 25 = 0$.

Solution $a = 9, \quad b = 30, \quad c = 25$

$$x = \frac{-30 \pm \sqrt{(30)^2 - 4(9)(25)}}{2(9)}$$

$$= \frac{-30 \pm \sqrt{900 - 900}}{2(9)}$$

$$= \frac{-30}{6(3)}$$

$$= -\tfrac{5}{3}$$

The solution set is $\{-\tfrac{5}{3}\}$.

Check: For $x = -\tfrac{5}{3}$,

$$9x^2 + 30x + 25 = 9(-\tfrac{5}{3})^2 + 30(-\tfrac{5}{3}) + 25$$

$$= 9(\tfrac{25}{9}) + 10(-5) + 25$$

$$= 25 - 50 + 25$$

$$= 0$$

In this case, $-\tfrac{5}{3}$ is a double root. Note that the discriminant $b^2 - 4ac = 0$:

$$b^2 - 4ac = (30)^2 - 4(9)(25) = 900 - 900 = 0$$

Note also that $9x^2 + 30x + 25 = (3x + 5)^2$, a perfect square trinomial.

EXERCISES 2.7 A

a. Write the equations in Exercises 1–5 as equivalent equations in the form $ax^2 + bx + c = 0$ *and state the value of a, b, and c in each equation.*

b. Find the value of the discriminant $b^2 - 4ac$ *and state whether the roots of the equation are real and unequal, imaginary, or real and equal.*

1. $2x^2 = 3x + 1$ **2.** $4x^2 = 12x - 9$

3. $\dfrac{5}{x} = x$ **4.** $(x + 2)(x + 3) = 6$

5. $4x - 8 = x^2$

Solve each of the equations in Exercises 6–20 by using the quadratic formula. Check each solution.

6. $3x^2 + 5x + 2 = 0$ **7.** $3x^2 - 5x - 2 = 0$

8. $x^2 - 3x + 1 = 0$ **9.** $x^2 + 3x - 1 = 0$

10. $x^2 - 3x + 3 = 0$ **11.** $x^2 + x + 1 = 0$

12. $y^2 - 2y - \frac{3}{4} = 0$ **13.** $3y^2 = 2(y - 1)$

14. $y + 2 = 3y(y + 1)$ **15.** $t^2 + 16 = 0$

16. $5t = 3t^2$ **17.** $4x - 36x^2 = \frac{1}{9}$

18. $80t - 16t^2 = 80$ **19.** $\frac{1}{2}d^2 - \frac{30}{8}d + \frac{36}{8} = 0$

20. $\dfrac{w}{w + 10} = \dfrac{w - 10}{10}$

Solve Exercises 21–28 for the variable indicated.

21. $x^2 + 2px + q = 0$ for x **22.** $y^2 - ny - n^2 = 0$ for y

23. $3x^2 + 8xy - 3y^2 = 0$ for x **24.** $y^2 - 4y + 2x = 6$ for y

25. $s = vt + \frac{1}{2}gt^2$ for t **26.** $s = (k - s)^2$ for s

27. $\dfrac{1}{R} + \dfrac{1}{a - R} = \dfrac{1}{b}$ for R **28.** $A = \pi r(r + s)$ for r

EXERCISES 2.7 B

a. Write the equations in Exercises 1–5 as equivalent equations in the form $ax^2 + bx + c = 0$ and state the value of a, b, and c in each equation.

b. Find the value of the discriminant $b^2 - 4ac$ and state whether the roots of the equation are real and unequal, imaginary, or real and equal.

1. $4x^2 - 2 = 2x$ **2.** $(x + 1)^2 = x + 3$

3. $9 = 6x + x^2$ **4.** $(5 - x)(4 + x) = 40$

5. $2x + \dfrac{1}{8x} = 1$

Solve each of the equations in Exercises 6–20 by using the quadratic formula. Check each solution.

6. $4x^2 + 7x + 3 = 0$ **7.** $4x^2 + 7x + 4 = 0$

8. $4x^2 + 7x - 2 = 0$ **9.** $4x^2 - 4x + 1 = 0$

10. $x^2 - 4x - 4 = 0$ **11.** $x^2 + 2x + 4 = 0$

12. $y^2 - y + \frac{1}{5} = 0$ **13.** $\dfrac{y^2 + 1}{6} = \dfrac{y}{2}$

14. $3(y + 1) = 5 - 2y^2$

15. $9 = 6t + t^2$

16. $\dfrac{t-1}{t} - \dfrac{t-1}{6} = \dfrac{1}{6}$

17. $25x - 100x^2 = 0$

18. $25u^2 = 40u - 16$

19. $\frac{1}{2}d^2 = 6d + 9$

20. $\dfrac{c^2}{0.01 - c} = 1.7 \times 10^{-5}$

Solve Exercises 21–28 for the variable indicated.

21. $cx^2 + 2bx + a = 0$ for x

22. $y^2 - 4ny + 2n^2 = 0$ for y

23. $4x^2 - 6xy + y^2 = 0$ for x

24. $\dfrac{W}{L} = \dfrac{L - W}{W}$ for W

25. $P = EI + RI^2$ for I

26. $F^2 = \dfrac{1}{LC} - \dfrac{R}{4L^2}$ for L

27. $y = mx - \dfrac{16(m^2 + 1)}{v^2} x^2$ for x

28. $V = \dfrac{h}{3}(a^2 + ab + b^2)$ for a

2.8 SUM AND PRODUCT OF ROOTS

Two important relationships can be established between the roots or solutions of a quadratic equation $ax^2 + bx + c = 0$ $(a \neq 0)$ and the coefficients a, b, and c.

Letting r and s designate the roots of the equation and using the quadratic formula, simple expressions can be obtained for the sum and product of the roots. If

$$r = \frac{-b + \sqrt{b^2 - 4ac}}{2a} \quad \text{and} \quad s = \frac{-b - \sqrt{b^2 - 4ac}}{2a}$$

then the sum

$$r + s = \frac{-b + \sqrt{b^2 - 4ac}}{2a} + \frac{-b - \sqrt{b^2 - 4ac}}{2a} = 2\left(\frac{-b}{2a}\right) = \frac{-b}{a}$$

and the product

$$rs = \left(\frac{-b + \sqrt{b^2 - 4ac}}{2a}\right)\left(\frac{-b - \sqrt{b^2 - 4ac}}{2a}\right)$$

$$= \frac{(-b)^2 - (b^2 - 4ac)}{4a^2}$$

$$= \frac{4ac}{4a^2} = \frac{c}{a}$$

Therefore, $r + s = -b/a$ and $rs = c/a$.

Now transforming the equation

$$ax^2 + bx + c = 0$$

to an equivalent equation with the coefficient of x^2 equal to 1,

$$x^2 + \frac{b}{a}x + \frac{c}{a} = 0$$

Comparing,

$$x^2 - (r + s)x + rs = 0$$

These results can be stated as the following theorem:

SUM AND PRODUCT OF ROOTS THEOREM

$\{r, s\}$ is the solution set of the equation

$$ax^2 + bx + c = 0 \qquad (a \neq 0)$$

if and only if

$$r + s = -\frac{b}{a} \quad \text{and} \quad rs = \frac{c}{a}$$

That is, the sum of the roots is equal to the negative ratio of the coefficient of the linear term to the coefficient of the second-degree term, and the product of the roots is equal to the ratio of the constant term to the coefficient of the second-degree term.

EXAMPLE 2.8.1 Without solving the equation, find the sum and the product of the roots of:

a. $3x^2 - 5x + 2 = 0$ b. $2x^2 + 6x - 7 = 0$

Solution

a. $3x^2 - 5x + 2 = 0$ is equivalent to $x^2 - \frac{5}{3}x + \frac{2}{3} = 0$

Comparing with

$$x^2 - (r + s)x + rs = 0$$

the sum of the roots, $r + s = \frac{5}{3}$
and the product of the roots, $rs = \frac{2}{3}$

b. $2x^2 + 6x - 7 = 0$ is equivalent to $x^2 + 3x - \frac{7}{2} = 0$
and to $x^2 - (-3)x - \frac{7}{2} = 0$
the sum of the roots, $r + s = -3$
the product of the roots, $rs = -\frac{7}{2}$

The sum and product of roots theorem provides an excellent check of the solution set of a quadratic equation, as illustrated in the next example.

EXAMPLE 2.8.2 Check that

$$\left\{\frac{5}{2} + \frac{\sqrt{2}}{2}, \frac{5}{2} - \frac{\sqrt{2}}{2}\right\}$$

is the solution set of $4x^2 - 20x + 23 = 0$.

Solution From the equation, the sum of the roots is $\frac{20}{4} = 5$, and the product of the roots is $\frac{23}{4}$. Adding the roots,

$$\left(\frac{5}{2} + \frac{\sqrt{2}}{2}\right) + \left(\frac{5}{2} - \frac{\sqrt{2}}{2}\right) = 5 \quad \text{(Check)}$$

Multiplying the roots,

$$\left(\frac{5}{2} + \frac{\sqrt{2}}{2}\right)\left(\frac{5}{2} - \frac{\sqrt{2}}{2}\right) = \frac{25}{4} - \frac{2}{4} = \frac{23}{4} \quad \text{(Check)}$$

Knowing the roots of a quadratic equation, it is possible to recapture the equation by retracing the steps in the solution process and by using the sum and product of roots theorem.

Let r and s be the roots of a quadratic equation.

Then $\qquad\qquad\qquad\qquad x = r \quad \text{or} \quad x = s$

and $\qquad\qquad\qquad\quad x - r = 0 \quad \text{or} \quad x - s = 0$

By the zero-product theorem, this set of equations is equivalent to

$$(x - r)(x - s) = 0$$

Performing the indicated multiplication,

$$x^2 - rx - sx + rs = 0$$

and $\qquad\qquad\qquad x^2 - (r + s)x + rs = 0$

EXAMPLE 2.8.3 Find a quadratic equation whose solution set is $\{2 + \sqrt{3}, 2 - \sqrt{3}\}$.

Solution Let $r = 2 + \sqrt{3}$ and $s = 2 - \sqrt{3}$.

Then adding, $\qquad r + s = (2 + \sqrt{3}) + (2 - \sqrt{3}) = 4$

and multiplying, $\quad rs = (2 + \sqrt{3})(2 - \sqrt{3}) = 4 - 3 = 1$

Using $x^2 - (r + s)x + rs = 0$,

$$x^2 - 4x + 1 = 0 \quad \text{(Answer)}$$

EXAMPLE 2.8.4   Find the value of c if 3 is one root of $x^2 + 5x + c = 0$.

Solution   Let r = the unknown root; then $r + s = r + 3$ and $rs = 3r$.

Comparing,
$$x^2 - (r + s)x + rs = 0$$
$$x^2 - (-5)x + c = 0$$
$$x^2 - (r + 3)x + 3r = 0$$

Then
$$r + 3 = -5 \quad \text{and} \quad c = 3r$$
$$r = -8 \quad \text{and} \quad c = 3(-8) = -24$$

EXERCISES 2.8 A

Without solving, state the sum and the product of the roots of Exercises 1–5.

1. $x^2 - 8x + 5 = 0$ **2.** $x^2 + 5x - 2 = 0$

3. $2x^2 - 3x + 4 = 0$ **4.** $3x^2 + 6x - 1 = 0$

5. $4x^2 - 5x - 7 = 0$

State the sum of the roots of Exercises 6–10.

6. $2x^2 - 6x + c = 0$ **7.** $4x^2 + 8x + c = 0$

8. $3x^2 + 12x + c = 0$ **9.** $5x^2 - 45x + c = 0$

10. $6x^2 + c = 0$

State the product of the roots of Exercises 11–15.

11. $2x^2 + bx + 8 = 0$ **12.** $3x^2 + bx - 42 = 0$

13. $6x^2 + bx = 0$ **14.** $5x^2 + bx - 30 = 0$

15. $x^2 + bx + 7 = 0$

Use the sum and product of roots theorem to check if the stated set is the solution set of the equations in Exercises 16–20.

16. $\{5 + \sqrt{7}, 5 - \sqrt{7}\}$ **17.** $\{1 - \sqrt{2}, 1 + \sqrt{2}\}$
$\qquad x^2 - 10x + 18 = 0$ $x^2 + 2x - 1 = 0$

18. $\left\{\dfrac{2 + \sqrt{5}}{2}, \dfrac{2 - \sqrt{5}}{2}\right\}$ **19.** $\left\{\dfrac{-2 + \sqrt{5}}{2}, \dfrac{-2 - \sqrt{5}}{2}\right\}$
$\qquad 2x^2 - 4x - 1 = 0$ $4x^2 + 8x - 1 = 0$

20. $\{-3 + i, -3 - i\}$
$\qquad x^2 + 6x + 10 = 0$

In Exercises 21–25, write a quadratic equation whose roots r and s have the given sum and product.

21. $r + s = 6$, $rs = 2$ **22.** $r + s = -3$, $rs = 0$

23. $r + s = 5\frac{2}{3}$, $rs = 3\frac{1}{3}$ **24.** $r + s = -\frac{2}{5}$, $rs = \frac{1}{25}$

25. $r + s = -\frac{3}{2}$, $rs = -\frac{5}{4}$

In Exercises 26–30, write a quadratic equation whose solution set is stated.

26. $\{3, -2\}$ **27.** $\{\sqrt{3}, -\sqrt{3}\}$

28. $\{3 + \sqrt{2}, 3 - \sqrt{2}\}$ **29.** $\{0, 5\}$

30. $\{1 + 2i, 1 - 2i\}$

Use the formula for the sum and product of the roots of a quadratic equation to find the value of k in each of the equations in Exercises 31–36.

31. $x^2 - 2x + k = 0$ (one root is 1)

32. $5x^2 + kx - 10 = 0$ (one root is -5)

33. $kx^2 - 19x + 6 = 0$ (one root is 6)

34. $x^2 + kx + 16 = 0$ (the roots are equal)

35. $4x^2 - 3x + k = 0$ (one root is 0)

36. $x^2 - kx - 25 = 0$ (one of the roots is the negative of the other)

Solve and check Exercises 37–40 by using the sum and product of roots theorem.

37. $x^2 - 4x - 2 = 0$ **38.** $x^2 + 4x + 8 = 0$

39. $5x = 2x^2 + 1$ **40.** $5x^2 = 4(2x - 1)$

EXERCISES 2.8 B

Without solving, state the sum and the product of the roots of Exercises 1–5.

1. $x^2 - 2x - 6 = 0$ **2.** $x^2 + 4x + 2 = 0$

3. $5x^2 - 3x + 4 = 0$ **4.** $2x^2 + 7x - 5 = 0$

5. $10x^2 + 5x + 2 = 0$

State the sum of the roots of Exercises 6–10.

6. $4x^2 - 12x + c = 0$ **7.** $7x^2 + c = 0$

8. $5x^2 + 40x + c = 0$ **9.** $2x^2 - 4nx + n^2 = 0$

10. $3x^2 + 7nx - n = 0$

State the product of the roots of Exercises 11–15.

11. $4x^2 + bx - 2 = 0$ **12.** $8x^2 - bx = 0$

13. $x^2 + bx + c = 0$ **14.** $2x^2 + nx - 4n = 0$

15. $3x^2 - 2nx + n^2 = 0$

Use the sum and product of roots theorem to check if the stated set is the solution set of the equations in Exercises 16–20.

16. $\{2\sqrt{2} - 1, 2\sqrt{2} + 1\}$ **17.** $\{2 - \sqrt{7}, 2 + \sqrt{7}\}$

   $x^2 - 4\sqrt{2}x + 7 = 0$ $x^2 - 4x - 3 = 0$

18. $\left\{\dfrac{1 + \sqrt{6}}{2}, \dfrac{1 - \sqrt{6}}{2}\right\}$ **19.** $\{2 - 3i, 2 + 3i\}$
                                 $x^2 - 4x - 5 = 0$

   $4x^2 + 4x - 5 = 0$

20. $\{3 + 5i, 3 - 5i\}$

   $x^2 - 6x + 34 = 0$

In Exercises 21–25, write a quadratic equation whose roots r and s have the given sum and product.

21. $r + s = -\frac{1}{4}, rs = -\frac{3}{4}$ **22.** $r + s = 2, rs = 5$

23. $r + s = 1, rs = -\frac{1}{9}$ **24.** $r + s = \sqrt{2}, rs = 0$

25. $r + s = -0.6, rs = 1.4$

In Exercises 26–30, write a quadratic equation whose solution set is stated.

26. $\{8, 3\}$ **27.** $\{-4 + \sqrt{2}, -4 - \sqrt{2}\}$

28. $\{\frac{1}{2}, -\frac{1}{4}\}$ **29.** $\left\{\dfrac{3 + 2i}{4}, \dfrac{3 - 2i}{4}\right\}$

30. $\{-\frac{1}{2}, 0\}$

Use the formula for the sum and product of the roots of a quadratic equation to find the value of k in each of the equations in Exercises 31–36.

31. $x^2 - 16x + k = 0$ (the roots are equal)

32. $2x^2 - 12x + k = 0$ (one root $= 3 - \sqrt{2}$)

33. $x^2 - kx - 72 = 0$ (one root is 14 more than the negative of the other)

34. $kx^2 - 12x + 18 = 0$ (the roots are equal)

35. $5x^2 + 7x - k = 0$ (one root is 0)

36. $x^2 - 6x + k = 0$ (one root is double the other)

Solve and check Exercises 37–40 by using the sum and product of roots theorem.

37. $3x^2 + 2x + 5 = 0$ **38.** $4x - 1 = 20x^2$
39. $(x + 2)^3 - x^3 = 2$ **40.** $2x = (2x - 1)^2$

2.9   EQUATIONS IN QUADRATIC FORM

It has been demonstrated that equations of the form $ax^2 + bx + c = 0$ ($a \neq 0$) can be solved by three methods: factoring, completing the square, or the quadratic formula. There are also many other equations which are not quadratic but which may be expressed in *quadratic form* and can then be solved by quadratic methods. For a polynomial equation, the degree of the polynomial determines the number of roots of the equation.

EXAMPLE 2.9.1   Find the solution set of the equation

$$x^4 - 11x^2 + 28 = 0$$

Solution   This equation is of the fourth degree and four roots must be found. By means of a substitution, it can be expressed in quadratic form. Let $u = x^2$; then $u^2 = x^4$, and the equation may now be written

$$u^2 - 11u + 28 = 0$$

Factoring,

$$(u - 4)(u - 7) = 0$$

$$u - 4 = 0 \quad \text{or} \quad u - 7 = 0$$

$$u = 4 \quad \text{or} \quad u = 7$$

If $u = 4$,

$$x^2 = 4 \quad \text{and} \quad x = +2 \quad \text{or} \quad x = -2$$

If $u = 7$,

$$x^2 = 7 \quad \text{and} \quad x = +\sqrt{7} \quad \text{or} \quad x = -\sqrt{7}$$

Therefore, the solution set is $\{2, -2, \sqrt{7}, -\sqrt{7}\}$.

EXAMPLE 2.9.2   Find the solution set of the equation

$$y^{-4} - 8y^{-2} + 15 = 0$$

Solution   This is not a polynomial equation, and the number of roots cannot be determined readily. Again, a substitution can be made to obtain an equation in quadratic form. Let $u = y^{-2}$; then $u^2 = y^{-4}$, so that

$$u^2 - 8u + 15 = 0$$

This equation can be solved by factoring:

$$(u - 3)(u - 5) = 0$$

$$u = 3 \quad \text{or} \quad u = 5$$

But $u = y^{-2} = 1/y^2$; therefore

$$\frac{1}{y^2} = 3 \qquad\qquad \text{or} \quad \frac{1}{y^2} = 5$$

$$y^2 = \tfrac{1}{3} \qquad\qquad \text{or} \quad y^2 = \tfrac{1}{5}$$

$$y = \pm\sqrt{\tfrac{1}{3}} = \pm\frac{\sqrt{3}}{3} \quad \text{or} \quad y = \pm\sqrt{\tfrac{1}{5}} = \pm\frac{\sqrt{5}}{5}$$

Thus the solution set is $\left\{ \pm\dfrac{\sqrt{3}}{3}, \pm\dfrac{\sqrt{5}}{5} \right\}$.

EXAMPLE 2.9.3   Solve for x: $x^{2/3} - x^{1/3} = 72$.

Solution   Let $u = x^{1/3}$; then $u^2 = x^{2/3}$.

$$x^{2/3} - x^{1/3} - 72 = 0$$

$$u^2 - u - 72 = 0$$

$$(u - 9)(u + 8) = 0$$

$$u = 9 \qquad \text{or} \qquad u = -8$$

$$x^{1/3} = 9 \qquad \text{or} \qquad x^{1/3} = -8$$

$$(x^{1/3})^3 = 9^3 \qquad \text{or} \quad (x^{1/3})^3 = (-8)^3$$

$$x = 729 \quad \text{or} \qquad x = -512$$

The solution set is $\{729, -512\}$.

EXAMPLE 2.9.4   Solve for x: $x^6 = 27 - 26x^3$.

Solution $x^6 + 26x^3 - 27 = 0$

Let $u = x^3$; then $u^2 = x^6$ and

$$u^2 + 26u - 27 = 0$$

$$(u + 27)(u - 1) = 0$$

$$u + 27 = 0 \quad \text{or} \quad u - 1 = 0$$

$$x^3 + 27 = 0 \quad \text{or} \quad x^3 - 1 = 0$$

$$(x + 3)(x^2 - 3x + 9) = 0 \quad \text{or} \quad (x - 1)(x^2 + x + 1) = 0$$

$$x + 3 = 0 \quad \text{or} \quad x^2 - 3x + 9 = 0 \quad \text{or} \quad x - 1 = 0 \quad \text{or} \quad x^2 + x + 1 = 0$$

$$x = -3 \quad \text{or} \quad x = \frac{3 \pm 3i\sqrt{3}}{2} \quad \text{or} \quad x = 1 \quad \text{or} \quad x = \frac{-1 \pm i\sqrt{3}}{2}$$

The solution set is

$$\left\{ -3, 1, \frac{3 + 3i\sqrt{3}}{2}, \frac{3 - 3i\sqrt{3}}{2}, \frac{-1 + i\sqrt{3}}{2}, \frac{-1 - i\sqrt{3}}{2} \right\}$$

EXERCISES 2.9 A

Find the solution set of each equation in Exercises 1–14 over the set of complex numbers. Check each solution.

1. $x^4 - 14x^2 + 45 = 0$

2. $2x^4 + x^2 = 1$

3. $(x - 3)^4 - 13(x - 3)^2 + 36 = 0$

4. $y^{-2} - y^{-1} = 20$

5. $9r^{-4} + 5r^{-2} = 4$

6. $x^{2/3} + 2x^{1/3} - 8 = 0$

7. $x^{-4} - 3x^{-2} + 2 = 0$

8. $(x^2 - 4x)^2 - 36 = 9(x^2 - 4x)$

9. $y^{1/2} - 7y^{1/4} + 12 = 0$

10. $2\left(x + \frac{1}{x}\right)^2 = \left(x + \frac{1}{x}\right) + 10$

11. $x^6 + 64 = 16x^3$

12. $(x^2 + 8x)^2 = 5(x^2 + 8x) + 36$

13. $16\left(\frac{x}{x + 1}\right)^4 + 9 = 25\left(\frac{x}{x + 1}\right)^2$

14. $(x^2 - 6x + 1)^{2/3} - (x^2 - 6x + 1)^{1/3} = 6$

EXERCISES 2.9 B

Find the solution set of each equation in Exercises 1–14 over the set of complex numbers. Check each solution.

1. $x^4 + x^2 = 6$

2. $10x^{-2} + 3x^{-1} = 1$

3. $4x^{-4} - 17x^{-2} + 4 = 0$

4. $\dfrac{6}{r^2} + \dfrac{1}{r^4} = 7$

5. $(x^2 - 6x)^2 - 2(x^2 - 6x) = 35$

6. $(x + 2)^4 + 24 = 10(x + 2)^2$

7. $y^{2/3} = 20 + y^{1/3}$

8. $t^8 - 12t^4 + 32 = 0$

9. $x^6 - 7x^3 - 8 = 0$

10. $\left(\dfrac{x}{x-1}\right)^2 = \dfrac{9x}{2x-2} + \dfrac{5}{2}$

11. $(8x + 41)^{1/2} - 8(8x + 41)^{1/4} + 15 = 0$

12. $28y^{-2/3} - 3y^{-1/3} = 1$

13. $(x^2 - 5x) + 12 = 8\sqrt{x^2 - 5x}$

14. $\sqrt{x^2 - 6x} - 3\sqrt[4]{x^2 - 6x} + 2 = 0$

2.10   IRRATIONAL EQUATIONS

An equation containing a variable in a radicand is called an irrational equation.

Some examples of irrational equations are:

$$\sqrt{x} = 4$$
$$\sqrt{x + 2} = 3$$
$$5\sqrt{x + 2} - \sqrt{x + 1} = 5$$

Solutions of irrational equations can often be found by applying the following theorem.

THEOREM

The solution set of the open equation $A = B$ is a **subset** of the solution set of $A^n = B^n$, where n is a natural number.

If two numbers are equal, then their squares are equal, their cubes are equal, and, in general, their nth powers are equal. However, if the nth powers of two numbers are equal, the numbers may not necessarily be equal. For example, $(-5)^2 = (+5)^2$ but $-5 \neq +5$.

EXAMPLE 2.10.1   Compare the solution sets of the equations $x = 3$ and $x^2 = 9$.

Solution   The solution set of $x = 3$ is $\{3\}$. The solution set of $x^2 = 9$ is $\{3, -3\}$. The solution set of $x = 3$ is a subset of the solution set of $x^2 = 9$. In symbols, $\{3\} \subset \{3, -3\}$.

EXAMPLE 2.10.2   Solve $\sqrt{x^2 - 9} = 4$.

Solution $\sqrt{x^2 - 9} = 4$

$$x^2 - 9 = 16 \quad \text{(Squaring each side)}$$
$$x^2 - 25 = 0$$
$$(x - 5)(x + 5) = 0$$
$$x = 5 \quad \text{or} \quad x = -5$$

The solution set of $x^2 - 9 = 16$ is $\{5, -5\}$. The solution set of $\sqrt{x^2 - 9} = 4$ is a subset of $\{5, -5\}$. Therefore, each member of $\{5, -5\}$ must be checked in the original equation to determine if it is a solution or not.

Check: $\sqrt{x^2 - 9} = 4$ for $x = 5$,
$$\sqrt{(5)^2 - 9} = \sqrt{25 - 9} = \sqrt{16} = 4$$

Therefore, 5 is a solution.

Check: $\sqrt{x^2 - 9} = 4$ for $x = -5$,
$$\sqrt{(-5)^2 - 9} = \sqrt{25 - 9} = \sqrt{16} = 4$$

Therefore, -5 is a solution.
The solution set of $\sqrt{x^2 - 9} = 4$ is $\{5, -5\}$.

EXAMPLE 2.10.3   Solve $\sqrt{x - 1} = 3 - x$.

Solution $(\sqrt{x - 1})^2 = (3 - x)^2$

$$x - 1 = 9 - 6x + x^2$$
$$x^2 - 6x + 9 = x - 1$$
$$x^2 - 7x + 10 = 0$$
$$(x - 2)(x - 5) = 0$$
$$x = 2 \quad \text{or} \quad x = 5$$

Check: $\sqrt{x - 1} = 3 - x$ for $x = 2$,
$$\sqrt{2 - 1} = 3 - 2$$
$$1 = 1$$

Thus 2 is a solution.

Check: $\sqrt{x - 1} = 3 - x$ for $x = 5$,
$$\sqrt{5 - 1} = 3 - 5$$
$$2 \neq -2$$

Thus 5 is not a solution. The solution set is $\{2\}$.

When a number is a root (solution) of the transformed equation $A^n = B^n$ and it is not a root of the original equation $A = B$, it is called an **extraneous root**. Actually, this term is misleading, since it is not a root at all.

Some equations involving radicals may require several applications of raising each side to an nth power. If an equation contains more than one radical, it is usually desirable to first find an equivalent equation containing a single radical on one side of the equation. This process is called isolating a radical.

EXAMPLE 2.10.4   Solve $\sqrt{x} - \sqrt{x-5} = 1$.

Solution $\sqrt{x} - \sqrt{x-5} = 1$
$$\sqrt{x} - 1 = \sqrt{x-5}$$
$$(\sqrt{x} - 1)^2 = (\sqrt{x-5})^2$$
$$x - 2\sqrt{x} + 1 = x - 5$$
$$-2\sqrt{x} + 1 = -5$$
$$-2\sqrt{x} = -6$$
$$\sqrt{x} = 3$$
$$(\sqrt{x})^2 = 3^2$$
$$x = 9$$

Check: $\sqrt{x} - \sqrt{x-5} = 1$
$$\sqrt{9} - \sqrt{9-5} = 1$$
$$3 - 2 = 1$$
$$1 = 1$$

The solution set is $\{9\}$.

EXAMPLE 2.10.5   Solve $\sqrt{2x - \sqrt{x-5}} = 4$.

Solution $\sqrt{2x - \sqrt{x-5}} = 4$
$$2x - \sqrt{x-5} = 16$$
$$2x - 16 = \sqrt{x-5}$$
$$4x^2 - 64x + 256 = x - 5$$
$$4x^2 - 65x + 261 = 0$$
$$(4x - 29)(x - 9) = 0$$
$$x = \tfrac{29}{4} \quad \text{or} \quad x = 9$$

Check: For $x = \frac{29}{4}$,

$$\sqrt{2x - \sqrt{x - 5}} = \sqrt{\frac{29}{2} - \sqrt{\frac{29}{4} - \frac{20}{4}}}$$

$$= \sqrt{\frac{29}{2} - \frac{3}{2}}$$

$$= \sqrt{13}$$

Since $\sqrt{13} \neq 4$, $\frac{29}{4}$ is not a solution.

Check: For $x = 9$,

$$\sqrt{2x - \sqrt{x - 5}} = \sqrt{18 - \sqrt{9 - 5}}$$

$$= \sqrt{18 - 2} = \sqrt{16} = 4$$

Therefore, 9 is a solution. The solution set is {9}.

It should be noted that checking is part of the solution process for an irrational equation. In other words, **checking is mandatory.**

In fact, some irrational equations may not have any solution, as the next example illustrates.

EXAMPLE 2.10.6   Solve $\sqrt{x - 3} - \sqrt{x} = 3$.

Solution $\sqrt{x - 3} = \sqrt{x} + 3$

$$(\sqrt{x - 3})^2 = (\sqrt{x} + 3)^2$$

$$x - 3 = x + 6\sqrt{x} + 9$$

$$-12 = 6\sqrt{x}$$

$$\sqrt{x} = -2$$

$$x = 4$$

Check: For $x = 4$,

$$\sqrt{x - 3} - \sqrt{x} = \sqrt{4 - 3} - \sqrt{4} = 1 - 2 = -1$$

Since $-1 \neq 3$, 4 is not a solution. The solution set is $\varnothing$, the empty set.

Note that $\sqrt{x} \geq 0$ for x a nonnegative real number; $\sqrt{x} = -2$ has no real solution.

EXERCISES 2.10 A

Determine the solution set of each equation in Exercises 1–20.

1. $\sqrt{2x + 3} = 5$ **2.** $\sqrt{2x} + 3 = 5$

3. $x + 2 + \sqrt{x + 8} = 0$ **4.** $x + 2 - \sqrt{x + 8} = 0$

5. $\sqrt{2x-3} - \sqrt{x+2} = 0$ **6.** $\sqrt{2x-3} + \sqrt{x+2} = 0$

7. $\sqrt[3]{x+3} = 3$ **8.** $1 + \sqrt{5x^2 - 5x - 1} = 2x$

9. $1 - \sqrt{5x^2 - 5x - 1} = 2x$ **10.** $\sqrt[3]{x^3 + 4x^2 - 2} = x + 2$

11. $x + 4 - \sqrt{2x^2 + 2x} = 0$ **12.** $x + 4 + \sqrt{2x^2 + 2x} + 2x = 0$

13. $\sqrt{y^2 + 3y + 2} - \sqrt{y+1} = 0$ **14.** $x - 1 + \sqrt{x-1} = 0$

15. $x - 1 - \sqrt{x-1} = 0$ **16.** $\sqrt{5y+1} - \sqrt{3y-5} = 2$

17. $\sqrt{7t+4} - \sqrt{3t+40} = 2$ **18.** $\sqrt{2x} + \sqrt{x+4} = 2$

19. $\sqrt{x} - \sqrt{x+3} = 3$ **20.** $\sqrt{2x} + \sqrt{2x+8} = 2$

EXERCISES 2.10 B

Determine the solution set of each equation in Exercises 1–20.

1. $\sqrt{x+8} = 3$ **2.** $\sqrt{x} + 8 = 3$

3. $\sqrt{2t+13} = t + 7$ **4.** $\sqrt{x} + \sqrt{2} = \sqrt{x+2}$

5. $2y = 9 + \sqrt{8y+9}$ **6.** $2y = 9 - \sqrt{8y+9}$

7. $r - \sqrt{r-5} = 5$ **8.** $2 + \sqrt[3]{x^2 + 15} = 6$

9. $\sqrt{3-x} + \sqrt{2+x} = 3$ **10.** $\sqrt{3-x} - \sqrt{2+x} = 3$

11. $2\sqrt{2x+5} = 1 + \sqrt{8x+1}$ **12.** $\sqrt{x-2} + \sqrt{x} - 2 = 0$

13. $\sqrt{x+5} = x + 5$ **14.** $\sqrt[3]{x^3 + 16x - 3} = x + 1$

15. $\sqrt{2y+45} - \sqrt{54-y} = 3$ **16.** $\sqrt{2x-5} + \sqrt{3x+1} = 3$

17. $\sqrt{y^2 - \sqrt{2y-2}} + y = 2$ **18.** $\sqrt{y+6} - \sqrt{y+2} = \sqrt{8y-1}$

19. $\sqrt{8x+25} - \sqrt{2x+5} = \sqrt{2x+8}$

20. $\sqrt{x} - \sqrt{x+3} = \sqrt{2x+7}$

2.11 VERBAL PROBLEMS

There are many practical applications that involve solving a quadratic equation. When finding the solutions to verbal problems that involve the application of quadratic equations, it is especially important to check each root of the equation in the statement of the problem to see if the necessary conditions are met. Often the equation will have two roots, but only one may apply to a given problem. For example, lengths of sides of rectangles, triangles, etc. are always positive numbers; ages of individuals are positive numbers; digits in a numeral cannot be fractions; and the number of people present at a certain gathering cannot be fractional or negative.

EXAMPLE 2.11.1 If the legs of a right triangle measure 5 inches and 12 inches, respectively, what is the length of the hypotenuse?

Solution The theorem of Pythagoras states that if a and b are the measures of the legs of a right triangle and c represents the measure of the hypotenuse, then

$$a^2 + b^2 = c^2$$

Let $a = 5$, $b = 12$; then

$$5^2 + 12^2 = c^2$$

$$25 + 144 = c^2$$

$$169 = c^2$$

$$\pm 13 = c$$

Since length is a positive number, the condition $c > 0$ is implied. Thus the common solution of $c = 13$ or $c = -13$ and $c > 0$ is $c = 13$.

EXAMPLE 2.11.2 If the hypotenuse of a right triangle is 25 inches long and one leg measures 24 inches, how long is the other leg?

Solution Applying the theorem of Pythagoras as in the preceding example, let $c = 25$, $a = 24$, and b designate the other leg:

$$a^2 + b^2 = c^2$$

$$(24)^2 + b^2 = (25)^2$$

$$b^2 = (25)^2 - (24)^2$$

$$b^2 = (25 - 24)(25 + 24)$$

$$b^2 = 49$$

$$b = \pm 7$$

Again we disregard the solution $b = -7$. Therefore, the other leg is 7 inches long.

EXAMPLE 2.11.3   A plane flies 300 miles with a tail wind of 10 mph and returns against a wind of 20 mph. What is the speed of the plane in still air if the total flying time is $4\frac{1}{2}$ hours?

Solution   Let x represent the speed of the plane in still air.

Then $x + 10 =$ speed with the tail wind

and $x - 20 =$ speed against the wind

Formula	r	$\cdot$	t	$=$	d
With wind	$x + 10$		$\dfrac{300}{x + 10}$		300
Against wind	$x - 20$		$\dfrac{300}{x - 20}$		300

Equation: Time going + time returning = total time = $4\frac{1}{2}$ hours = $\frac{9}{2}$ hours.

$$\frac{300}{x + 10} + \frac{300}{x - 20} = \frac{9}{2}$$

Multiplying both sides of this equation by the L.C.D.,

$$2(x + 10)(x - 20)$$

yields

$$600(x - 20) + 600(x + 10) = 9(x + 10)(x - 20)$$

Dividing both sides by 3,

$$200(x - 20) + 200(x + 10) = 3(x + 10)(x - 20)$$

$$200x - 4000 + 200x + 2000 = 3x^2 - 30x - 600$$

$$3x^2 - 430x + 1400 = 0$$

$$(3x - 10)(x - 140) = 0$$

$$3x - 10 = 0 \quad \text{or} \quad x - 140 = 0$$

$$x = 3\tfrac{1}{3} \quad \text{or} \quad x = 140$$

Since speed is always a positive number, $x \neq 3\frac{1}{3}$ because

$$x - 20 = 3\tfrac{1}{3} - 20$$

which is a negative number.

Therefore, the speed of the plane in still air is 140 mph.

EXAMPLE 2.11.4   A square flower bed has a 3-foot walk surrounding it. If the walk were to be replaced and planted with flowers, the new flower bed would have 4 times the area of the original bed. What is the length of one side of the original bed?

Solution   Let x = length of a side of the original flower bed.

Then x^2 = area of original bed

$x + 6$ = length of a side of the new flower bed

and $(x + 6)^2$ = area of the new flower bed

$$(x + 6)^2 = 4x^2$$
$$x^2 + 12x + 36 = 4x^2$$
$$3x^2 - 12x - 36 = 0$$
$$3(x^2 - 4x - 12) = 0$$
$$x^2 - 4x - 12 = 0$$
$$(x + 2)(x - 6) = 0$$
$$x + 2 = 0 \quad \text{or} \quad x - 6 = 0$$
$$x = -2 \quad \text{or} \quad x = 6$$

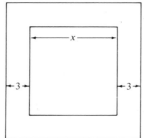

FIGURE 2.11.1

Again we discard the negative answer, and the length of the original side is 6 feet.

Check: Area of original bed = $x^2 = 6^2 = 36$ square feet

Area of new bed = $(x + 6)^2 = (6 + 6)^2 = 12^2 = 144$ square feet

Area of new bed = 4 times area of old bed

$$144 = 4 \times 36$$
$$144 = 144$$

EXAMPLE 2.11.5 Working alone, a carpenter can make a set of cabinets in 3 hours less time than his helper can. Working together, they can make the set of cabinets in 6 hours. Find the time (correct to the nearest minute) that each requires to make the set alone.

Solution

Let $x = $ time it takes helper alone

Then $x - 3 = $ time it takes carpenter alone

   Formulas:

   $tr = w$, and w of first $+ w$ of second $= w$ of both $= 1$

where $t = $ time, $r = $ rate, and $w = $ amount of work done.

		Working Alone				Working Together (*whole job*)	
	t	$\cdot$ r	$= w$	t	$\cdot$ r	$= w$
Carpenter	$x - 3$	$\dfrac{1}{x-3}$	1	6	$\dfrac{1}{x-3}$	$\dfrac{6}{x-3}$
Helper	x	$\dfrac{1}{x}$	1	6	$\dfrac{1}{x}$	$\dfrac{6}{x}$

   Equation: Work of helper + work of carpenter = whole job.

$$\frac{6}{x} + \frac{6}{x-3} = 1 \qquad \text{with restrictions } x > 0 \text{ and } x - 3 > 0$$

$$6(x - 3) + 6x = x(x - 3)$$

$$12x - 18 = x^2 - 3x$$

$$x^2 - 15x + 18 = 0$$

Since $x^2 - 15x + 18$ is not readily factorable, and since $b/a = -15$ is not an even integer, the method using the quadratic formula is recommended for solving the equation.

   Solution of equation:

$$ax^2 + bx + c = 0 \text{ if and only if } x = \frac{-b \pm \sqrt{b^2 - 4ac}}{2a}$$

If $x^2 - 15x + 18 = 0$, then $a = 1$, $b = -15$, $c = 18$.

Thus

$$x = \frac{-(-15) \pm \sqrt{(-15)^2 - 4(18)}}{2} = \frac{15 \pm \sqrt{225 - 72}}{2}$$

$$x = \frac{15 + \sqrt{153}}{2} \quad \text{or} \quad x = \frac{15 - \sqrt{153}}{2}$$

Approximating (by using the tables), $\sqrt{153} = 12.37$. Thus, to the nearest hundredth,

$$x = \frac{15 + 12.37}{2} \quad \text{or} \quad x = \frac{15 - 12.37}{2}$$

$$x = \frac{27.37}{2} \quad \text{or} \quad x = \frac{2.63}{2}$$

$$x = 13.69 \quad \text{or} \quad x = 1.32$$

and

$$x - 3 = 10.69 \quad \text{or} \quad x - 3 = -1.68$$

(This solution rejected, since $x - 3 > 0$)

Thus $x = 13.69$ and $x - 3 = 10.69$, correct to the nearest hundredth. Since there are 60 minutes in an hour,

0.69 hour = 0.69(60) minutes = 41.4 minutes

Therefore, correct to the nearest minute,

$x = 13$ hours 41 minutes, time of helper alone
$x - 3 = 10$ hours 41 minutes, time of carpenter alone

EXERCISES 2.11 A

1. The hypotenuse of a right triangle is 3 units and the legs are equal in length. Find the length of a leg of the triangle.
2. The length of a rectangle exceeds 3 times its width by 1 inch. The area is 52 square inches. Find the dimensions of the rectangle.
3. A certain number exceeds its reciprocal by $\frac{15}{4}$. Find the number.
4. It takes John 3 hours longer to do a certain job than it does his brother Bob. For 3 hours they worked together; then John left and Bob finished the job in 1 hour. How many hours would it have taken Bob to do the whole job by himself?
5. A rectangular piece of sheet metal is twice as long as it is wide. From each of its four corners a square piece 2 inches on a side is cut out. The

flaps are then turned up to form an uncovered metal box. If the volume of this box is 320 cubic inches, find the dimensions of the original piece of sheet metal.

6. The sum of two numbers is 7 and the difference of their reciprocals is $\frac{1}{12}$. Find the numbers.

7. The roof line of a certain house has a pitch of 3 to 12. This means that for each vertical rise of 3 feet, there is a horizontal run of 12 feet. To find the length of lumber he must cut for the roof, the carpenter has to calculate the length of the hypotenuse of a right triangle whose legs are the height of the roof and the half-span of the roof. Find the length of this hypotenuse, correct to the nearest inch, for this house having a half-span of 13 feet.

8. A fisherman trolled upstream in a motorboat to a spot 6 miles from his camp site and then returned to camp. If the rate of the current was $1\frac{1}{2}$ mph and if the round trip took 3 hours, find the rate of the motorboat in still water.

9. The span s of a circular arch is related to its height h and its radius r by the formula

$$s^2 = 8rh - 4h^2$$

Find the height of a circular arch whose span is 80 feet and whose radius is 50 feet. (Assume the circular arch is less than a semicircle.)

10. A circular hole has a radius of 5 inches. How much larger should the radius be so that a new circular hole will have a cross-sectional area twice as large as the original area? (Area of circle, $A = \pi r^2$.)

11. The ionization constant K of a weak acid is related to the hydrogen ion concentration x and the molarity M of the acid solution by the equation

$$\frac{x^2}{M - x} = K$$

Find the hydrogen ion concentration for a certain solution of acetic acid having $M = 0.0002$ and $K = 1.8 \times 10^{-5}$.

12. A private plane flew from San Francisco to Lake Tahoe, a distance of 180 miles, with a tail wind and then returned against the same wind. If the total flying time was $2\frac{1}{2}$ hours and if the speed of the plane in still air was 150 mph, find the speed of the wind.

13. One man can do a job in 8 days less time than another man, but he charges $50 a day, whereas the slower man charges $20 a day. When the two men work together, they take 3 days to complete the job. Which would cost the less, to have the faster man do the job alone, to have the slower man do the job alone, or to have both work together? State the cost for each case.

14. A fisherman in a boat on a small lake sees some plants growing in the water with their roots at the bottom of the lake. To find the depth of the lake, he pushes a plant extending 8 inches out of the water so that the plant is completely submerged with its tip just touching the water. He measures the distance from the point where the plant first emerged from the water to the new position of its tip and finds this to be 32 inches. Find the depth of the lake.

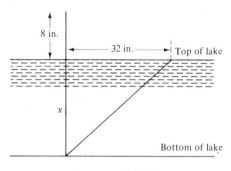

FIGURE 2.11.2

15. A dive bomber traveling 480 feet per second vertically downward releases a bomb 1024 feet above the earth. If s is the distance of the bomb below the point where it was dropped and t is the time in seconds after the bomb was dropped, then

$$s = 480t + 16t^2$$

How long does the bomb take to reach the earth? With what velocity v does the bomb hit the earth if

$$v = 480 + 32t$$

16. For a certain electric motor having a mechanical output of 25,000 watts, a resistance in the armature of 0.04 ohm, and a line voltage of 110 volts, the armature current I in amperes is given by

$$25,000 = 110I - 0.04I^2$$

Find the current I.

EXERCISES 2.11 B

1. A concrete walk of uniform width extends around a rectangular lawn having dimensions of 20 feet by 80 feet. Find the width of the walk if the area of the walk is 864 square feet.

2. The hypotenuse of a right triangle is 13 inches. If one leg is 7 inches longer than the other, how long are the legs of the triangle?

3. If 4 times a number is added to 3 times its square, the sum is 95. Find the number.

4. Find two consecutive odd integers the sum of whose squares is 514.

5. A 36-inch length of copper tubing is bent to form a right triangle having a 15-inch hypotenuse. Find the lengths of the other two sides of the triangle.

6. A wire is stretched from the top of a 4-foot fence to the top of a 20-foot vertical pole. If the fence and the pole are 30 feet apart, find the length of the wire.

7. One of two outlets can fill a swimming pool in 6 hours. The time for the other outlet to fill the pool is 2 hours longer than the two outlets together. Find the time it takes for the two outlets together to fill the tank, correct to the nearest minute.

8. One inlet pipe takes 12 minutes longer than another inlet pipe to fill a certain tank. An outlet pipe can empty the tank in 45 minutes. When all three pipes are open, it takes 15 minutes to fill the tank. Find the time it takes to fill the tank if only the larger inlet pipe is open.

9. A jet plane flying against a head wind of 20 mph takes 20 minutes longer to fly a distance of 2610 miles than a plane with the same still air speed flying in the opposite direction. Find the still air speed of the plane.

10. A boat that travels 12 mph in still water takes 2 hours less time to go 45 miles downstream than to return the same distance upstream. Find the rate of the current.

11. A baseball diamond has the shape of a square with each side 90 feet long. The pitcher's mound is 60.5 feet from home plate on the line joining home plate to second base. Find the distance from the pitcher's mound to second base.

12. The bending moment M of a beam fixed at one end and simply supported at the other is related to its length L, its uniform load distribution w, and the distance x from the fixed end by the relation

$$M = \frac{-w}{8}(4x^2 - 5Lx + L^2)$$

A certain beam has a length L of 12 feet and a uniform load distribution w of 100 pounds per foot.

a. For what position x on the beam is the bending moment 0?

b. For what position x on the beam is the bending moment 100 foot-pounds?

13. If I is the intensity of illumination in lumens, c is the candlepower of the source of light, and s is the distance in feet from the source, then

$$I = \frac{c}{s^2}$$

A 20-candlepower light is 3 feet to the left of a 45-candlepower light. Find the distance x from the 20-candlepower source on the line joining the two sources so that the illumination is the same from each source. Use

$$\frac{20}{x^2} = \frac{45}{(3 - x)^2}$$

14. A mechanized division 2 miles long is moving at the rate of 30 mph. A messenger from the rear rushes to the head of the division and immediately returns to the rear, arriving $7\frac{1}{2}$ minutes after he left. Find the speed of the messenger.

15. The number of grams x of barium sulfate that is dissolved when 500 milliliters of 0.0001 M ammonium sulfate is mixed with an excess of solid barium sulfate, using 1.00×10^{-10} as the solubility product and 233 as the molecular weight of barium sulfate, is given by

$$\frac{2x}{233}\left(0.0001 + \frac{2x}{233}\right) = 1.00 \times 10^{-10}$$

Find x. (*Hint:* Let $y = 2x/233$ and first solve for y.)

16. If an object is dropped a distance of s feet and if the sound is heard t seconds later, then

$$s = 16\left(t - \frac{s}{1120}\right)^2$$

where 1120 feet per second is used as the speed of sound. A stone is dropped from the top of a building and is heard hitting the ground 7.7 seconds later. Find the height of the building. (Find s for $t = 7.7$.)

SUMMARY

Complex Numbers

☐ **Definition.** $i = \sqrt{-1}$ and $i^2 = -1$.

☐ **Definition.** $\sqrt{-a} = i\sqrt{a}$ if a is a positive real number.

☐ **Standard form:** Every complex number can be expressed in the form $a + bi$, called the standard form, where a and b are real numbers and i is the imaginary unit such that $i^2 = -1$.

☐ **Definition.** Two complex numbers are equal if and only if their real parts are equal and their imaginary parts are equal—that is,

$$a + bi = c + di \quad \text{if and only if} \quad a = c \text{ and } b = d$$

☐ **Definition.** The conjugate of $a + bi$ is $a - bi$.

The conjugate of $a - bi$ is $a + bi$.

☐ **Definition.** Sum: $(a + bi) + (c + di) = (a + c) + (b + d)i$.

Product: $(a + bi)(c + di) = (ac - bd) + (ad + bc)i$.

☐ **The quadratic formula:** The quadratic equation $ax^2 + bx + c = 0$ $(a \neq 0)$ has the solutions

$$x = \frac{-b \pm \sqrt{b^2 - 4ac}}{2a}$$

☐ **Sum and product of roots theorem:** $\{r, s\}$ is the solution set of the equation $ax^2 + bx + c = 0$ $(a \neq 0)$ if and only if $r + s = -b/a$ and $rs = c/a$.

☐ The discriminant of a quadratic equation is $b^2 - 4ac$.

If $b^2 - 4ac > 0$, then the roots are real and unequal.

If $b^2 - 4ac = 0$, then the roots are real and equal.

If $b^2 - 4ac < 0$, then the roots are imaginary.

☐ The solution set of $A = B$ is a subset of the solution set of $A^n = B^n$, where $A = B$ is an open equation and n is a natural number.

REVIEW EXERCISES

1. Write each of the following complex numbers in the standard form:

a. $2 + \sqrt{-16}$ b. $(3 + 5i) - (2 - 3i)$

c. $\dfrac{4 + \sqrt{-12}}{2}$ d. $(2 + 3i)(3 - 4i)$

e. $2i(3 + 4i)^2$ f. $\dfrac{1 - i}{2 + 3i}$

2. Determine the real numbers x and y for which each equation is true:

a. $x + yi = 3 + 2i$ b. $(x + 2y + 2) + (2x + y)i = 0$

3. Two forces represented by the vectors $7 - 8i$ and $5 + 3i$ act on an object. Find the resultant force vector, its magnitude, and its direction. Illustrate geometrically.

4. Show that

$$\sqrt{\frac{9}{2} + \frac{9i\sqrt{3}}{2}} = \frac{3\sqrt{3}}{2} + \frac{3i}{2}$$

by showing that their squares are equal.

5. Determine the solution set of each of the following equations over the complex numbers by the factoring method; check each solution:

  a. $x^2 - 6x - 40 = 0$ b. $12x^2 + 32x + 5 = 0$

  c. $3x^2 - 4x = 0$ d. $x^2 + 9 = 0$

6. Solve the following equations by completing the square; check each solution:

  a. $x^2 + 8x + 15 = 0$ b. $2x - x^2 - 3 = 0$ c. $3x^2 = 12x + 3$

  d. $(x + 4)(x + 6) = 7$

7. Solve the following equations by the quadratic formula; check each solution:

  a. $2x^2 + 5x - 1 = 0$ b. $2 = 2x^2 - x$ c. $6x^2 - 7x - 20 = 0$

  d. $16x = 3x^2 + 8$

8. Solve the following equations by any method you wish; check each solution:

  a. $\frac{3}{4}x^2 = \frac{7}{8}$ b. $x^2 + 4x + 8 = 0$ c. $(2x + 3)(x - 2) = 4$

  d. $3x^2 + 5x = -1$

9. Find the value of k for which the roots of each equation are equal:

  a. $x^2 + 6x + k = 0$ b. $x^2 + 2kx + 4 = 0$

  c. $kx^2 + 5x + 5 = 0$ d. $kx^2 + 8x + k = 0$

10. State a quadratic equation whose roots have the given sum and product:

  a. $r + s = -2$; $rs = 4$ b. $r + s = \frac{1}{2}$; $rs = \frac{3}{4}$

11. Solve each of the following for y:

  a. $10x^2 = y^2 - 3xy$ b. $y^2 + 2x^2y = x^4$

12. State the sum of the roots of each equation:

  a. $x^2 - 6x + c = 0$ b. $6x^2 + 12x + c = 0$

  c. $5x^2 + k^2 = 0$

13. State the product of the roots of each equation:

    a. $x^2 + bx + 4 = 0$ b. $3x^2 - bx - 6 = 0$ c. $4x^2 + 12 = 0$

    d. $5x^2 + 10x = 0$

14. Find the solution set of each of the following:

    a. $x^{2/3} - 6x^{1/3} + 5 = 0$ b. $\sqrt{y + 2} - \sqrt[4]{y + 2} = 6$

    c. $5y^{-2} + 9y^{-1} - 2 = 0$ d. $\sqrt{x + 15} = x - 5$

    e. $17 - \sqrt{x - 3} = 10 + \sqrt{32 + x}$

    f. $\sqrt{y - 2} + \sqrt{2y - 2} = \sqrt{3y + 20}$

15. John can mow his law in 20 minutes less time with his power mower than with his hand mower. One day his power mower broke down 15 minutes after he started mowing, and he had to complete the job with his hand mower. It took him 25 minutes to finish mowing by hand. How long does it take John to do the complete job with the power mower?

16. A pilot left a Chicago airport and flew 200 miles south to a town T with a tail wind of 20 mph. From T he flew back to Chicago against a head wind of 30 mph. If his total flying time was $2\frac{1}{3}$ hours, what was the average speed of the plane in still air?

17. A farmer has 90 feet of fencing he can use to enclose a rectangular piece of land. Find the dimensions of the rectangle if the area is to be 450 square feet.

18. One leg of a right triangle is 9 inches longer than the other leg. The hypotenuse is 45 inches long. Find the lengths of the legs of the triangle.

19. The height H of a projectile at the end of t seconds is given by

$$H = cvt - \tfrac{1}{2}gt^2$$

Solve for t.

20. The total surface area T of a right circular cylinder of radius r and height h is given by

$$T = 2\pi r(r + h)$$

Solve for r.

21. The ancient Greeks considered the most beautiful rectangle to be one in which the ratio of the length to the width was equal to the ratio of width to the length minus the width.

$$\frac{L}{W} = \frac{W}{L - W}$$

Find the exact numerical value of L/W, the "Golden Ratio." Compare L/W with W/L.

22. A manufacturer finds that the number N of lamps demanded daily by the public is given by $N = 2000/x$, where x is the price per lamp. The number of lamps he can supply daily is given by $N = 30x - 500$. Find the price, called the equilibrium price, when the demand equals the supply—that is, when

$$\frac{2000}{x} = 30x - 500$$

CHAPTER **3**

FACTORING, INEQUALITIES, ABSOLUTE VALUE

3.1 FACTORING BY GROUPING AND SUBSTITUTION

The following types of factoring have been discussed:

1. $AB + AC = A(B + C)$	Distributive axiom
2. $A^2 - B^2 = (A - B)(A + B)$	Difference of squares
3. $X^2 + 2AX + A^2 = (X + A)^2$	Perfect square trinomial
4. $X^2 + (A + B)X + AB = (X + A)(X + B)$	Simple trinomial
5. $ACX^2 + (AD + BC)X + BD =$	
$\quad (AX + B)(CX + D)$	General trinomial
6. $A^3 - B^3 = (A - B)(A^2 + AB + B^2)$	Difference of cubes
7. $A^3 + B^3 = (A + B)(A^2 - AB + B^2)$	Sum of cubes

Since these are basic types or forms, they can be used to factor more complicated expressions.

EXAMPLE 3.1.1   Factor $3(x + 2)^2 - (x + 2) - 2$.

Solution   If the substitution $y = x + 2$ is made, the expression can be written

$$3y^2 - y - 2$$

which factors as

$$(3y + 2)(y - 1)$$

Replacing y by $x + 2$ yields

$$[3(x + 2) + 2][(x + 2) - 1]$$

which simplifies as

$$(3x + 8)(x + 1)$$

Therefore, $3(x + 2)^2 - (x + 2) - 2 = (3x + 8)(x + 1)$.

EXAMPLE 3.1.2   Factor $4(a + n)^2 - x^2$.

Solution   Recognizing the form,

$$X^2 - A^2 = (X - A)(X + A)$$

$$4(a + n)^2 - x^2 = (2[a + n])^2 - x^2 = (2[a + n] - x)(2[a + n] + x)$$

$$= (2a + 2n - x)(2a + 2n + x)$$

EXAMPLE 3.1.3   Factor $x^2 - y^2 - 10y - 25$.

Solution   The presence of more than three terms suggests that the terms should be *grouped*. The recognition of one or more perfect squares indicates *how* the terms should be grouped. Recognizing

$$-y^2 - 10y - 25 = -(y^2 + 10y + 25) = -(y + 5)^2$$

then

$$x^2 - y^2 - 10y - 25 = x^2 - (y^2 + 10y + 25)$$

$$= x^2 - (y + 5)^2 \qquad\qquad = X^2 - A^2$$

$$= (x - [y + 5])(x + [y + 5]) = (X - A)(X + A)$$

$$= (x - y - 5)(x + y + 5)$$

EXAMPLE 3.1.4   Factor $a^4 - 7a^2b^2 + 9b^4$.

Solution   It is not possible to use the form $X^2 + (A + B)X + AB = (X + A)(X + B)$ with $X = a^2$, $A + B = -7b^2$, and $AB = 9b^4$ since the only factors of 9 are (9 and 1) or (3 and 3), and neither of these combinations will produce a middle term with the numerical coefficient -7.

However, *noting the two squares*, a^4 and $9b^4$, this suggests the perfect square trinomial form,

$$X^2 + 2AX + A^2 = (X + A)^2$$

Now
$$(a^2 + 3b^2)^2 = a^4 + 6a^2b^2 + 9b^4$$

and
$$(a^2 - 3b^2)^2 = a^4 - 6a^2b^2 + 9b^4$$

Rewriting the given expression to try to obtain one of these forms, there are two possibilities:

$$1. \ a^4 - 7a^2b^2 + 9b^4 = a^4 + 6a^2b^2 + 9b^4 - 13a^2b^2$$
$$= (a^2 + 3b^2)^2 - 13a^2b^2$$

or

$$2. \ a^4 - 7a^2b^2 + 9b^4 = a^4 - 6a^2b^2 + 9b^4 - a^2b^2$$
$$= (a^2 - 3b^2)^2 - (ab)^2$$

The difference of squares form can be recognized for possibility (2). Finally

$$a^4 - 7a^2b^2 + 9b^4 = a^4 - 6a^2b^2 + 9b^4 - a^2b^2$$
$$= (a^2 - 3b^2)^2 - (ab)^2$$
$$= (a^2 - 3b^2 - ab)(a^2 - 3b^2 + ab)$$

Some polynomials can be factored by grouping pairs of terms; for example, $ax + ay + bx + by$ can be grouped by the associative axiom:

$$(ax + ay) + (bx + by)$$

The first two terms have a common factor, a, and the next two terms have a common factor, b:

$$a(x + y) + b(x + y)$$

The expression now consists of two terms, each of which has a common factor, $(x + y)$.

Now making the substitution $N = x + y$,

$$a(x + y) + b(x + y) = a(N) + b(N)$$
$$= (a + b)(N)$$
$$= (a + b)(x + y)$$

EXAMPLE 3.1.5 Factor $x^3 + 5x^2 - 4x - 20$.

Solution Since a basic factoring form cannot be recognized readily, the terms are grouped to try to find a form that can be recognized. Since no two terms are perfect squares, the terms are grouped in pairs:

$$x^3 + 5x^2 - 4x - 20 = (x^3 + 5x^2) - (4x + 20)$$
$$\text{(Grouping in pairs)}$$

$$= x^2(x + 5) - 4(x + 5)$$
$$\text{(Factoring each group)}$$

$$= x^2(N) - 4(N)$$
$$\text{(Letting } N = x + 5)$$
$$= (x^2 - 4)(N)$$
$$\text{(Using \quad the \quad distributive}$$
$$\text{axiom)}$$

$$= (x^2 - 4)(x + 5)$$

$$= (x - 2)(x + 2)(x + 5)$$
$$\text{(Difference of squares)}$$

It should be recalled that the process of factoring is a relative one— that is, the coefficients of the factors may be restricted to be members of a particular set of numbers. For example, to factor completely over the integers means to express the given polynomial as the product of polynomials of smallest degree with integral coefficients. Similarly, to factor completely over the reals means that the factors should be those polynomials of smallest degree having real coefficients. In solving quadratic equations, it was seen that it was sometimes necessary to factor over the set of complex numbers.

EXAMPLE 3.1.6 Factor over the reals $x^2 - 6$.

Solution $x^2 - 6 = x^2 - (\sqrt{6})^2$
$$= (x - \sqrt{6})(x + \sqrt{6})$$

EXAMPLE 3.1.7 Factor over the complex numbers $x^2 + 4$.

Solution $x^2 + 4 = x^2 - (-4)$
$$= x^2 - (\sqrt{-4})^2$$
$$= x^2 - (2i)^2$$
$$= (x - 2i)(x + 2i)$$

EXAMPLE 3.1.8 Factor $x^4 + 4x^2 - 45$ over: (a) the integers, (b) the reals, (c) the complex numbers.

Solution

a. $x^4 + 4x^2 - 45 = (x^2 - 5)(x^2 + 9)$

b. $x^4 + 4x^2 - 45 = (x - \sqrt{5})(x + \sqrt{5})(x^2 + 9)$

c. $x^4 + 4x^2 - 45 = (x - \sqrt{5})(x + \sqrt{5})(x - 3i)(x + 3i)$

It is often necessary to do several factoring processes before a polynomial is completely factored. For example, factor completely $3ax^4 - 3ay^4$.

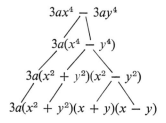

Summarizing these steps,

1. $3ax^4 - 3ay^4 = 3a(x^4 - y^4)$ (Removal of greatest common monomial factor by distributive principle)

2. $= 3a(x^2 + y^2)(x^2 - y^2)$ (Factor difference of two squares)

3. $= 3a(x^2 + y^2)(x + y)(x - y)$
                        (Factor difference of two squares)

4. $= 3a(x + y)(x - y)(x^2 + y^2)$
                        (Commutative property of multiplication)

5. $3ax^4 - 3ay^4 = 3a(x + y)(x - y)(x^2 + y^2)$

It is desirable to arrange the factors in such an order that monomials are written first, then linear factors, followed by quadratic factors, and so on.

PROCEDURE FOR COMPLETE FACTORIZATION

Steps in Order	*Formulas*
1. Remove greatest common monomial factor:	$ab + ac = a(b + c)$
2. Factor binomials, if present:	$a^2 - b^2 = (a - b)(a + b)$
	$a^3 - b^3 = (a - b)(a^2 + ab + b^2)$
	$a^3 + b^3 = (a + b)(a^2 - ab + b^2)$
3. Factor trinomials, if present:	$a^2 + 2ab + b^2 = (a + b)^2$
	$x^2 + (p + q)x + pq = (x + p)(x + q)$
	$rsx^2 + (rq + ps)x + pq = (rx + p)(sx + q)$
4. Factor by grouping, if necessary:	$ax + bx + ay + by = (x + y)(a + b)$
	$a^2 + 2ab + b^2 - x^2 - 2xy - y^2$
	$= (a + b)^2 - (x + y)^2$
	$= (a + b - x - y)(a + b + x + y)$

EXERCISES 3.1 A

Factor Exercises 1–40 completely over the integers.

1. $x(x + 4) + 5(x + 4)$
2. $y(y - 2) - (y - 2)$
3. $(x + 7)^2 - 6(x + 7) + 5$
4. $3(y - 4)^2 + 14(y - 4) - 5$
5. $(t + 3)^4 - 3(t + 3)^2 - 4$
6. $(x^2 + 6)^2 - y^2$
7. $x^4 - (y - 9)^2$
8. $(x - 4)^2 - (y + 6)^2$
9. $x^2(x + 2) + 3(x + 2)$
10. $4x^2(x - 8) - (x - 8)$
11. $(2x + 5)x^3 - 8(2x + 5)$
12. $x^3 + 6x^2 - 9x - 54$
13. $x^4 - x^3 + 27x - 27$
14. $x^2 + 12x + 36 - 4y^2$
15. $x^2 - y^2 - 16y - 64$
16. $x^4 - x^2 + 18x - 81$
17. $y^4 + 4y^2 + 4 - 25x^2$
18. $(x + 1)^3 - (x - 1)^3$
19. $x^4 - 15x^2 - 250$
20. $(t - 1)^4 - 35(t - 1)^2 - 36$
21. $c^2 + 6c + 9 - 16x^2$
22. $3x^3 + 2x^2 - 3xy^2 - 2y^2$
23. $4a^3 - 12a^2 - a + 3$
24. $(x + 2)^2 - (x - 7)^2$
25. $a^3(x - 3)^2 + b^3(x - 3)^2$
26. $(x^2 - 1)^3 + 1$
27. $-(x + 2)^2 + 14(x + 2) - 49$
28. $2p^3 - 3p^2 - 2p - 3$
29. $x^8 - y^8$
30. $t^6 - 64$
31. $p^3(r - s)^3 + p^3$
32. $(2x + 1)^2 - 3(2x + 1) - 10$
33. $x(x + 3)(4x + 9) - 5(x + 3)$
34. $r^2 + 4rs + 4s^2 - r - 2s$

35. $x^2 + 6xy + 9y^2 - 9$

36. $r^4 + 4t^4$ (*Hint:* Add and subtract $4r^2t^2$.)

37. $a^4 + a^2 + 1$ **38.** $x^2 + 2x - 9y^2 + 1$

39. $(x - y)^5 - 4(x - y)^3$ **40.** $x^3 + x^2y + x^2 - xy^2 - y^2 - y^3$

Factor Exercises 41–45 over (a) the integers, (b) the reals, (c) the complex numbers.

41. $x^4 - 2x^2 - 3$ **42.** $2x^6 - 20x^4 + 50x^2$

43. $x^2 - 8x + 10$ **44.** $y^2 - y + 1$

45. $4t^3 + 4t^2 - 25t - 25$

EXERCISES 3.1 B

Factor Exercises 1–40 completely over the integers.

1. $x(x - 5) + (x - 5)$ **2.** $t^2(2t + 7) - 16(2t + 7)$

3. $(r - 6)^2 + 3(r - 6) - 18$ **4.** $5(y + 8)^2 - 3(y + 8) - 2$

5. $(s - 10)^2 - 20(s - 10) + 100$ **6.** $(x^2 - 4)^2 - 4x^2$

7. $y^4 - (y^2 + 5)^2$ **8.** $(s + 3t)^2 - (s - t)^2$

9. $(y - 4)y^2 - (y - 4)$ **10.** $(9x^2 - 1)x^2 - 25(9x^2 - 1)$

11. $(6x^2 - 7x)x^3 + 6x^2 - 7x$ **12.** $x^3 + 2x^2 - 25x - 50$

13. $3x^3 - 2x^2 - 12x + 8$ **14.** $r^2 - s^2 - 6s - 9$

15. $16u^2 - 8u + 1 - 49v^2$ **16.** $t^4 + 12t^2 - 64$

17. $(t - 2)^4 - 34(t - 2)^2 - 72$ **18.** $(x + 1)^3 + (x - 1)^3$

19. $y^4 + 2y^3 - 1000y - 2000$ **20.** $x^4 - 10x^2 + 25 - 49x^2$

21. $x - x^2 - x^3 + x^4$ **22.** $2x^2 - yz^2 - x^2y + 2z^2$

23. $a(1 - b)^2 - c(b - 1)$ **24.** $4x^2 - 4xy + y^2 - 36$

25. $(x + 3)^2 + 3(x + 3) + 2$ **26.** $p^3 - (r - s)^3$

27. $b^6 - 1$ **28.** $8x^6 - 38x^3 + 35$

29. $a^6 + 9a^3 + 18$ **30.** $3ax^2 - 3ay^2 - x^2 + y^2$

31. $1 - x^2 - 2xy - y^2$ **32.** $9 - 4a^2 - 4a - 1$

33. $-x^2 - y^2 + 2xy + z^2$ **34.** $(x^2 - 1)^2 + 3(x^2 - 1) + 2$

35. $(a^2 + 2a - 3)^2 - 4$ **36.** $2x^4 - x^3 - 4x + 2$

37. $x^2 - 9y^2 + 5x + 15y$ **38.** $a^4 - 9(2a - 3)^2$

39. $2x^2 + y^2 + 3xy + 3x + 3y$

40. $25x^2 - x^2y^2 + 10xy - a^2 - 2axy + y^2$

Factor Exercises 41–45 over (a) *the integers,* (b) *the reals,* (c) *the complex numbers.*

41. $x^4 + 14x^2 - 32$ **42.** $5x^5 + 30x^3 + 45x$

43. $x^2 - 10x + 29$ **44.** $y^2 - y - 1$

45. $16z^3 - 16z^2 + 49z - 49$

3.2 SYNTHETIC DIVISION

The notation $P(x)$ is used to indicate a polynomial in the variable x. For example, $P(x) = x^4 + 3x^2 + 2x + 1$ is read "P of x is the polynomial $x^4 + 3x^2 + 2x + 1$." Similarly, the symbols $Q(x)$ or $R(x)$ may represent polynomials in x.

If a given polynomial $P(x)$ is divided by a **linear polynomial of the form** $(x - a)$, the process may be done by long division. Since this particular type of quotient is frequently used in algebra, and since the long-division algorithm is often tedious, a shorter method called *synthetic division* has been devised.

To illustrate, let us first divide $x^4 + 3x^2 + 2x + 1$ by $x - 2$, using the long-division algorithm:

$$
\begin{array}{r}
x^3 + 2x^2 + 7x + 16 \\
x - 2\overline{)x^4 \qquad\; + 3x^2 + 2x + 1} \\
\underline{x^4 - 2x^3} \qquad\qquad\qquad\; \\
+ 2x^3 + 3x^2 \qquad\quad \\
\underline{+ 2x^3 - 4x^2} \qquad\quad \\
+ 7x^2 + 2x \quad \\
\underline{+ 7x^2 - 14x} \quad \\
+ 16x + 1 \\
\underline{+ 16x - 32} \\
+ 33 \quad \text{(Remainder)}
\end{array}
$$

Notice that both polynomials were arranged in descending powers of x, and that a space was left for the missing power of x whose coefficient was zero, $(0x^3)$.

If $P(x)$ is the dividend polynomial and $Q(x)$ the quotient polynomial, the coefficients of the terms of $P(x)$ are 1, 0, 3, 2, 1, and the coefficients of the terms of $Q(x)$ are 1, 2, 7, 16. The remainder is 33. Thus

$$\frac{x^4 + 3x^2 + 2x + 1}{x - 2} = x^3 + 2x^2 + 7x + 16 + \frac{33}{x - 2}, \quad x \neq 2$$

and

$$x^4 + 3x^2 + 2x + 1 = (x - 2)(x^3 + 2x^2 + 7x + 16) + 33 \text{ even if } x = 2$$

Therefore $P(x) = (x - 2)Q(x) + 33$

Rewriting the division problem, but this time omitting the variable and using only the numerical coefficients,

$$
\begin{array}{r}
1 \quad\;\; 2 \quad\; 7 \quad\; 16 \\
1 - 2)\overline{1 \quad\;\; 0 \quad\; 3 \quad\; 2 \quad\;\; 1} \\
(1)\;-2 \\
\hline
+2 \quad (3) \\
(+2)\;-4 \\
\hline
+7 \quad (2) \\
(+7)\;-14 \\
\hline
+16 \quad (1) \\
(+16)\;-32 \\
\hline
+33
\end{array}
$$

Notice that the numerals in parentheses are repetitions of the numerals directly above them. Omitting these repetitions, the problem may be condensed as follows:

$$
\begin{array}{r}
1 \quad\;\; 2 \quad\;\; 7 \quad\;\; 16 \\
1 \;\; -2)\overline{1 \quad\;\; 0 \quad\;\; 3 \quad\;\; 2 \quad\;\;\; 1} \\
-2 \quad -4 \quad -14 \quad -32 \\
\hline
1 \quad\;\; 2 \quad\;\; 7 \quad\;\; 16 \;\big|\; 33
\end{array}
$$

Except for the last term (the remainder 33), the bottom row is identical to the top row, and therefore the top row may be omitted. The coefficient 1 of the divisor is not needed for the subtractions involved, so it may be omitted. Moreover, the subtractions may be changed to additions by using the definition of subtraction—that is, $a - b = a + (-b)$. Replacing -2 by $+2$, the division may be expressed as follows:

$$
\begin{array}{r}
2\big|\;1 \quad 0 \quad 3 \quad 2 \quad 1 \\
2 \quad 4 \quad 14 \quad 32 \\
\hline
1 \quad 2 \quad 7 \quad 16 \;\big|\; 33
\end{array}
\quad \text{or} \quad
\begin{array}{r}
1 \quad 0 \quad 3 \quad 2 \quad 1\;\big|\;2 \\
2 \quad 4 \quad 14 \quad 32 \\
\hline
1 \quad 2 \quad 7 \quad 16 \;\big|\; 33
\end{array}
$$

A detailed discussion of the application of synthetic division to this problem now follows.

Divide $x^4 + 3x^2 + 2x + 1$ by $x - 2$.

1. Arrange the terms in dividend and divisor in descending powers of the variable.

$$x^4 + 0x^3 + 3x^2 + 2x + 1$$
$$x - 2$$

2. Write the coefficients of the terms of the dividend in this order, and be sure to use 0 for every missing power of the variable. On the right, write c from the divisor $(x - c)$.

$$+1 \quad 0 \quad +3 \quad +2 \quad +1 \,\big|\, +2$$

3. Bring down the first term to third row ($+1$ in the example) and multiply by c ($+2$ in the example). Write this product under second coefficient and add. Multiply this sum by c, and repeat the process until last column has been totaled.

$$
\begin{array}{rrrrr}
+1 & 0 & +3 & +2 & +1 \,\big|\, +2 \\
 & +2 & +4 & +14 & +32 \\
\hline
+1 & +2 & +7 & +16 & +33
\end{array}
$$

4. The last entry represents the remainder, and the other numbers are the coefficients of the quotient. Since the dividend is a polynomial of degree n, and the divisor is a first-degree polynomial, the degree of the quotient is $n - 1$,

$$1(x^3) + 2(x^2) + 7(x) + 16 \,\big|\, +33$$

$$x^3 + 2x^2 + 7x + 16$$
Remainder: 33

$$\left(\frac{ax^n}{x} = ax^{n-1} \right)$$

Therefore,

$$(x^4 + 3x^2 + 2x + 1) \div (x - 2) = x^3 + 2x^2 + 7x + 16 + \frac{33}{x - 2}$$

EXAMPLE 3.2.1   Use synthetic division to find the quotient

$$(x^5 - 1) \div (x - 1)$$

Solution $x^5 - 1 = x^5 + 0x^4 + 0x^3 + 0x^2 + 0x - 1$

$$
\begin{array}{rrrrrr}
+1 & 0 & 0 & 0 & 0 & -1 \,\big|\, +1 \\
 & +1 & +1 & +1 & +1 & +1 \\
\hline
+1 & +1 & +1 & +1 & +1 & \big|\, 0
\end{array}
$$

Quotient: $x^4 + x^3 + x^2 + x + 1$, with no remainder.
The result can be checked by multiplying $(x^4 + x^3 + x^2 + x + 1)(x - 1)$.

EXAMPLE 3.2.2   Use synthetic division to find the quotient

$$(x^3 + 3x^2 + 4) \div (x + 2)$$

Solution $\begin{array}{rrrr|r} +1 & +3 & 0 & +4 & -2 \\ & -2 & -2 & +4 & \\ \hline +1 & +1 & -2 & +8 \end{array}$

Quotient: $x^2 + x - 2$; remainder: $+8$.

Check: If $P(x) = x^3 + 3x^2 + 4$ and $Q(x) = x^2 + x - 2$,   then

$$P(x) = (x + 2) \cdot Q(x) + 8$$

(The dividend is equal to the product of the divisor and the quotient, plus the remainder.)

$$(x + 2) \cdot Q(x) = (x + 2)(x^2 + x - 2) = x^3 + 3x^2 - 4$$

$$(x + 2) \cdot Q(x) + 8 = x^3 + 3x^2 - 4 + 8 = x^3 + 3x^2 + 4 = P(x)$$

The answer may be written

$$x^2 + x - 2 + \frac{8}{x + 2}$$

EXERCISES 3.2 A

Use synthetic division to determine the quotient and remainder for Exercises 1–25. Check by using $P(x) = (x - c)Q(x) + R$.

1. $(2x^3 - 3x^2 + 4x - 5) \div (x - 2)$ 2. $(x^4 + 3x^2 + 2x - 3) \div (x + 1)$
3. $(x^4 + x^3 - x^2 - x + 2) \div (x + 2)$
4. $(x^3 - x^2 - 11x + 15) \div (x - 3)$ 5. $(x^4 + x^2 - 6) \div (x + 3)$
6. $(2x^4 - 200x + 6) \div (x - 5)$
7. $(x^3 - 4x^2 + 8x - 32) \div (x - 4)$
8. $(5y^4 + 12y^3 - 30y^2 - 32) \div (y + 4)$
9. $(x^3 - 27) \div (x - 3)$ 10. $(x^3 + 8) \div (x + 2)$
11. $(3x^3 + x^2 - 7) \div (x - 2)$
12. $(3y^3 + 4y^2 - 7y - 14) \div (y + 3)$
13. $(x^5 + 32) \div (x + 2)$ 14. $(y^6 - 64) \div (y - 2)$
15. $(4x^3 - 10x^2 + x - 1) \div (x - \frac{1}{2})$
16. $(3x^3 + 4x^2 - 5x - 2) \div (x + \frac{1}{3})$

17. $(5x^4 - x^3 - 2x^2 + 7x - 1) \div (x - \frac{1}{5})$

18. $(8x^3 - 10x^2 + 7) \div (4x + 3)$ (*Hint:* Use synthetic division to divide by $x + \frac{3}{4}$. Then divide result by 4.)

19. $(8x^3 - 10x^2 + 9) \div (4x + 3)$

20. $(3x^3 - 7x^2 - x + 1) \div (3x - 1)$

21. $(5x^4 + 8x^3 - 4x^2 + 10x - 6) \div (5x - 2)$

22. $[x^3 + (a + 1)x^2 + (a - 2a^2)x - 2a^2] \div (x - a)$

23. $(x^3 - bx^2 - 4b^2x - 2b^3) \div (x + b)$

24. $(x^3 - x^2 + 4x - 4) \div (x + 2i)$

25. $(x^3 + 2x^2 + 9x + 18) \div (x - 3i)$

26. Use synthetic division to divide $x^4 - 14x^2 + 5x + 30$ by $(x - 2)(x - 3)$. (*Hint:* First divide by $x - 2$ and then divide by $x - 3$.)

27. Repeat Exercise 26 for $(x^4 + 4x^3 + 20x - 25) \div (x + 5)(x - 1)$.

28. Determine p so that the remainder is 0 when $3x^2 + 2x - p$ is divided by $x - 4$.

29. In Exercise 28, $P(x) = 3x^2 + 2x - p$. Substitute the value of p for which the remainder in Exercise 16 is 0, then evaluate $P(4)$.

30. Let $P(x) = x^3 - 3x + 6$. Find the remainder R when $P(x)$ is divided by $x - 2$. Compare R with $P(2)$.

EXERCISES 3.2 B

Use synthetic division to determine the quotient and remainder for Exercises 1–25. Check by using $(Px) = (x - c)Q(x) + R$.

1. $(2x^4 - 4x^3 + x^2 - 1) \div (x - 1)$ **2.** $(x^3 - 5x^2 + 2x + 3) \div (x + 1)$

3. $(x^4 - x^3 + x^2 - x + 2) \div (x - 2)$

4. $(x^3 + 125) \div (x + 5)$ **5.** $(x^3 - 64) \div (x - 4)$

6. $(y^3 - 11y + 6) \div (y - 3)$ **7.** $(2x^4 - 17x^2 - 4) \div (x + 3)$

8. $(x^2 - 4x + 2) \div (x + 2)$ **9.** $(y^4 - y^2 + 6) \div (y + 2)$

10. $(x^5 + 1) \div (x + 1)$

11. $(5x^5 + 2x^4 - 50x^3 - x + 1) \div (x - 3)$

12. $(64x^3 - 1) \div (x + \frac{1}{4})$ **13.** $(ix^3 + 2x^2 + 1) \div (x - i)$

14. $(y^4 + ky^3 - k^2y^2 + k^3y + 2) \div (y - k)$

15. $(x^4 - 2x^3 - 2x - 1) \div (x + i)$

16. $(8x^3 - 6x^2 - 3x + 1) \div (x - \frac{1}{4})$

17. $(8x^3 - 6x^2 - 3x + 1) \div (4x - 1)$

18. $(9x^4 - 9x^3 - 10x^2 + x + 1) \div (3x + 1)$

19. $(6x^3 - 11x^2 - 7x + 10) \div (6x - 5)$

20. $(x^3 + 2ix^2 - x - 2i) \div (x + 2i)$

21. $\left(2k^2x^3 - kx^2 - [k + 1]x + 1\right) \div (kx - 1)$

22. $\left(ax^3 + [2a + b]x^2 + [3a + 2b]x + 3b\right) \div (ax + b)$

23. $(x^2 - 2x + 4) \div (x - 1 - i\sqrt{3})$ **24.** $(x^2 - 2i) \div (x - 1 - i)$

25. $(x^5 - 4x^3 - x^2 + 4) \div (x - 1)(x + 2)$ (*Hint:* See Exercise 26, Exercises 3.2 A.)

26. Repeat Exercise 25 for $(x^4 + 2x^3 - 8x^2 - 18x - 9) \div (x + 1)(x - 3)$.

27. Determine k so that the remainder is 0 when $2x^3 - 15x + k$ is divided by $x - 3$.

28. Determine k so that the remainder is 5 when $4x^4 - 3x^3 + k$ is divided by $x + 2$.

29. Let $P(x) = x^4 - 4x^3 + 6x^2 - 10$.

    a. Find the remainder R when $P(x)$ is divided by $x - 5$.

    b. Find $P(5)$ and compare with R.

30. Let $P(x) = x^3 + 7x^2 + 10x - 8$.

    a. Find the remainder R when $P(x)$ is divided by $x + 4$.

    b. Find $P(-4)$ and compare with R.

3.3 REMAINDER AND FACTOR THEOREMS

Let $P(x)$ represent any polynomial. Then $P(2)$ is the value of the polynomial when x is replaced by 2; $P(0)$ is the value of the polynomial when x is replaced by zero; and in general, $P(c)$ is the value of the polynomial when x is replaced by c.

To illustrate, let

$$P(x) = 5x^2 + 3x + 2$$

Then

$$P(2) = 5(2)^2 + 3(2) + 2 = 28$$

and

$$P(0) = 5(0)^2 + 3(0) + 2 = 2$$

Also

$$P(c) = 5c^2 + 3c + 2$$

Let $P(x) = 5x^2 + 3x + 2$; divide $P(x)$ by $x + 3$. By synthetic division,

$$
\begin{array}{rrr|l}
5 & 3 & 2 & -3 \\
 & -15 & +36 & \\
\hline
5 & -12 & \multicolumn{1}{|l}{38} & \text{(Remainder)}
\end{array}
$$

$$(5x^2 + 3x + 2) \div (x + 3) = 5x - 12 + \frac{38}{x + 3}$$

$$P(-3) = 5(-3)^2 + 3(-3) + 2 = 38$$

Let $Q(x) = 2x^3 + 7x - 1$; divide $Q(x)$ by $x + 2$:

$$
\begin{array}{rrrr|l}
2 & 0 & 7 & -1 & -2 \\
 & -4 & +8 & -30 & \\
\hline
2 & -4 & +15 & \multicolumn{1}{|l}{-31} & \text{(Remainder)}
\end{array}
$$

$$(2x^3 + 7x - 1) \div (x + 2) = 2x^2 - 4x + 15 + \frac{-31}{x + 2}$$

$$Q(-2) = 2(-2)^3 + 7(-2) - 1 = -31$$

Note that the remainder, after dividing $P(x)$ by $x + 3$, is $P(-3)$, and that the remainder, after dividing $Q(x)$ by $x + 2$, is $Q(-2)$. This is not just coincidence but can be proved by the following theorem:

THE REMAINDER THEOREM

If c is any complex number, and if a polynomial $P(x)$ is divided by a divisor of the form $(x - c)$, then the remainder R is equal to $P(c)$.

Proof: Let $P(x)$ be any polynomial. Let $Q(x)$ represent the quotient and R the remainder.

$$P(x) \div (x - c) = Q(x) + \frac{R}{x - c} \qquad x \neq c \qquad (1)$$

or

$$P(x) = Q(x) \cdot (x - c) + R \qquad (2)$$

Statement (2) is true for all values of x; in particular, it is true for $x = c$. Then

$$
\begin{aligned}
P(c) &= Q(c) \cdot (c - c) + R \\
&= Q(c) \cdot 0 + R \\
&= 0 + R \\
&= R
\end{aligned}
$$

We can now confirm the results of the two examples worked above by applying the remainder theorem.

EXAMPLE 3.3.1 Find the remainder when $5x^2 + 3x + 2$ is divided by $x + 3$.

Solution $P(x) = 5x^2 + 3x + 2$

$$(x + 3) = [x - (-3)]$$

Therefore $c = -3$

$$P(-3) = 5(-3)^2 + 3(-3) + 2 = 38$$

Therefore, the remainder is 38.

EXAMPLE 3.3.2 Find the remainder when $2x^3 + 7x - 1$ is divided by $x + 2$.

Solution $Q(x) = 2x^3 + 7x - 1,\quad c = -2$

$$Q(-2) = 2(-2)^3 + 7(-2) - 1 = -31$$

Therefore, the remainder is -31.

Examples 3.3.1 and 3.3.2 were checked by synthetic division in the illustration and are shown to be correct.

Another useful theorem results as a direct consequence of the remainder theorem.

THE FACTOR THEOREM

Let $P(x)$ be a polynomial. Let c be any complex number. Then $x - c$ is a *factor* of $P(x)$ if and only if $P(c) = 0$.

The factor theorem is an "if and only if" theorem, and it therefore consists of two separate statements which must be proved.

a. If $x - c$ is a factor of $P(x)$, then $P(c) = 0$.

b. If $P(c) = 0$, then $x - c$ is a factor of $P(x)$.

Proof: a. Assume $x - c$ is a factor of $P(x)$; prove $P(c) = 0$. If $x - c$ is a factor of $P(x)$, then there exists a polynomial $Q(x)$ such that

$$P(x) = Q(x) \cdot (x - c)$$

and also $P(c) = Q(c) \cdot (c - c) = Q(c) \cdot 0 = 0$

b. Assume, for the polynomial $P(x)$, $P(c) = 0$. Prove that $(x - c)$ is a factor of $P(x)$.

By the remainder theorem,

$$P(x) = Q(x)\cdot(x - c) + R$$
$$= Q(x)\cdot(x - c) + P(c)$$

(Since $R = P(c)$)

$$= Q(x)\cdot(x - c) + 0 \qquad (P(c) = 0 \text{ by assumption})$$
$$= Q(x)\cdot(x - c)$$

Since $P(x)$ is shown to be the product of two polynomials, each of these polynomials is a factor of $P(x)$; in particular, therefore, $x - c$ is a factor of $P(x)$. This proves the factor theorem.

EXAMPLE 3.3.3 Show that $x + 2$ is a factor of the polynomial

$$P(x) = x^4 + 5x^3 + 8x^2 + 5x + 2$$

Solution $P(-2) = (-2)^4 + 5(-2)^3 + 8(-2)^2 + 5(-2) + 2$

$$= 16 - 40 + 32 - 10 + 2$$
$$= 0$$

Since $P(-2) = 0$, by the factor theorem, $x + 2$ is a factor of

$$x^4 + 5x^3 + 8x^2 + 5x + 2$$

EXAMPLE 3.3.4 Determine if $x - 6$ is a factor of

$$2x^5 - x^4 - 16x^3 + 5x^2 - 1832x + 12$$

Solution Use synthetic division to find $P(6)$:

$$
\begin{array}{rrrrrr|r}
2 & -1 & -16 & 5 & -1832 & 12 & \underline{6} \\
  & 12 & 66 & 300 & 1830 & -12 & \\
\hline
2 & 11 & 50 & 305 & -2 & 0 &
\end{array}
$$

$P(6) = R = 0$. Thus $x - 6$ is a factor.

EXAMPLE 3.3.5 Determine if $x + 5$ is a factor of

$$x^3 + 8x^2 + 16x + 10$$

Solution Use synthetic division to find $P(-5)$:

$$
\begin{array}{rrrr|r}
1 & 8 & 16 & 10 & \underline{-5} \\
  & -5 & -15 & -5 & \\
\hline
1 & 3 & 1 & 5 &
\end{array}
$$

$P(-5) = R = 5$. Since $R \neq 0$, $x + 5$ is *not* a factor.

EXAMPLE 3.3.6   Find k so that $x - 3$ is a factor of
$$kx^3 - 6x^2 + 2kx - 12$$

Solution   By the factor theorem, $x - 3$ is a factor of
$$P(x) = kx^3 - 6x^2 + 2kx - 12 \text{ if } P(3) = 0$$
$$P(3) = k(3)^3 - 6(3)^2 + 2k(3) - 12$$
$$= 27k - 54 + 6k - 12 = 33k - 66$$

Let $P(3) = 33k - 66 = 0$
Then $k = 2$

Therefore, when $k = 2$, $x - 3$ is a factor of $kx^3 - 6x^2 + 2kx - 12$.

EXERCISES 3.3 A

Use the factor theorem to determine if $Q(x)$ is a factor of $P(x)$ for Exercises 1–5.

1. $P(x) = x^3 + 3x^2 + 2x - 24$ **2.** $P(x) = 2x^3 - 11x^2 + 12x + 9$
   $Q(x) = x - 2$ $Q(x) = x - 3$
3. $P(x) = x^4 - x^2 + 2x + 6$ **4.** $P(x) = 2x^3 - 7x^2 - 21x + 54$
   $Q(x) = x + 2$ $Q(x) = x + 3$
5. $P(x) = x^6 - 1$
   $Q(x) = x - 1$

By using the remainder theorem, find the remainder for each indicated division in Exercises 6–10.

6. $(x^{27} - 4x^{10} + 5x^2 - 7) \div (x - 1)$
7. $(3x^{12} + x^3 - x + 2) \div (x + 1)$ **8.** $(x^{15} - 1) \div (x - 1)$
9. $(x^{15} + 1) \div (x + 1)$ **10.** $(x^5 - 3x^3 - 8) \div (x - 2)$

For $P(x) = x^4 - 6x^2 - 7x - 6$, find each of Exercises 11–16 by using (a) the remainder theorem and (b) synthetic division.

11. $P(1)$ **12.** $P(-1)$
13. $P(2)$ **14.** $P(-2)$
15. $P(3)$ **16.** $P(-3)$

17. Using the results of Exercises 11–16, factor $P(x)$ over the integers.

For $P(x) = x^4 + 3x^3 - 6x^2 + 12x - 40$, find each of Exercises 18–25.

18. $P(1)$ **19.** $P(-1)$
20. $P(2)$ **21.** $P(-2)$
22. $P(5)$ **23.** $P(-5)$
24. $P(2i)$ **25.** $P(-2i)$

26. Using the results of Exercises 18–25, factor $P(x)$ over the set of complex numbers.

27. For $P(x) = 2x^3 - 3x^2 - 3x + 2$, (a) find $P(1)$, $P(-1)$, $P(2)$, $(P-2)$, $P(\frac{1}{2})$, $P(-\frac{1}{2})$ and (b) factor $P(x)$ as a product of linear factors.

28. If $P(x) = 3x^2 + 4x - k$, find k so that $P(3) = 6$.

29. For $P(x)$ in Exercise 28, find k so that $P(3) = 0$.

30. If $Q(x) = 2x^3 - kx^2 + 2x - k$, find k so that $Q(4) = 0$.

31. Find k so that $x + 2$ is a factor of $x^4 + 2x^3 + x + k$.

32. Find k so that $x - 3$ is a factor of $2x^3 - 3x^2 + x + k$.

33. Find m so that when $y^3 - m^2y^2 - my - 4$ is divided by $y - 3$ the remainder is 23.

34. Find k if $x^3 - 4kx^2 + kx + 6$ is exactly divisible by $x - 2$.

35. Find the remainder when $2x^4 - 3x^3 + 5x - 1$ is divided by $x - 4$ (a) by long division, (b) by synthetic division, and (c) by the remainder theorem.

36. Use the factor theorem to check if $2 + 4i$ is a root of the equation $y^4 - 4y^3 + 18y^2 + 8y - 40 = 0$.

37. Use the factor theorem to show that $2i$ is a root of the equation $x^3 - x^2 + 4x - 4 = 0$.

EXERCISES 3.3 B

Use the factor theorem to determine if $Q(x)$ is a factor of $P(x)$ for Exercises 1–5.

1. $P(x) = x^4 - 12x^2 - 5x + 12$
$Q(x) = x + 3$

2. $P(x) = 2x^3 - 3x^2 - 2x + 6$
$Q(x) = x + 1$

3. $P(x) = x^3 + 3x^2 + 3x - 7$
$Q(x) = x - 1$

4. $P(x) = x^3 - 3x^2 - 9x - 5$
$Q(x) = x - 5$

5. $P(x) = x^8 + 1$
$Q(x) = x + 1$

By using the remainder theorem, find the remainder for each indicated division in Exercises 6–10.

6. $(x^{35} + 2x^{16} + 3x + 2) \div (x + 1)$

7. $(5x^{20} - 7x^{15} + 3x - 1) \div (x - 1)$

8. $(x^{20} - 1) \div (x - 1)$ **9.** $(x^{20} - 1) \div (x + 1)$

10. $(x^7 - 4x^5 + x^4 - x^2 + 2) \div (x + 2)$

For $P(x) = x^3 - 26x + 5$ find each of Exercises 11–16 by using (a) the remainder theorem and (b) synthetic division.

11. $P(1)$ **12.** $P(-1)$

13. $P(5)$ **14.** $P(-5)$

15. $P(2)$ **16.** $P(-2)$

17. Using the results of Exercises 11–16, factor $P(x)$ over the integers.

For $P(x) = x^5 - x^4 - 4x^3 + 4x^2 - 5x + 5$, find each of Exercises 18–25.

18. $P(1)$ **19.** $P(-1)$

20. $P(5)$ **21.** $P(-5)$

22. $P(\sqrt{5})$ **23.** $P(-\sqrt{5})$

24. $P(i)$ **25.** $P(-i)$

26. Using the results of Exercises 18–25, factor $P(x)$ over the set of complex numbers.

27. For $P(x) = 3x^3 - 23x^2 + 13x + 7$, (a) find $P(1)$, $P(-1)$, $P(7)$, $P(-7)$, $P(\frac{1}{3})$, $P(-\frac{1}{3})$ and (b) factor $P(x)$ as a product of linear factors.

28. If $P(x) = x^3 + 3x^2 - 2x + k$, find k so that $P(2) = 6$.

29. For $P(x)$ in Exercise 28, find k so that $P(2) = 0$.

30. If $Q(x) = 2x^3 + kx^2 - 5x + 4k$, find k so that $Q(3) = 0$.

31. Find m so that $x - 3$ is a factor of $2x^3 - x^2 + 3x + m$.

32. Find p so that $a + 2$ is a factor of $a^4 - 3a^2 - a + p$.

33. Find p so that when $2x^3 + x^2p - 2xp + 5$ is divided by $x + 2$ the remainder is -11.

34. Find k so that 5 is a root of the equation $2x^2 - 3x + k = 0$.

35. Find the remainder when $3x^5 - 46x^3 + x - 9$ is divided by $x + 4$ (a) by long division, (b) by synthetic division, and (c) by the remainder theorem.

36. Use the factor theorem to check if $1 + 2i$ is a root of the equation $x^4 + 3x^2 - 15x + 1 = 0$. Justify your answer.

37. Find m so that $Q(i) = 0$ if $Q(x) = x^4 + mx^2 - 2$.

3.4 SOLUTION OF POLYNOMIAL EQUATIONS BY FACTORING

The techniques of factoring and the factor theorem provide useful methods for solving certain polynomial equations.

If $P(x) = (x - r_1)(x - r_2)\cdots(x - r_n) = 0$, then the zero-product theorem implies that

$$x - r_1 = 0 \quad \text{or} \quad x - r_2 = 0 \quad \text{or} \quad \cdots \quad \text{or} \quad x - r_n = 0$$

EXAMPLE 3.4.1   Solve $(x - 1)(x - 2)(x + 5) = 0$.

Solution   By the zero-product theorem,

$$x - 1 = 0 \quad \text{or} \quad x - 2 = 0 \quad \text{or} \quad x + 5 = 0$$

Thus $x = 1$ or $x = 2$ or $x = -5$, and the solution set is $\{1, 2, -5\}$.

EXAMPLE 3.4.2   Solve $x^3 - 3x^2 - 4x + 12 = 0$.

Solution   Factoring by grouping,

$$(x^3 - 3x^2) - (4x - 12) = x^2(x - 3) - 4(x - 3)$$
$$= (x^2 - 4)(x - 3)$$
$$= (x + 2)(x - 2)(x - 3)$$

Thus $x^3 - 3x^2 - 4x + 12 = (x + 2)(x - 2)(x - 3) = 0$ if and only if $x + 2 = 0$ or $x - 2 = 0$ or $x - 3 = 0$. The solution set is $\{-2, 2, 3\}$.

Note in Example 3.4.2 that the number of solutions in the solution set is 3, the same as the degree of the polynomial $x^3 - 3x^2 - 4x + 12$ (the degree of a polynomial in one variable is the greatest exponent that occurs on the variable). In general, **a polynomial of degree n has n linear factors, and therefore n roots must be accounted for when solving a polynomial equation.**

EXAMPLE 3.4.3   Solve $x^3 - 8 = 0$.

Solution   Factoring $x^3 - 8$,

$$x^3 - 8 = (x - 2)(x^2 + 2x + 4) = 0$$
$$x - 2 = 0 \quad \text{or} \quad x^2 + 2x + 4 = 0$$
$$x = 2 \quad \text{or} \quad x = -1 \pm i\sqrt{3}$$

The solution set is $\{2, -1 + i\sqrt{3}, -1 - i\sqrt{3}\}$.

EXAMPLE 3.4.4   Solve $x^4 - 4x^3 + 4x^2 - (x^2 - 4x + 4) = 0$.

Solution   Factoring,

$$x^2(x^2 - 4x + 4) - (x^2 - 4x + 4) = 0$$
$$(x^2 - 4x + 4)(x^2 - 1) = 0$$
$$x^2 - 4x + 4 = 0 \quad \text{or} \quad x^2 - 1 = 0$$
$$(x - 2)^2 = 0 \quad \text{or} \quad (x + 1)(x - 1) = 0$$
$$x = 2 \quad \text{or} \quad x = 2 \quad \text{or} \quad x = -1 \quad \text{or} \quad x = 1$$

The solution set is $\{2, -1, 1\}$, where 2 is called a *double root* or a root of *multiplicity 2*. With the root 2 counted twice, it may then be said that the number of roots (4) is equal to the degree of the polynomial (4).

In general, if a polynomial $P(x)$ has exactly k factors of the form $x - r$ (that is, $P(x) = (x - r)^k Q(x)$) then r is said to be a **root of multiplicity** k of the equation $P(x) = 0$.

EXAMPLE 3.4.5 Solve $x^3 - 5x - 12 = 0$.

Solution Using the factor theorem to find r so that $P(r) = 0$:

$$\text{If } x = 1, \quad \text{then} \quad P(1) = 1 - 5 - 12 \neq 0$$
$$\text{If } x = 2, \quad \text{then} \quad P(2) = 8 - 10 - 12 \neq 0$$
$$\text{If } x = 3, \quad \text{then} \quad P(3) = 27 - 15 - 12 = 0$$

Thus $x - 3$ is a factor.

Now by using synthetic division to find the quotient polynomial,

$$
\begin{array}{rrrr|r}
1 & 0 & -5 & -12 & \underline{3} \\
  & 3 & 9 & 12 & \\
\hline
1 & 3 & 4 & 0 &
\end{array}
$$

Thus

$$x^3 - 5x - 12 = (x - 3)(x^2 + 3x + 4) = 0$$

and

$$x - 3 = 0 \quad \text{or} \quad x^2 + 3x + 4 = 0$$

$$x = 3 \quad \text{or} \quad x = \frac{-3 \pm i\sqrt{7}}{2}$$

The solution set is

$$\left\{ 3, \frac{-3 + i\sqrt{7}}{2}, \frac{-3 - i\sqrt{7}}{2} \right\}$$

The work involved in finding an integral root of a polynomial equation may be reduced by the following theorem.

THEOREM ON INTEGRAL ROOTS

If $x - r$ is a factor of $P(x) = a_n x^n + a_{n-1} x^{n-1} + \cdots + a_1 x + a_0$, where r is an integer and the coefficients of $P(x)$ are integers, then r is a factor of a_0.

Proof: The factor theorem states that $P(r) = 0$ if $x - r$ is a factor of $P(x)$. Thus

$$P(r) = a_n r^n + \cdots + a_1 r + a_0 = 0$$
$$= r(a_n r^{n-1} + \cdots + a_1) + a_0 = 0$$
$$r(a_n r^{n-1} + \cdots + a_1) = -a_0$$

Thus r is a factor of a_0, the constant term of the polynomial.

In other words, an examination of the constant term of a polynomial can quickly limit the number of trials to see what value of r will yield

$P(r) = 0$. For example, if $P(x) = x^2 + 5x - 3$, the only numbers to try for integral factors of 3 are ± 1 and ± 3.

EXAMPLE 3.4.6   Solve $x^4 - 8x^2 + 5x + 6 = 0$.

Solution   The integral factors of the constant term 6 are $\pm 1, \pm 2, \pm 3, \pm 6$.

1. Apply the factor theorem:

$$\text{If } x = 1, \quad \text{then} \quad P(1) = 1 - 8 + 5 + 6 \neq 0$$

$$\text{If } x = 2, \quad \text{then} \quad P(2) = 16 - 32 + 10 + 6 = 0$$

and $x - 2$ is a factor.

2. Use synthetic division to find the quotient polynomial:

$$
\begin{array}{rrrrr|r}
1 & 0 & -8 & 5 & 6 & \,2 \\
  & 2 & 4 & -8 & -6 & \\
\hline
1 & 2 & -4 & -3 & 0 &
\end{array}
$$

$$x^4 - 8x^2 + 5x + 6 = (x - 2)(x^3 + 2x^2 - 4x - 3)$$

3. Now apply the factor theorem to the cubic polynomial; the integral factors of the constant term -3 are $\pm 1, \pm 3$:

$$\text{If } x = 1, \quad \text{then} \quad P(1) = 1 + 2 - 4 - 3 \neq 0$$

$$\text{If } x = 3, \quad \text{then} \quad P(3) = 27 + 18 - 12 - 3 \neq 0$$

$$\text{If } x = -1, \quad \text{then} \quad P(-1) = -1 + 2 + 4 - 3 \neq 0$$

$$\text{If } x = -3, \quad \text{then} \quad P(-3) = -27 + 18 + 12 - 3 = 0$$

and $x + 3$ is a factor.

4. Use synthetic division to find the quotient polynomial:

$$
\begin{array}{rrrr|r}
1 & 2 & -4 & -3 & \,-3 \\
  & -3 & 3 & 3 & \\
\hline
1 & -1 & -1 & 0 &
\end{array}
$$

Thus $x^3 + 2x^2 - 4x - 3 = (x + 3)(x^2 - x - 1)$

5. Apply the zero-product theorem to the original equation:

$$x^4 - 8x^2 + 5x + 6 = (x - 2)(x + 3)(x^2 - x - 1) = 0$$

$$x - 2 = 0 \quad \text{or} \quad x + 3 = 0 \quad \text{or} \quad x^2 - x - 1 = 0$$

The solution set is

$$\left\{ 2, \; -3, \; \frac{1 + \sqrt{5}}{2}, \; \frac{1 - \sqrt{5}}{2} \right\}$$

EXERCISES 3.4 A

One root of each of the equations in Exercises 1–5 is given. Find the other roots and state the solution set.

1. $2x^3 - x^2 - 13x - 6 = 0$; 3   **2.** $x^3 + 2x^2 - 5x + 12 = 0$; -4

3. $3x^3 + 2x^2 - 4x + 1 = 0$; $\frac{1}{3}$

4. $x^4 - x^3 - 5x^2 - x - 6 = 0$; -2

5. $2x^4 - 9x^3 - 7x^2 + 9x + 5 = 0$; $-\frac{1}{2}$

In Exercises 6–20, solve for x.

6. $x^3 + 3x^2 - x - 3 = 0$ **7.** $x^3 - 26x + 5 = 0$

8. $9x^4 - 13x^2 + 4 = 0$ **9.** $3x^4 - x^3 - 27x^2 + 9x = 0$

10. $x^5 - 9x^3 + 8x^2 - 72 = 0$ **11.** $2x^3 + 7x^2 - 9 = 0$

12. $3x^4 + 2x^3 - 9x^2 + 4 = 0$ **13.** $x^3 + 64 = 0$

14. $x^3 - 1 = 0$ **15.** $x^4 - 4x^3 - 6x^2 + 4x + 5 = 0$

16. $x^3 - 3x^2 = 16$ **17.** $x^4 - 14x^2 + 49 = 0$

18. $x^4 - x^3 - 6x - 36 = 0$ **19.** $x^3 + x^2 - 12 = 0$

20. $x^3 - 19x + 30 = 0$

21. The depth x to which a floating sphere of radius r and specific gravity s will sink in water is given by

$$x^3 - 3rx^2 + 4r^3s = 0$$

  a. Find x for $r = 2$ inches and $s = \frac{27}{32}$ inches

  b. Find x for $r = 2$ inches and $s = 1$ inch

  c. Find x for $r = 2$ inches and $s = 0.5$ inch

22. The sides of a rectangular box measure 4 centimeters, 4 centimeters, and 6 centimeters. In order to triple the volume of the box, each side is to be increased by the same amount. Find this amount.

23. The total revenue R in dollars that can be received from the sale of x television sets is given by

$$R = x(280 - 8x - 2x^2)$$

Find x for $R = \$960$. (*Hint:* The answer must be an integer.)

24. The depth x in inches to which a certain bullet will penetrate a block of wood is given by

$$v_o^2 x^2(12 - x) = 1728 v^2$$

where v_o = initial velocity in feet per second and v = final velocity in feet per second. Find x for $v_o = 2400$ feet per second and $v = 900$ feet per second.

EXERCISES 3.4 B

One root of each of the equations in Exercises 1–5 is given. Find the other roots and state the solution set.

1. $x^4 - 5x^2 + 2x = 0; 2$ **2.** $5x^3 - 7x^2 - 3x + 2 = 0; \frac{2}{5}$

3. $x^3 - 6x^2 + 25 = 0; 5$ **4.** $3x^4 + 2x^3 - 9x^2 + 4 = 0; -\frac{2}{3}$

5. $4x^5 - 29x^3 - 24x^2 + 7x + 6 = 0; 3$

In Exercises 6–20, solve for x.

6. $x^3 + 5x^2 - 2x - 10 = 0$ **7.** $x^3 - 48x - 7 = 0$

8. $4x^4 - 15x^2 - 4 = 0$ **9.** $x^5 + x^4 - 25x^3 - 25x^2 = 0$

10. $x^5 - x^3 - 27x^2 + 27 = 0$ **11.** $x^3 - 7x^2 + 36 = 0$

12. $x^3 + 6x^2 + 4x - 5 = 0$ **13.** $8x^3 - 125 = 0$

14. $27x^3 + 1 = 0$

15. $x^4 - 6x^3 + 5x^2 + 24x - 36 = 0$

16. $x^3 + 6x^2 + 12x + 8 = 0$ **17.** $x^4 - 4x^3 + 6x^2 - 4x + 1 = 0$

18. $x^4 - 4x^3 - 2x^2 + 12x + 9 = 0$ **19.** $x^4 - 4x^2 + 12x - 9 = 0$

20. $x^5 + 2x^2 - x - 2 = 0$

21. The radius r of a cone having a height of 24 inches and a lateral surface area of 550 square inches is given by

$$\pi^2 r^2(r^2 + 576) = (550)^2$$

Find r if $\pi = \frac{22}{7}$.

22. The sides of a rectangular box measure 3 inches, 4 inches, and 5 inches. Each side is to be increased by the same amount in order to double the volume of the box. Find the amount each side should be increased.

23. The bending moment M of a certain beam having a length L and carrying a certain weight distribution of W pounds is given by

$$M = \frac{Wx(L^2 - x^2)}{3L^2}$$

where x is the distance from one end of the beam. Find x for $M = \frac{375}{4}$, $W = 100$ pounds, and $L = 12$ feet. (*Hint:* There are two possible solutions.)

24. Warehouse A is 9 miles north of a point P on a road running east–west. Warehouse B is 7 miles south of a point Q on the road that is 8 miles east of P. A store is to be located at a point R on the road between P

and Q so that the cost of constructing the broken-line road ARB is a minimum (see Figure 3.4.1).

If the cost of constructing road AR is 3 times the cost for road BR, then the value of x that produces the minimum cost may be obtained by solving

$$x^4 - 16x^3 + 109x^2 + 162x - 648 = 0$$

Find the value of x that produces the minimum cost.

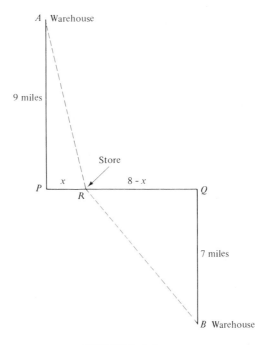

FIGURE 3.4.1

3.5 INEQUALITIES

The trichotomy axiom states: If a and b are real numbers, then exactly one of the following statements is true:

$$a < b \quad \text{or} \quad a = b \quad \text{or} \quad a > b$$

On the number line drawn horizontally with its positive direction to the right, it is seen that $a < b$ if and only if the point whose coordinate is a is to the left of the point whose coordinate is b, and $a > b$ if and only if the point whose coordinate is a is located to the right of the point whose coordinate is b.

FIGURE 3.5.1   FIGURE 3.5.2

From Figures 3.5.1. and 3.5.2. it may be observed that $a < b$ if and only if $b > a$.

A formal definition of the "less than" relation between two real numbers is stated below.

DEFINITIONS

If a and b are real numbers, then $a < b$ (a is less than b) if and only if there exists a positive real number p such that $a + p = b$.

$$a > b \text{ if and only if } b < a$$

For example, $3 < 5$ because $3 + 2 = 5$, and $5 > 3$ because $3 < 5$.

The relations "less than" and "greater than" ($<$ and $>$) are called order relations. A statement involving an order relation is called an inequality. Examples of inequalities are:

$$3 < 5, \quad 2x + 1 < 4, \quad y \geq 2x + 1$$

The inequalities $a < b$ and $c < d$ are called inequalities of the same order or inequalities having the same sense, while the inequalities $a < b$ and $c > d$ are called inequalities of opposite order or inequalities having the opposite sense.

Since all numbers to the right of 0 on the number line are said to be positive, it follows that "$a > 0$" is another way of writing "a is a positive number." Similarly, "$a < 0$" is another way of writing "a is a negative number."

A property of the set of positive real numbers is that it is closed with respect to addition and multiplication.

THE CLOSURE AXIOM FOR POSITIVE REAL NUMBERS

Addition: If $a > 0$ and $b > 0$, then $a + b > 0$.

Multiplication: If $a > 0$ and $b > 0$, then $ab > 0$.

In words, this axiom states that the sum of two positive numbers is positive and the product of two positive numbers is positive.

Inequalities and the Number Line

The number line is often a useful graphic aid in the solution of inequalities. For example, the set $\{x \mid x > 3\}$ is graphed as shown in Figure 3.5.3.

The circle above the numeral 3 indicates that 3 is excluded from the solution set; the solution set is indicated by the half-line starting at 3 (but not including 3), and all values greater than 3 as shown by the direction of the line.

FIGURE 3.5.3   **FIGURE 3.5.4**

The set $\{x \mid x \le -2\}$ is graphed as shown in Figure 3.5.4.

This time a solid dot over the -2 coordinate indicates that -2 is included in the solution set, as well as all points to the left of -2, since x is less than or equal to -2.

The statement $x \le a$ means "x is less than a or x equals a," and the statement $x \ge a$ means "x is greater than a or x equals a."

It is often convenient to graph intersections and unions of inequalities. The statement $a < x$ and $x < b$ may be expressed as $a < x < b$, read "x is between a and b," whenever $a < b$.

DEFINITION

$$a < x < b \text{ means } a < x \text{ and } x < b \text{ for } a < b$$

Unions and Intersections

$$\{x \mid a < x\} \cap \{x \mid x < b\} = \{x \mid a < x \text{ and } x < b\} = \{x \mid a < x < b\}$$

$$\{x \mid x < a\} \cup \{x \mid x > b\} = \{x \mid x < a \text{ or } x > b\}$$

$$\{x \mid x > a\} \cap \{x \mid x > b\} = \{x \mid x > a\} \text{ if } a > b$$

$$\{x \mid x < a\} \cap \{x \mid x < b\} = \{x \mid x < a\} \text{ if } a < b$$

The words *and* and *or* as used in the first two statements above are very significant and must be used correctly. Intersection, ∩, corresponds to the word *and*, whereas union, ∪, corresponds to the word *or*. Thus $3 < x$ or $x < -2$ cannot be expressed in a *single* statement, but $3 < x$ and $x < 5$ is equivalent to the statement $3 < x < 5$.

EXAMPLE 3.5.1 The graph of $\{x \mid -2 < x < 1\}$ is illustrated in Figure 3.5.5.

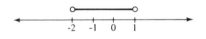

FIGURE 3.5.5

In words, the graph illustrates $x > -2$ and $x < 1$. It also represents $\{x \mid x < 1\} \cap \{x \mid x > -2\}$.

EXAMPLE 3.5.2 The graph of $\{x \mid x < -2$ or $x \geq 1\}$ is illustrated in Figure 3.5.6. The circle over -2 indicates that -2 is not included in the set, and the solid dot over 1 indicates that 1 is included in the set.

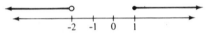

FIGURE 3.5.6

In set symbols, Figure 3.5.6 is the graph of $\{x \mid x < -2\} \cup \{x \mid x \geq 1\}$.

EXAMPLE 3.5.3 Graph on a number line and write a verbal statement to describe the set $\{x \mid -2 \leq x < -1\} \cup \{x \mid x > 0\}$.

Solution The union of the two stated sets is needed; thus *all* the numbers in both sets will be in the union set. In words, "x is greater than or equal to -2 and x is less than -1" or "x is greater than zero."

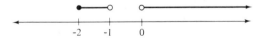

FIGURE 3.5.7

EXAMPLE 3.5.4   Graph on a number line and write a verbal statement to describe the set $\{x \mid -4 < x \le 1\} \cap \{x \mid -1 < x \le 2\}$.

Solution   This time the intersection of the two sets is needed, so the final set will contain the numbers which are in both sets. From Figure 3.5.8,

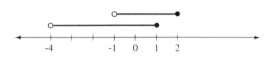

FIGURE 3.5.8

it is clear that the intersection is the set $\{x \mid -1 < x \le 1\}$ (see Figure 3.5.9). In words, "x is greater than -1" and "x is less than or equal to 1."

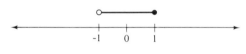

FIGURE 3.5.9

EXERCISES 3.5 A

Graph Exercises 1–20 on the number line and write a verbal statement describing the set.

1. $\{x \mid x < 3\}$ **2.** $\{x \mid x > -2\}$

3. $\{x \mid x \ge 3\}$ **4.** $\{x \mid x \le -2\}$

5. $\{x \mid 2 > x\}$ **6.** $\{x \mid -5 \ge x\}$

7. $\{x \mid x < 0\}$ **8.** $\{x \mid x \ge 0\}$

9. $\{x \mid x > -4\} \cap \{x \mid x < 4\}$ **10.** $\{x \mid x \ge -4\} \cup \{x \mid x \le 4\}$

11. $\{x \mid -2 \le x\} \cup \{x \mid x \ge 3\}$ **12.** $\{x \mid -2 \ge x\} \cup \{x \mid x \ge 3\}$

13. $\{x \mid x > -2\} \cap \{x \mid x \le 3\}$ **14.** $\{x \mid 1 < x < 5\}$

15. $\{x \mid -2 \le x < 3\}$ **16.** $\{x \mid 0 < x < 4\}$

17. $\{x \mid -2 < x \le 0\} \cup \{x \mid 2 < x < 3\}$

18. $\{x \mid 0 < x < 1\} \cup \{x \mid 1 \le x < 3\}$

19. $\{x \mid -1 < x < 2\} \cap \{x \mid 1 \le x < 3\}$

20. $\{x \mid -2 \le x \le \frac{1}{2}\} \cap \{x \mid 0 < x \le 15\}$

EXERCISES 3.5 B

Graph Exercises 1–20 on the number line and write a verbal statement describing the set.

1. $\{x \mid x > 2\}$ 2. $\{x \mid x \leq 1\}$
3. $\{x \mid x < -4\}$ 4. $\{x \mid x \geq -1\}$
5. $\{x \mid 1 > x\}$ 6. $\{x \mid -2 \geq x\}$
7. $\{x \mid 1 \leq x\}$ 8. $\{x \mid -3 \leq x\}$
9. $\{x \mid -5 < x < 5\}$ 10. $\{x \mid -5 \leq x\} \cap \{x \mid x \geq 5\}$
11. $\{x \mid x \leq -1\} \cap \{x \mid x < 2\}$ 12. $\{x \mid -5 \geq x\} \cup \{x \mid x \geq 5\}$
13. $\{x \mid -5 < x\} \cup \{x \mid x < 5\}$ 14. $\{x \mid x < -1\} \cup \{x \mid x \leq 2\}$
15. $\{x \mid 2 \leq x < 4\}$ 16. $\{x \mid -3 < x \leq 0\}$
17. $\{x \mid -1 < x \leq 1\} \cup \{x \mid 3 < x\}$
18. $\{x \mid -2 \leq x < 0\} \cup \{x \mid 0 < x < 2\}$
19. $\{x \mid -2 \leq x \leq -1\} \cap \{x \mid -\frac{3}{2} \leq x < 5\}$
20. $\{x \mid -4 < x \leq 1\} \cap \{x \mid x \geq 2\}$

3.6   LINEAR INEQUALITIES IN ONE VARIABLE

A **solution of an inequality in one variable** is a number from a specified set of numbers that makes the inequality true when the variable is replaced by the name of this specified number.

The **solution set of an inequality** is the set of all solutions of the inequality.

Two inequalities are equivalent if and only if they have the same solution set.

To **solve an inequality in one variable** means to find the solution set of an equivalent inequality having the form $x < a$, $x \leq a$, $x > a$, or $x \geq a$ or unions or intersections of sets whose corresponding inequalities have these forms.

The theorems for inequalities provide the technique for finding these equivalent inequalities. Solving an inequality is similar to solving an equation, with the important exception that multiplication (or division) by a negative number changes the order of the inequality.

It should also be pointed out at this time that the order relation is *not symmetric*. Thus, interchanging the two members of an inequality also changes the sense. For example, if $4 < x$, then $x > 4$.

The following theorems on inequalities are stated without proof. Let a, b, and c be real numbers.

THE TRANSITIVITY THEOREM

If $a < b$ and $b < c$, then $a < c$.

For example, if $3 < 5$ and $5 < 8$, then $3 < 8$. Also, if $-5 < -1$ and $-1 < 7$, then $-5 < 7$.

THE ADDITION THEOREM

If c is any real number and if $a < b$, then $a + c < b + c$.

For example, if $3 < 5$, then $3 + 2 < 5 + 2$, or $5 < 7$. If $(x - 2) < 7$ then $(x - 2) + 2 < 7 + 2$, or $x < 9$.

Since c is *any* real number, it could also be a negative number; thus

if $x + 3 < 5$

then $(x + 3) + (-3) < 5 + (-3)$

or $x < 2$

THE MULTIPLICATION THEOREMS

Theorem 1   If $a < b$ and $c > 0$, then $ac < bc$.

Theorem 2   If $a < b$ and $c < 0$, then $ac > bc$.

Theorem 1 states that if $a < b$, then multiplication of each side of the inequality by the same positive number c preserves the order of the inequality. In other words, the product of a and c is also less than the product of b and c.

Theorem 2 states, however, that if $a < b$ and a and b are each multiplied by the same negative number, the order of the resulting inequality is opposite to $a < b$.

For example, if $3 < 5$, then $3(2) < 5(2)$ or $6 < 10$

and $3(\frac{1}{2}) < 5(\frac{1}{2})$ or $\frac{3}{2} < \frac{5}{2}$

However, if $3 < 5$, then $3(-2) > 5(-2)$ or $-6 > -10$

and $3(-\frac{1}{2}) > 5(-\frac{1}{2})$ or $-\frac{3}{2} > -\frac{5}{2}$

A proof of the addition theorem is stated below. The other theorems are proved in a similar manner.

THE ADDITION THEOREM

If c is any real number and if $a < b$, then $a + c < b + c$.

Proof:

1. $a < b$ 1. Given
2. $a + p = b$ and $p > 0$ 2. Definition of $<$
3. $a + p + c = b + c$ 3. Addition theorem of equality
4. $a + c + p = b + c$ 4. Commutative axiom, addition
5. $a + c < b + c$ 5. Definition of $<$

EXAMPLE 3.6.1   Solve the inequality $2x + 3 < x + 5$ and graph the solution set.

Solution $2x + 3 < x + 5$

$\qquad\qquad 2x < x + 2$ (Addition theorem—adding -3)

$\qquad\qquad\quad x < 2$ (Addition theorem—adding $-x$)

Solution set $= \{x \mid x < 2\}$.

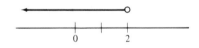

FIGURE 3.6.1

EXAMPLE 3.6.2   Solve for x and graph the solution set:

$$2 - x \le 4 + 3x$$

Solution $2 - x \le 4 + 3x$

$\qquad\quad 2 - 4x \le 4$ (Addition theorem—adding $-3x$)

$\qquad\qquad -4x \le 2$ (Addition theorem—adding -2)

$\qquad\qquad\quad x \ge -\tfrac{1}{2}$ (Multiplication theorem—note that multiplication by $-\tfrac{1}{4}$ reversed the order of the inequality)

Solution set: $\{x \mid x \ge -\tfrac{1}{2}\}$.

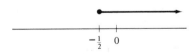

FIGURE 3.6.2

EXAMPLE 3.6.3   Solve $\dfrac{x}{x+1} > 2$.

Solution   Since the inequality has a variable in the denominator, it is necessary to consider two cases, $x + 1 > 0$ and $x + 1 < 0$. The solution set of the inequality is the union of the solution sets for the two cases.

Case 1: $x + 1 > 0$ and thus $x > -1$. Since $x + 1 > 0$, multiplying each side of the inequality by $x + 1$ does not change the order.

$$x > 2x + 2$$

$$-x > 2 \qquad \text{(Subtracting } 2x \text{ from both sides)}$$

$$x < -2 \qquad \text{(Multiplication by } -1 \text{ reversed the order)}$$

The solution set is $\{x \mid x > -1\} \cap \{x \mid x < -2\} = \varnothing$.

Case 2: $x + 1 < 0$ and thus $x < -1$. Since $x + 1 < 0$, multiplying each side of the inequality by $x + 1$ *does change the order.*

$$x < 2x + 2$$

$$-x < 2$$

$$x > -2$$

The solution set is $\{x \mid x < -1\} \cap \{x \mid x > -2\} = \{x \mid -2 < x < -1\}$. The union of the solution sets for the two cases is

$$\{x \mid -2 < x < -1\} \cup \varnothing = \{x \mid -2 < x < -1\}$$

Note that the third possibility, $x + 1 = 0$, was not considered, since this would have made $x/(x + 1) = -\frac{1}{0}$, and division by zero is not defined.

A verbal statement of the solution set $\{x \mid -2 < x < -1\}$ is "all real numbers x such that x is greater than -2 *and* x is less than -1"; or "x is any real number between -2 and -1."

EXERCISES 3.6 A

Solve Exercises 1–20, represent the solution set on a number line, and write a symbolic statement for the solution set.

1. $3x - 4 > 0$ **2.** $3x - 2 > 1 + 2x$

3. $3x + 1 < x - 5$ **4.** $4 - 2x \le 0$

5. $1 < 2x - 3$ **6.** $3(4 - 5x) < 4(x + 3)$

7. $-2(3 + x) \ge 4(2x + 1)$ **8.** $5 - x \le 3 - (x - 2)$

9. $3(x + 2) - 16 < x - 6$ **10.** $x + 4 - 3x \le 2x + 4$

11. $\dfrac{3(4 - 5x)}{2} + 5x > 1$ **12.** $\dfrac{4}{x} > 3$

13. $\dfrac{3}{x} \le 2$ **14.** $\dfrac{x + 5}{x} \ge 6$

15. $\dfrac{1}{x - 1} < 1$ **16.** $\dfrac{-2}{x + 3} \ge 1$

17. $2 \le \dfrac{x}{x + 5}$ **18.** $\dfrac{2 - x}{8} > \dfrac{1}{4}$

19. $\dfrac{3}{x + 1} - 2 < 0$ **20.** $\dfrac{x}{x + 2} + 1 \ge 0$

21. Find all values for x which satisfy *both* statements:

$$5(2 - x) > 3x - 8$$
$$2(3x + 4) + 7 < 5 + x$$

22. A student must have an average of 90 to 100 percent inclusive on five tests in a course to receive an A grade. If his grades on the first four tests were 93 percent, 86 percent, 82 percent, and 96 percent, what grade must he achieve on the fifth test in order to qualify for the A?

23. A baseball team wins 40 of its first 50 games. What is the smallest number of games the team must win in the remaining 40 games in order to have an average of at least 60 percent wins for the complete season?

24. What temperatures in degrees centigrade correspond to 50 degrees $\le$ $F \le 77$ degrees, where F is the temperature in degrees Fahrenheit? [Use $F = (9C + 160)/5$].

25. If a 12-year-old child has an I.Q. between 110 and 140, what are the possible mental ages of the child? (Use I.Q. $= 100M/C$, where M is the mental age and C is the chronological age.)

★ **26.** If $a > b$ and $b > c$, prove that $a > c$.

★ **27.** If $a > b$ and $ab > 0$, prove that $(1/a) < (1/b)$.

EXERCISES 3.6 B

Solve Exercises 1–20, represent the solution set on a number line, and write a symbolic statement for the solution set.

1. $3x + 4 < x + 7$ **2.** $5 - 3x > 0$

3. $11 - 3x \le \dfrac{2x}{3}$ **4.** $3x - 2 > 1 + 2x$

5. $\dfrac{x - 6x}{2} < -20$ **6.** $2(x + 3) \geq 8(2 - x)$

7. $-3(2 - 3x) < 15 - (x + 1)$ **8.** $x - 2 < 2 - x$

9. $\dfrac{x}{2} + 7 < \dfrac{x}{3} - x$ **10.** $\frac{3}{4}(2x - 4) - \frac{1}{2}(4x + 3) \geq 0$

11. $\dfrac{x - 2}{3} \leq \dfrac{2 - x}{2}$ **12.** $\dfrac{3}{x} < 0$

13. $\dfrac{2 + x}{x} < 5$ **14.** $\dfrac{4}{x} \leq \dfrac{8}{x}$

15. $\dfrac{2}{x + 1} < 4$ **16.** $\dfrac{10}{x - 3} \geq 2$

17. $10 \leq \dfrac{6x}{x + 2}$ **18.** $\dfrac{x}{2x + 1} + 4 < 0$

19. $x + \dfrac{5}{x} \leq 7 + x$ **20.** $x + \dfrac{5}{x} \leq x - 7$

21. Find all values for x which satisfy *both* statements:

$$x + 2(x + 2) > 10 + x$$

$$5x - 2 < 2x + 13$$

22. A student must have an average of 80 percent to 89 percent inclusive on five tests in a course to receive a B grade. His grades on the first four tests were 98 percent, 76 percent, 86 percent, and 92 percent. Find the range of grades on the fifth test which would qualify him for a B in the course.

23. If a basketball team wins 30 of its first 35 games, what is the largest number of games it can lose in the remaining 45 games to have an average of at least 70 percent wins for the complete season?

24. Write an inequality that expresses the number of degrees A in angle A of a triangle if angle $B = 40$ degrees and angle C is such that 110 degrees $\leq C < 140$ degrees. (Use $A + B + C = 180$ degrees.)

25. How many quarts of milk containing 5 percent butterfat can be mixed with cream containing 80 percent butterfat so that 100 quarts of the resulting mixture will have a butterfat content between 20 percent and 26 percent?

★ **26.** Prove: If $a > 1$ and $b > 0$, then $ab > b$.

★ **27.** Prove: If $a \geq 0$ and $b \geq 0$, and $a \leq b$, then $a^2 \leq b^2$ and $a^3 \leq b^3$.

3.7 QUADRATIC INEQUALITIES IN ONE VARIABLE

Statements having any one of the forms

$$ax^2 + bx + c > 0 \qquad ax^2 + bx + c < 0$$

$$ax^2 + bx + c \geq 0 \qquad ax^2 + bx + c \leq 0$$

where x is a variable and a, b, and c are constants with $a \neq 0$ are called *quadratic inequalities* in the variable x. The following discussion shows how quadratic inequalities are solved.

The multiplication of two real numbers can be separated into the following four cases:

1. Both factors are positive: $a > 0$ and $b > 0$, then $ab > 0$.

2. Both factors are negative: $a < 0$ and $b < 0$, then $ab > 0$.

3. The first factor is positive, the second is negative: $a > 0$ and $b < 0$, then $ab < 0$.

4. The first factor is negative, the second is positive: $a < 0$ and $b > 0$, then $ab < 0$.

Since these are the only cases possible, the following theorems may be stated.

THE POSITIVE PRODUCT THEOREM

If $ab > 0$, then either $(a > 0$ and $b > 0)$ or $(a < 0$ and $b < 0)$.

THE NEGATIVE PRODUCT THEOREM

If $ab < 0$, then either $(a > 0$ and $b < 0)$ or $(a < 0$ and $b > 0)$.

The positive product theorem states that if a product of two factors is positive, then both factors are positive or both factors are negative.

The negative product theorem states that if a product of two factors is negative, then one of the factors is positive and the other is negative.

EXAMPLE 3.7.1   Find the solution set of $(x + 2)(x + 3) > 0$.

Solution   By the positive product theorem, there are two cases:

Case 1: $x + 2 > 0$ and $x + 3 > 0$

Case 2: $x + 2 < 0$ and $x + 3 < 0$

In set notation, these two cases are written

         Case 1   *or*   *Case 2*

$(\{x \mid x + 2 > 0\} \cap \{x \mid x + 3 > 0\}) \cup (\{x \mid x + 2 < 0\} \cap \{x \mid x + 3 < 0\})$

     Case 1: If $x + 2 > 0$, then $x > -2$.

           If $x + 3 > 0$, then $x > -3$.

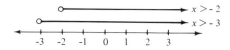

FIGURE 3.7.1

$$\{x \mid x + 2 > 0\} \cap \{x \mid x + 3 > 0\} = \{x \mid x > -2\}$$

Case 2: If $x + 2 < 0$, then $x < -2$.

          If $x + 3 < 0$, then $x < -3$.

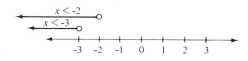

FIGURE 3.7.2

$$\{x \mid x + 2 < 0\} \cap \{x \mid x + 3 < 0\} = \{x \mid x < -3\}$$

Therefore, $\{x \mid x > -2\} \cup \{x \mid x < -3\}$ is the solution set—that is, $\{x \mid x < -3 \text{ or } x > -2\}$.

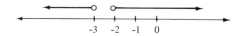

FIGURE 3.7.3

EXAMPLE 3.7.2   Solve $(x + 2)(x + 3) < 0$.

Solution   By the trichotomy axiom, if r is any real number, then $r < 0$ or $r = 0$ or $r > 0$.

1. If $(x + 2)(x + 3) > 0$, then $x < -3$ or $x > -2$ by Example 3.7.1.

2. If $(x + 2)(x + 3) = 0$, then $x = -3$ or $x = -2$.

3. If $(x + 2)(x + 3) < 0$, then $x \nleq -3$ and $x \ngeq -2$.

In other words, the solution set consists of those real numbers not in the solution set of $(x + 2)(x + 3) \geq 0$.
Therefore, $\{x \mid -3 < x < -2\}$ is the solution set (see Figure 3.7.4.)

FIGURE 3.7.4

EXAMPLE 3.7.3   Find the solution set of the inequality $2x^2 - 5x < 3$ and represent the solution set on a line graph.

Solution $2x^2 - 5x < 3$

$$2x^2 - 5x - 3 < 0 \qquad \text{(Addition theorem)}$$

$$(2x + 1)(x - 3) < 0 \qquad \text{(Factoring the polynomial)}$$

Thus

$[2x + 1 < 0 \quad$ and $\quad x - 3 > 0]$ or $[2x + 1 > 0 \quad$ and $\quad x - 3 < 0]$

$\quad 2x < -1 \quad$ and $\quad x > 3 \qquad\qquad\qquad 2x > -1 \quad$ and $\quad x < 3$

$\quad x < -\frac{1}{2} \quad$ and $\quad x > 3 \qquad\qquad\qquad x > -\frac{1}{2} \quad$ and $\quad x < 3$

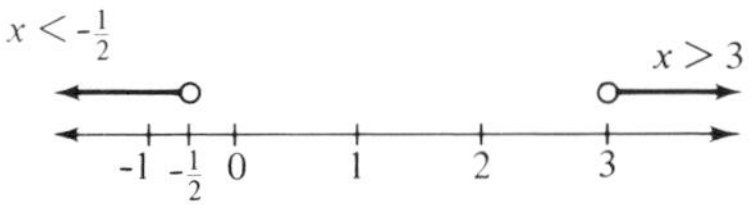

$x < -\frac{1}{2}$

$x > 3$

FIGURE 3.7.5 $\{x \mid x < -\frac{1}{2}\} \cap \{x \mid x > 3\} = \varnothing$.

$x < 3$

$x > -\frac{1}{2}$

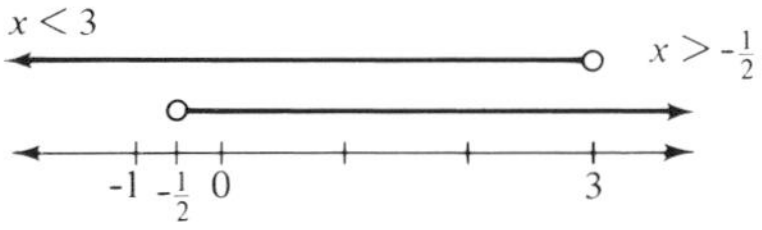

FIGURE 3.7.6 $\{x \mid x > -\frac{1}{2}\} \cap \{x \mid x < 3\} = \{x \mid -\frac{1}{2} < x < 3\}$.

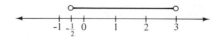

FIGURE 3.7.7 $\{x \mid -\frac{1}{2} < x < 3\}$.

Since division can be defined as multiplication

$$\frac{a}{b} = a \cdot \frac{1}{b} \qquad (b \neq 0)$$

the product theorems for inequalities apply to the solution of expressions such as

$$\frac{f(x)}{g(x)} > 0 \quad \text{or} \quad \frac{f(x)}{g(x)} < 0$$

where $f(x)$ and $g(x)$ are polynomials in x, and $g(x) \neq 0$.

EXAMPLE 3.7.4   Find the solution set for

$$\frac{x + 2}{x - 3} \leq 5$$

and graph the solution set.

Solution   (This is an alternate method of solution to the one presented in the preceding section.)

$$\text{If} \quad \frac{x + 2}{x - 3} \leq 5, \quad \text{then} \quad \frac{x + 2}{x - 3} - 5 \leq 0$$

Expressing the left side of the inequality as a single fraction,

$$\frac{x + 2 - 5(x - 3)}{x - 3} \leq 0$$

$$\frac{-4x + 17}{x - 3} \leq 0$$

The quotient is negative only if the numerator is positive and the denominator is negative, or if the numerator is negative and the denominator is positive. That is,

$$(\{x \mid -4x + 17 \geq 0\} \cap \{x \mid x - 3 < 0\}) \cup$$
$$(\{x \mid -4x + 17 \leq 0\} \cap \{x \mid x - 3 > 0\})$$

Thus

$$(-4x + 17 \geq 0 \quad \text{and} \quad x - 3 < 0) \quad \text{or} \quad (-4x + 17 \leq 0 \quad \text{and} \quad x - 3 > 0)$$

$17 \geq 4x$ and	$x < 3$	$-4x \leq -17$ and	$x > 3$
$\frac{17}{4} \geq x$ and	$x < 3$	$x \geq \frac{17}{4}$ and	$x > 3$

$$\{x \mid x \leq \tfrac{17}{4}\} \cap \{x \mid x < 3\} = \{x \mid x < 3\}$$

or

$$\{x \mid x \geq \tfrac{17}{4}\} \cap \{x \mid x > 3\} = \{x \mid x \geq \tfrac{17}{4}\}$$

FIGURE 3.7.8

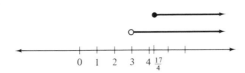

FIGURE 3.7.9

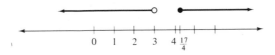

FIGURE 3.7.10

$$(\{x \mid -4x + 17 \geq 0\} \cap \{x \mid x - 3 < 0\}) \cup$$

$$(\{x \mid -4x + 17 \leq 0\} \cap \{x \mid x - 3 < 0\})$$

$$= \{x \mid x < 3 \quad \text{or} \quad x \geq \tfrac{17}{4}\}$$

The chart on the following page summarizes the properties of inequalities and compares them with the properties of equalities.

COMPARISON OF EQUATIONS AND INEQUALITIES

	Equations	*Inequalities*
1. Reflexivity	$x = x$	
Trichotomy		If $x \neq y$, then
		$x < y$ or $x > y$
2. Symmetry	If $x = y$, then $y = x$	
Nonsymmetry		If $x < y$, then $y > x$
3. Transitivity	If $x = y$ and $y = z$, then	If $x < y$ and $y < z$, then
	$x = z$	$x < z$
4. Addition	If $x = y$, then	If $x < y$, then
	$x + z = y + z$	$x + z < y + z$
5. Multiplication	If $x = y$, then $xz = yz$	If $x < y$ and $z > 0$, then
		$xz < yz$
		If $x < y$ and $z < 0$, then
		$xz > yz$
6. Zero product	$ab = 0$ if and only if	
	$a = 0$ or $b = 0$	
Positive product		$ab > 0$ if and only if
		either $a > 0$ and $b > 0$
		or $a < 0$ and $b < 0$
Negative product		$ab < 0$ if and only if
		either $a > 0$ and $b < 0$
		or $a < 0$ and $b > 0$

Examples

Equations	*Inequalities*
1. $5 = 5$	$5 < 7$ and $5 > 3$
2. If $5 = x$, then $x = 5$	If $5 < x$, then $x > 5$
3. If $x = 3 + 2$ and $3 + 2 = 5$, then $x = 5$	If $x < \sqrt{8}$ and $\sqrt{8} < 3$, then $x < 3$
4. If $x - 2 = 5$, then $x = 7$	If $x - 2 < 5$, then $x < 7$
5. If $\dfrac{x}{2} = 5$, then $x = 10$	If $\dfrac{x}{2} < 5$, then $x < 10$
If $\dfrac{x}{-2} = 5$, then $x = -10$	If $\dfrac{x}{-2} < 5$, then $x > -10$
6. If $(x - 3)(x + 2) = 0$, then $x - 3 = 0$ or $x + 2 = 0$	If $(x - 3)(x + 2) > 0$, then $x - 3 > 0$ and $x + 2 > 0$ or $x - 3 < 0$ and $x + 2 < 0$
	If $(x - 3)(x + 2) < 0$, then $x - 3 > 0$ and $x + 2 < 0$ or $x - 3 < 0$ and $x + 2 > 0$

EXERCISES 3.7 A

Solve Exercises 1–25, represent the solution on a line graph, and write a symbolic statement describing the solution set.

1. $(x + 3)(3x + 2) > 0$

2. $(x - 2)(2x + 3) \leq 0$

3. $x(x + 4) \geq 0$

4. $x^2 - 4 \geq 0$

5. $x^2 > 2x$

6. $x^2 - 3x - 4 < 0$

7. $x^2 + 2x + 1 \geq 0$

8. $(x - 2)^2 \geq 0$

9. $(x + 1)^2 < 0$

10. $(x + 1)^2 \leq 0$

11. $x^2 + 1 \geq 2x$

12. $x^2 + 4 > 4x$

13. $x^2 < 4x + 12$

14. $2x^2 - 35 > 9x$

15. $(x - 4)^2 \leq 5$

16. $x^2 - 2x > 1$

17. $x^2 + 5 < 4x$

18. $(x - k)(x - k + 2) \geq 0$

19. $(x - a)^2 \leq b^2 \quad (b > 0)$

20. $x + \dfrac{1}{x} > 2$

21. $\dfrac{1 - x}{x - 4} > 0$

22. $\dfrac{3x - 2}{3} \geq \dfrac{x^2 - 4}{x}$

23. $\dfrac{3}{3x + 1} < \dfrac{4}{4x - 5}$

24. $\dfrac{x}{x + 4} > \dfrac{x - 1}{x + 3}$

25. $\dfrac{1}{x} < \dfrac{1}{x - 5}$

For what values of x does each of the expressions in Exercises 26–28 designate a real number? Illustrate on a line graph.

26. $\sqrt{5x + 10}$

27. $\sqrt{x^2 - 25}$

28. $\sqrt{25 - x^2}$

★ **29.** If $a > 0$ and $b > 0$ and $a \neq b$, prove that

$$a + b > \frac{4a}{a + b}$$

★ **30.** If $a > 0$ and $a \neq 1$, prove that

$$\frac{1}{a} + a > 2$$

★ **31.** The braking distance s for a car decelerating at 22 ft/sec² from an initial velocity v is given by

$$s = \frac{v^2}{44}$$

Find the possible initial velocities v in (a) feet per second and (b) mph so that the car may stop in less than 99 feet. (*Hint:* 44 feet per second = 30 mph.)

★ **32.** If a cubical box is lined with an insulating material t inches thick and if s is the length of a side of the interior of the box before insulation, then the loss of volume V due to the insulation may be approximated to within 1 unit by

$$V = 6ts^2 - 12t^2s \quad (0 < t < \tfrac{1}{2})$$

If $s = 6$ inches, for what values of t will V be between 64 cubic inches and 90 cubic inches?

EXERCISES 3.7 B

Solve Exercises 1–25, represent the solution on a line graph, and write a symbolic statement describing the solution set.

1. $(2x + 1)(x - 3) \geq 0$ **2.** $(x - 1)(x + 5) < 0$

3. $x(2x + 3) > 0$ **4.** $9 - x^2 \geq 0$

5. $x^2 < 4x$ **6.** $x^2 > 25$

7. $x^2 - 5x \leq -6$ **8.** $(x - 1)^2 \geq 0$

9. $(x + 3)^2 < 0$ **10.** $(x + 3)^2 \leq 0$

11. $x^2 + 9 \geq 6x$ **12.** $x^2 + 100 \geq 20x$

13. $x^2 > x + 12$ **14.** $4 - 5x^2 \leq 19x$

15. $(x - 6)^2 \geq 2$ **16.** $x^2 - 4x < 1$

17. $x^2 - 3 \geq 6x$

18. $x^2 - (a + b)x + ab \leq 0 \quad (a < b)$

19. $x^2 - n^2 \geq 0 \quad (n > 0)$ **20.** $x + \dfrac{9}{x} < 6$

21. $\dfrac{x - 3}{3 + x} \leq 0$ **22.** $\dfrac{3x^2 - 4}{3x} \geq x + 2$

23. $x^2 + 1 \leq 2x$ **24.** $\dfrac{x - 1}{x + 4} < \dfrac{x - 4}{x + 1}$

25. $\dfrac{1}{x + 3} \geq \dfrac{1}{x - 3}$

*For what values of x does each of the expressions in Exercises 26–28 designate a
real number? Illustrate on a line graph.*

26. $\sqrt{12 - 3x}$　　　　　　　　　　**27.** $\sqrt{4x^2 - 9}$

28. $\sqrt{75 - x^2}$

★ **29.** If $0 < a < 1$, prove that $a^2 < a$ and $a^3 < a$.

★ **30.** If a and b are real numbers such that $a \le b$ and $b \le a$, prove that $a = b$.

★ **31.** A traffic flow study produced the empirical relation

$$n = 5(60v - v^2)$$

where n = the number of vehicles per hour traveling on a certain bridge
and v = the average speed in miles per hour maintained on the bridge.
Find the values of v for which n is at most 4420 vehicles per hour.

★ **32.** If a missile is fired directly upward from the ground with an initial
velocity of 800 feet per second, then its distance s in feet above the
ground t seconds after the missile was fired may be approximated by

$$s = 800t - 16t^2$$

For what values of t is the missile more than 9600 feet above the ground?

3.8　ABSOLUTE VALUE: EQUALITIES

　　　The number line has been used as a geometric model for the set of real
numbers, and points on the number line correspond to real numbers. Any
point x and its opposite, or additive inverse, $-x$ are the same distance from
the origin but on opposite sides of the origin. The algebraic sign of a number
indicates on which side of the origin the corresponding point is located.
The distance of the point from the origin, regardless of the side on which it
is located, is called the absolute value of the number and is denoted by the
symbol $|x|$.

DEFINITION

The absolute value of a real number $|x|$:

$$|x| = x \text{ if } x \ge 0$$

$$|x| = -x \text{ if } x < 0$$

　　　The definition indicates that the absolute value of a number is never negative, $|x| \geq 0$. In particular, $|0| = 0$. Figure 3.8.1 illustrates that the graphs of the number 3 and the number -3 are the same distance from the origin but on opposite sides.

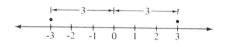

FIGURE 3.8.1

If $x = 3$, then $x > 0$, and $|x| = x$ implies that $|3| = 3$.

If $x = -3$, then $x < 0$, and $|x| = -x$ implies that $|-3| = -(-3) = 3$.

If $P(x)$ is a polynomial in the variable x and a is a nonnegative real number, then the solution set of $|P(x)| = a$ is the **union** of the solution sets of $P(x) = a$ and $-P(x) = a$. Thus there are two cases to be considered.

EXAMPLE 3.8.1   Solve for x: $|x| = 4$.

Solution

Case 1: $x \geq 0$, $|x| = x$, and $x = 4$

Case 2: $x < 0$, $|x| = -x$, $-x = 4$, and $x = -4$

The solution set is $\{4, -4\}$.

EXAMPLE 3.8.2   Solve for x: $|x - 3| = 5$.

Solution

Case 1: $x - 3 \geq 0; \quad |x - 3| = x - 3$

$$x - 3 = 5$$

$$x = 8$$

Case 2: $x - 3 < 0; \quad |x - 3| = -(x - 3)$

$$-(x - 3) = 5$$

$$x - 3 = -5$$

$$x = -2$$

The solution set is $\{-2, 8\}$.

If a and b are any two real numbers, the distance between a and b is defined in terms of absolute value.

DEFINITION

If A and B are two points on a number line with coordinates a and b, respectively, then the distance $|AB|$ between $A:(a)$ and $B:(b)$ is

$$|AB| = |b - a|$$

If $a < b$, then by definition there exists a positive real number d so that $a + d = b$ and thus $d = b - a$. If $A:(a)$ and $B:(b)$ are two points on the number line such that A is to the left of B, then $a < b$ and $d = b - a$ is the length of the line segment AB.

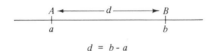

$$d = b - a$$

FIGURE 3.8.2

In some problems it is not known whether A is to the left of B or not. Since $-(a - b) = b - a$, subtraction in the opposite order would produce a negative number. However, $|a - b| = |b - a|$, and thus a positive number can be obtained by using the concept of absolute value. In the special case that A is the same point as B, then $a = b$ and $a - b = b - a = 0$.

Examples:

$$A:(3),\ B:(5),\ |AB| = |5 - 3| = 2$$

$$A:(3),\ B:(-5),\ |AB| = |-5 - 3| = |-8| = 8$$

$$A:(-3),\ B:(-5),\ |AB| = |-5 - (-3)| = |-5 + 3| = |-2| = 2$$

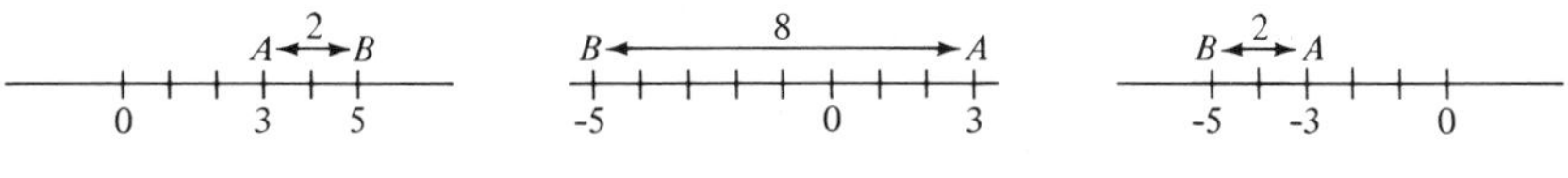

FIGURE 3.8.3

EXAMPLE 3.8.3 Solve for x and graph the solution set on a number line: $|x + 2| + 1 = 4$.

Solution If $|x + 2| + 1 = 4$

then $$|x + 2| = 3$$

$$Case\ 1:\ x + 2 \geq 0;\quad |x + 2| = x + 2$$

$$x + 2 = 3$$

$$x = 1$$

$$Case\ 2:\ x + 2 < 0;\quad |x + 2| = -(x + 2)$$

$$-(x + 2) = 3$$

$$-x - 2 = 3$$

$$-x = 5$$

$$x = -5$$

The solution set is $\{-5, 1\}$, and its graph is shown in Figure 3.8.4.

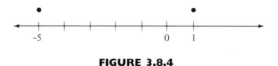

FIGURE 3.8.4

The following theorems are useful when working with absolute values and will be stated without proof.

THEOREMS

For every real number a and b,

$$|a||b| = |ab|$$

For every real number a and for every nonzero real number b,

$$\frac{|a|}{|b|} = \left|\frac{a}{b}\right|$$

Examples: $|3x| = 3|x|$, since $|3| = 3$

$$\frac{|-5|}{|2|} = \left|\frac{-5}{2}\right|$$

EXERCISES 3.8 A

Solve Exercises 1–20 for all values of x.

1. $|x| = 2$

2. $|x| = 5$

3. $|x + 2| = 1$

4. $|3x + 1| = 4$

5. $|x - 2| = 3$

6. $|3 - 2x| = 4$

7. $|x - 4| = 3$

8. $|4 - x| = 3$

9. $|x - 4| = -3$

10. $\left|\dfrac{2x - 1}{4}\right| = 2$

11. $3 - |x| = 4$

12. $3 = \left|4 - \dfrac{x}{2}\right|$

13. $\dfrac{3}{|x + 1|} = 4$

14. $\dfrac{2}{|x - 2|} = 1$

15. $\dfrac{|x|}{x} = 1$

16. $\dfrac{|x|}{x} = -1$

17. $|x - a| = b \quad (b \geq 0)$

18. $|cx| = a \quad (c > 0, a \geq 0)$

19. $|ax + b| = c \quad (a \neq 0, c \geq 0)$

20. $\left|b - \dfrac{x}{2}\right| = a \quad (a \geq 0)$

In Exercises 21–26, find the distance between A and B.

21. $A:(7), B:(3)$

22. $A:(3), B:(7)$

23. $A:(-7), B:(-2)$

24. $A:(-2), B:(-7)$

25. $A:(-3), B:(4)$

26. $A:(4), B:(-3)$

In Exercises 27–28, state in symbols that the distance between A: (a) and B: (x) is d for the values indicated.

27. $a = 4, b = x, d = 3$

28. $a = -2, b = x, d = 5$

Find the values of x for which each of Exercises 29–35 is true.

29. $|x| = |-3|$

30. $|2x| = 2x$

31. $|x| = -3$

32. $|x - 2| = 2 - x$

33. $|2x - 5| = -3$

34. $|x - 2| = |2 - x|$

35. $|x - 2| = x - 2$

EXERCISES 3.8 B

Solve Exercises 1–20 for all values of x.

1. $|x| = 3$ 2. $|x| = \frac{1}{2}$
3. $|x + 4| = 1$ 4. $|2x + 1| = 3$
5. $|x - 3| = 2$ 6. $|2 - 3x| = 4$
7. $|x - 2| = 5$ 8. $|2 - x| = 5$

9. $|x - 2| = -5$ 10. $\left|\dfrac{3x + 1}{2}\right| = 1$

11. $4 - |x + 1| = 3$ 12. $1 = \left|3 - \dfrac{x}{4}\right|$

13. $\dfrac{2}{|x - 3|} = 4$ 14. $\dfrac{1}{|x + 3|} = 4$

15. $2|x - 2| = 5$ 16. $|2x + \frac{1}{2}| - \frac{1}{4} = 0$
17. $|a - x| = b \quad (b \geq 0)$ 18. $|px| = a \quad (p < 0, a > 0)$
19. $|b - ax| = c \quad (a \neq 0, c > 0)$ 20. $|a - \frac{1}{4}x| = b \quad (b > 0)$

In Exercises 21–26, find the distance between A and B.

21. $A:(5), B:(-2)$ 22. $A:(2), B:(-2)$
23. $A:(-2), B:(-2)$ 24. $A:(2), B:(5)$
25. $A:(-6), B:(0)$ 26. $A:(0), B:(4)$

27. State in symbols that the distance between $A:(3)$ and $B:(x)$ is less than or equal to 2.
28. State in symbols that the distance between $A:(-6)$ and $B:(x)$ is greater than or equal to 5.

Find the values of x for which each of Exercises 29–35 is true.

29. $|x| = |-4|$ 30. $|2x - 1| = 2x - 1$
31. $|2x - 1| = |1 - 2x|$ 32. $|2x - 1| = 1 - 2x$
33. $|3x| = 9$ 34. $|x + 5| = -1$
35. $|2 - x| = x - 2$

3.9 ABSOLUTE VALUE: INEQUALITIES

The solution of inequalities involving absolute value involves a closer look at what is meant by expressions such as $|x| > 3$ or $|x + 2| < 1$.

$|x| = 3$ means that $x = 3$ or $x = -3$—that is, x is three units to the right of the origin, or x is three units to the left of the origin on a horizontal number line.

$|x| > 3$ means that x is *more* than three units from the origin. This again raises two possibilities:

Case 1: x is *more* than three units to the *right* of the origin—that is $x > 3$.

Case 2: x is *more* than three units to the *left* of the origin—that is $x < -3$.

Thus $|x| > 3$ indicates that

$$x > 3 \quad \text{or} \quad x < -3$$

$|x + 2| < 1$ means that the distance between x and -2 is less than one unit. (Recall that $|a - b|$ is the distance between a and b, and that $|x + 2| = |x - (-2)|$.)

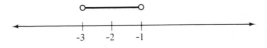

FIGURE 3.9.I

Thus, according to Figure 3.9.1,

$$-3 < x < -1$$

The above intuitive arguments are formalized in the following theorems:

THEOREMS

Let $P(x)$ be an algebraic expression in the variable x and a be a real number such that $a \geq 0$.

1. $|P(x)| \leq a$ if and only if $-a \leq P(x) \leq a$.

2. $|P(x)| \geq a$ if and only if $P(x) \geq a$ or $P(x) \leq -a$.

EXAMPLE 3.9.1   Solve for x: $|x| < 5$ and graph the solution set.

Solution   By theorem 1, $|x| < 5$ means

$$-5 < x < 5$$

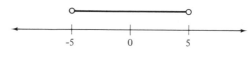

FIGURE 3.9.2

EXAMPLE 3.9.2   Solve for x: $|x| \geq 5$ and graph the solution set.

Solution   By theorem 2, $|x| \geq 5$ means that

$$x \geq 5 \quad \text{or} \quad x \leq -5$$

FIGURE 3.9.3

EXAMPLE 3.9.3   Solve for x and graph the solution set: $|x - 2| \leq 3$.

Solution   By theorem 1, $|x - 2| \leq 3$ means

$$-3 \leq x - 2 \leq 3$$

Thus $$-1 \leq x \leq 5 \qquad \text{(Addition theorem, } +2)$$

FIGURE 3.9.4

The solution set is $\{x \mid -1 \leq x \leq 5\}$. A verbal description of $|x - 2| \leq 3$ is that x is less than three units from 2, or x is exactly three units from 2.

EXAMPLE 3.9.4   Solve $|x + 2| > 4$.

Solution $x + 2 > 4$ or $x + 2 < -4$ (Using theorem 2)

$\qquad\qquad\quad x > 2$ or $x < -6$

The solution set is $\{x \mid x > 2 \text{ or } x < -6\}$.

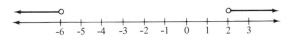

FIGURE 3.9.5

EXAMPLE 3.9.5   Solve for x: $\dfrac{3}{|x|} < 5$.

Solution $|x|$ is never negative, so multiplication by $|x|$ does not change the sense of the inequality. Since $|x|$ is in the denominator, $x \neq 0$.

$$\frac{3}{|x|} < 5$$

$$3 < 5|x|$$

$$|x| > \tfrac{3}{5}$$

$$x > \tfrac{3}{5} \text{ or } x < -\tfrac{3}{5} \text{ (by theorem 2)}$$

EXERCISES 3.9 A

In Exercises 1–20, solve and represent the solution set on a line graph.

1. $|x| > 2$

2. $|x| \leq 3$

3. $|x| < \tfrac{1}{2}$

4. $|x - 4| < 3$

5. $|2x + 3| \leq 5$

6. $3 \geq |2x - 1|$

7. $|2x - 1| \geq 3$

8. $2|x - 2| < 5$

9. $|2x| + 3 > 4$

10. $\left|\dfrac{x - 2x}{3}\right| > 2$

11. $\dfrac{2}{|2 - x|} < 3$

12. $\dfrac{4}{|x|} > 2$

13. $|x| \leq -4$

14. $3 - |2 - x| < 6$

15. $|x - 3| > -2$

16. $|x - 3| < -2$

17. $2|2x - 8| > 6$

18. $\left|\dfrac{3 - 2x}{x + 5}\right| < -4$

19. $\left|\dfrac{4}{x}\right| \geq 3$

20. $|x + 2||x - 3| > 6$

EXERCISES 3.9 B

In Exercises 1–20, solve and represent the solution set on a line graph.

1. $|x| < 2$

2. $|x| > 3$

3. $|x| \geq \frac{1}{4}$

4. $|x - 2| > 3$

5. $|x + 4| < 2$

6. $|3 - 2x| > 4$

7. $2 > |1 - x|$

8. $|1 - x| > 2$

9. $\left|\dfrac{2x - 1}{4}\right| < 2$

10. $|x| + 2 \leq 3$

11. $3 \leq \left|4 - \dfrac{x}{2}\right|$

12. $\dfrac{3}{|x + 1|} > 4$

13. $|x| \leq -2$

14. $2 + |x - 3| < 5$

15. $|x + 3| > -3$

16. $|x + 3| < -3$

17. $3|4x + 1| \leq 6$

18. $\left|\dfrac{-2}{x}\right| > 3$

19. $\left|\dfrac{2 - 3x}{x}\right| > 1$

20. $|x - 2| < |x + 3|$

SUMMARY

Basic Factoring Forms

☐ $ax + ay = a(x + y)$ Common monomial factor
                                      (Distributive principle)

☐ $x^2 + 2ax + a^2 = (x + a)^2$ Perfect square trinomial

☐ $x^2 - a^2 = (x + a)(x - a)$ Difference of two squares

☐ $x^2 + (a + b)x + ab = (x + a)(x + b)$ Quadratic trinomial

☐ $x^3 + a^3 = (x + a)(x^2 - ax + a^2)$ Sum of two cubes

☐ $x^3 - a^3 = (x - a)(x^2 + ax + a^2)$ Difference of two cubes

Theorems

☐ **The remainder theorem.** If c is any constant, and if a polynomial $P(x)$ is divided by a divisor of the form $(x - c)$, then the remainder R is equal to $P(c)$.

☐ **The factor theorem.** Let $P(x)$ be a polynomial. Then $(x - c)$ is a factor of $P(x)$ if and only if $P(c) = 0$.

☐ **Definition.** $a < b$ if and only if there exists a positive real number, p, such that $a + p = b$.

☐ **Definition.** $a > b$ if and only if $b < a$.

☐ **The trichotomy axiom.** Exactly one of the following is true: $a < b$ or $a = b$ or $a > b$.

☐ **The transitivity theorem.** If $a < b$ and $b < c$, then $a < c$.

☐ **The addition theorem.** If $a < b$, then $a + c < b + c$.

☐ **The multiplication theorems.** If $a < b$ and $c > 0$, then $ac < bc$. If $a < b$ and $c < 0$, then $ac > bc$.

☐ **Definition.** $a < x < b$ if and only if $a < x$ and $x < b$.

☐ **The positive product theorem.** $ab > 0$ if and only if either $(a > 0$ and $b > 0)$ or $(a < 0$ and $b < 0)$.

☐ **The negative product theorem.** $ab < 0$ if and only if either $(a > 0$ and $b < 0)$ or $(a < 0$ and $b > 0)$.

☐ **Definition of absolute value.**

$$|x| = x \quad \text{if} \quad x \geq 0$$
$$|x| = -x \quad \text{if} \quad x < 0$$

☐ **The theorems for solving absolute-value inequalities.** Let $P(x)$ be an algebraic expression in the variable x. Let a be a nonnegative real number, $a \geq 0$.

$$|P(x)| < a \quad \text{if and only if} \quad -a < P(x) < a$$

$$|P(x)| > a \quad \text{if and only if} \quad P(x) > a \text{ or } P(x) < -a$$

☐ **Distance.** $|AB|$ between two points, $A:(a)$ and $B:(b)$, on the real number line:

$$|AB| = |b - a|$$

REVIEW EXERCISES

Factor completely Exercises 1–10.

1. $x^3 y^2 + y^2$
2. $7(x + a)^2 - 11y(x + a) - 6y^2$
3. $r^2 - 25s^2 + r - 5s$
4. $ax + cx + a^2 b + abc$
5. $x^3 - 27$
6. $x^4 + x^2 + 1$
7. $x^3 - 2x^2 - 5x + 6$
8. $(b - 1)^3 + 8$
9. $x^3 + 125 + x^4 + 125x$
10. $(m^2 + 2m + 1)^2 - 16$

11. Use synthetic division to determine the quotient and remainder:

    a. $(2x^5 - 30x^3 + 9x^2 + 5x - 3) \div (x + 4)$

    b. $(3x^3 + 2x^2 - x + 4) \div (x - 3)$

12. Check your answers to Exercise 11 by using the remainder theorem.

13. Use the factor theorem to determine m so that $(x - 3)$ is a factor of $P(x) = x^4 - 3x^3 + 2x^2 - 3x + m$.

14. Find k so that 3 is a root of the equation $4x^3 - 8x^2 - 2x + k = 0$.

Determine the solution set of each equation in Exercises 15–18.

15. $x^4 - x^2 + 4x - 4 = 0$ **16.** $x^5 - 3x^3 + 8x^2 - 24 = 0$

17. $x^3 + 2x^2 - 19x - 20 = 0$ **18.** $x^4 - x^3 - 2x - 4 = 0$

Solve Exercises 19–36 and represent the solution set on a line graph.

19. $2x - 1 > x + 3$ **20.** $x^2 + 3x > 0$

21. $2x - 1 \le 3x + 5$ **22.** $\dfrac{x}{x - 2} > 0$

23. $\dfrac{2x + 3}{x - 2} < 0$ **24.** $3 \le \dfrac{x + 6}{x}$

25. $x^2 - x > 12$ **26.** $x^2 \le 2x + 15$

27. $\dfrac{x - 1}{x + 3} \le 5$ **28.** $|2x + 5| = 1$

29. $|5 - 3x| = 2$ **30.** $|x + 4| > 5$

31. $|3 - 2x| \le 6$ **32.** $|2x - 6| \ge 4$

33. $\dfrac{3}{|x + 2|} \le 5$ **34.** $4 + \dfrac{|x - 3|}{2} < 6$

35. $|4x - 7| \ge 0$ **36.** $|5x + 4| < 0$

CHAPTER 4

RELATIONS
AND FUNCTIONS

One of the fundamental concepts of concern to the mathematician is how numbers, or their "real-world" applications, are related. For instance, profit from the sale of a certain commodity is related somehow to the number of items produced. The distance an object travels is related to the time it travels.

The mathematical term for these relationships is *relation*. A special type of relation, and one that is of great importance in mathematics, is called a *function*. Profit is a function of the number of items produced, and distance is a function of time. Not all relations are such that the term *function* can be applied. It is the purpose of this chapter to define the relation and function concepts and to introduce some special mathematical relations and functions.

4.1 ORDERED PAIRS AND RELATIONS

An **ordered pair** is an expression having the form (a, b), where a is called the **first component** (or first member) of the ordered pair and b is called the **second component** (or second member) of the ordered pair.

The order in which the components of an ordered pair are written is important. For example, the ordered pair $(3, 5)$ is not the same as the ordered pair $(5, 3)$.

Besides the set operations ∪ (union) and ∩ (intersection), there is another operation on sets that is very useful in the study of algebra. This is the operation × (cross), which is also called the Cartesian product of two sets.

DEFINITION OF CARTESIAN PRODUCT

If A and B are nonempty sets, then

$$A \times B = \{(x, y) | x \in A \text{ and } y \in B\}$$

This is read "A cross B is the set of all ordered pairs (x, y) such that x is an element of A and y is an element of B." Thus the Cartesian product of two sets is again a set, but its elements are ordered pairs.

EXAMPLE 4.1.1   Let $A = \{a, b, c\}$ and $B = \{1, 2\}$; then

$$A \times B = \{(a, 1), (a, 2), (b, 1), (b, 2), (c, 1), (c, 2)\}$$
$$B \times A = \{(1, a), (2, a), (1, b), (2, b), (1, c), (2, c)\}$$

Since the order of the elements in the ordered pair is important, the pair $(a, 1)$ and the pair $(1, a)$ are not the same. Therefore, $A \times B \neq B \times A$, and the Cartesian product of sets is not a commutative operation.

DEFINITION

A **relation** is a set whose elements are ordered pairs.

In this text, the universal set for all relations, unless otherwise specified, shall be $R \times R$, the Cartesian product of the real numbers, R, with the real numbers, R. That is to say, the ordered pairs shall consist of components that are real numbers, and every relation will be a subset of $R \times R$ unless otherwise specified.

DEFINITIONS

The **domain** of a relation is the set of all first components of the relation.
    The **range** of a relation is the set of all second components of the relation.

For the set $A \times B$ in Example 4.1.1, set A is the domain of $A \times B$ and set B is the range of $A \times B$.

Specifying a Relation by an Equation

There are several methods for specifying particular relations. One method commonly used is to state an equation relating the first and second components of the relation. This involves the concept of the solution set of an equation in two variables.

DEFINITION

The **solution set of an open equation in two variables** x and y is the set of ordered pairs of the form (a, b) such that the equation becomes true when x is replaced by a and y is replaced by b.

EXAMPLE 4.1.2   Let $A = \{-3, -2, -1, 0, 1, 2, 3\}$; let r be a subset of $A \times A$, where r is defined by

$$r = \left\{ (x, y) \mid y = \frac{x + 2}{x} \right\}$$

a. List the elements in r.
b. State the domain and range of r.

Solution

a. Find solutions of $y = \dfrac{x + 2}{x}$ where x is in A and y is in A.

x	y	
-3	$-$	For $x = -3$, $\dfrac{x + 2}{x} = \dfrac{-3 + 2}{-3} = \dfrac{1}{3}$, but $\dfrac{1}{3}$ is not in A.
-2	0	For $x = -2$, $\dfrac{x + 2}{x} = \dfrac{-2 + 2}{-2} = 0$.
-1	-1	For $x = -1$, $\dfrac{x + 2}{x} = \dfrac{-1 + 2}{-1} = -1$.
0	$-$	For $x = 0$, $\dfrac{x + 2}{x} = \dfrac{0 + 2}{0}$, which is undefined.
1	3	For $x = 1$, $\dfrac{x + 2}{x} = \dfrac{1 + 2}{1} = 3$.
2	2	For $x = 2$, $\dfrac{x + 2}{x} = \dfrac{2 + 2}{2} = 2$.
3	$-$	For $x = 3$, $\dfrac{x + 2}{x} = \dfrac{3 + 2}{3} = \dfrac{5}{3}$, but $\dfrac{5}{3}$ is not in A.

Thus $r = \{(-2, 0), (-1, -1), (1, 3), (2, 2)\}$.
b. The domain of $r = \{-2, -1, 1, 2\}$, the set of all first components.
    The range of $r = \{0, -1, 3, 2\}$, the set of all second components.

Specifying a Relation by a Table

A table of values may specify a relation, as shown in the following example.

EXAMPLE 4.1.3   At Capital City Community College, a survey was taken of the number of students enrolled in certain courses. The courses were coded by number, and the results are tabulated below:

Course C	1031	1032	1034	1035	1036	1037	1038
Number of Students n	120	148	315	46	216	15	156

Let $r = \{(C, n) \mid C$ and n are entries in the table$\}$.

a. List the elements in r.
b. State the domain and range of r.

Solution

a. $r = \{(1031,\ 120),\ (1032,\ 148),\ (1034,\ 315),\ (1035,\ 46),\ (1036,\ 216),\ (1037,\ 15),\ (1038,\ 156)\}$
b. Domain of $r = \{1031,\ 1032,\ 1034,\ 1035,\ 1036,\ 1037,\ 1038\}$
    Range of $r = \{120,\ 148,\ 315,\ 46,\ 216,\ 15,\ 156\}$

Specifying a Relation by a Graph

Since there is a one-to-one correspondence between the set of ordered pairs in $R \times R$ and the geometric points on a rectangular (Cartesian) number plane, the points on a graph may be used to define the ordered pairs in a relation.

A rectangular coordinate system is shown in Figure 4.1.1.

A **rectangular coordinate system** is obtained by taking two perpendicular number lines, called the **axes**, intersecting at their origins. The point of intersection of the axes is called the **origin** of the coordinate system.

It is customary to select one axis horizontal, called the **x-axis**, with its positive direction to the right, and the other axis vertical, called the **y-axis**, with its positive direction upward.

The axes separate the plane into four regions, called **quadrants**, that are numbered consecutively starting with the upper right quadrant and proceeding counterclockwise.

With each ordered pair (a, b) in $R \times R$ is associated a unique point P, the point of intersection of a vertical line through point a on the x-axis and a horizontal line through point b on the y-axis. The numbers of the ordered

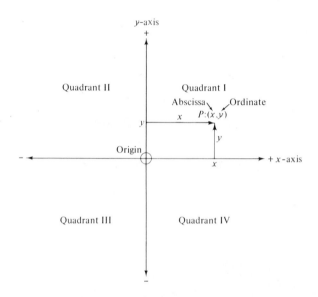

FIGURE 4.1.1   Rectangular coordinate system.

pair (a, b) are called the **coordinates** of P, with the first component a called the **abscissa** and the second component b called the **ordinate**.

If the axes on a graph are labeled by letters other than x and y, it is customary to take the domain elements from values on the horizontal axis and the range elements from values on the vertical axis.

EXAMPLE 4.1.4   For the relation specified by the graph shown in Figure 4.1.2,

a. List the ordered pairs in the relation for $a \in \{3000, 6000, 9000, 12000\}$
b. Specify the relation by an equation

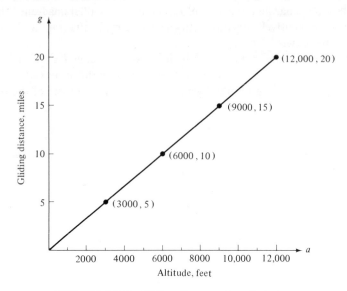

FIGURE 4.1.2   Gliding distances of small airplanes.

Solution

a. (3000, 5), (6000, 10), (9000, 15), (12000, 20)

b. $\left\{(a, g) \mid g = \dfrac{a}{600} \text{ where } a \geq 0\right\}$

It should be noted that there is *not* exactly one answer to the problem of determining an equation from a graph or a list of ordered pairs. In some cases, however, one possibility may appear more obvious than others.

EXERCISES 4.1 A

If $A = \{1, 2, 3\}$ and $B = \{2, 4\}$, list the elements in each of Exercises 1–4.

1. $A \times B$ **2.** $B \times A$

3. $A \times A$ **4.** $B \times B$

For each of the relations in Exercises 5–8, state the domain and range.

5. $\{(1, 2), (3, 2), (5, 6), (2, 1)\}$ **6.** $\{(-1, -2), (0, 0), (3, -1)\}$

7. $\{(1, 1), (1, -1), (4, 2), (4, -2), (0, 0)\}$

8. $\{(10, 20), (12, 20), (14, 20), (15, 30)\}$

Which of 9–14 are solutions of $y = 2x + 3$?

9. $(0, 3)$ **10.** $(2, 6)$
11. $(7, 2)$ **12.** $(3, 0)$
13. $(1, 5)$ **14.** $(5, 13)$

Specify each of the relations in Exercises 15–18 as (a) a table of values and (b) a graph.

15. $\{(x, y) \mid y = 3x, 0 < x \le 5, x$ is a natural number, y is a real number$\}$

16. $\left\{(x, y) \mid y = \dfrac{1}{x}, 2 < x < 6, x$ is an integer, y is a real number$\right\}$

17. $\{(r, s) \mid r + s = 5, r$ and s are natural numbers$\}$
18. $\{(a, b) \mid a - b = 0, a$ and b are natural numbers$\}$

Specify each of the relations in Exercises 19–22 as (a) an equation and (b) a graph.

19. $\{(1, 1), (2, 2), (3, 3), (4, 4), (5, 5)\}$
20. $\{(1, 1), (2, 4), (3, 9), (4, 16), (5, 25)\}$
21. $\{(-1, 2), (-2, 4), (-3, 6), (-4, 8), (-5, 10), (-6, 12)\}$
22. $\{(1, 0), (2, 0), (3, 0), (4, 0), (5, 0)\}$

Find the domain for the relation defined by each of the equations in Exercises 23–26 if the domain is a subset of the set of real numbers.

23. $y = \dfrac{x - 2}{x - 3}$ **24.** $y = \dfrac{2}{3x - x^2}$

25. $y = \sqrt{16 - x}$ **26.** $y = \sqrt{4 + x}$

In Exercises 27–30, let $A = \{-2, -1, 0, 1, 2\}$. *Find a relation that is a subset of* $A \times A$ *and that is specified by the given equation. List the ordered pairs in each relation and graph each relation.*

27. $y = 2x$ **28.** $y < x$
29. $y \ge x$ **30.** $y = |x|$

31. Given the relation $\{(t, s) \mid s = 16t^2, 0 \le t \le 6\}$, which describes the distance s (in feet) that an object falls in t seconds if air resistance is disregarded.

a. Make a table of values for this relation for domain values $\{0, 1, 2, 3, 4, 5, 6\}$.
b. Graph the relation for the specified values of t. (Let the horizontal axis be the t-axis.)

c. Join the points of the graph by a smooth curve.

d. From the graph, determine the distance the object falls in $3\frac{1}{2}$ seconds.

e. From the graph, find the time it takes the object to fall 324 feet.

32. A relation is specified by the table below, giving the speed records for cars and the year in which they were made.

	Piston Engine				Jet Engine			
Date d	1898	1904	1927	1935	1963	1964	1965	1970
Speed record s (in mph)	39	104	204	301	407	526	601	622

a. List the ordered pairs having the form (d, s) in this relation.

b. Graph this relation.

33. The graph in Figure 4.1.3 relates the profit P in dollars for a certain commodity to the number of items x that are sold.

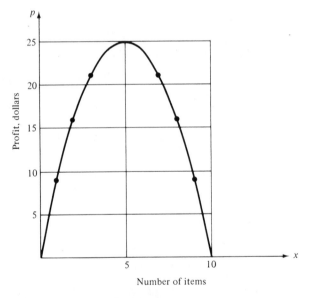

FIGURE 4.1.3

a. Using the graph, list the ordered pairs in the relation

$$\{(x, P) \mid 0 \leq x \leq 10 \text{ and } x \text{ is an integer}\}$$

b. For what value of x is the profit the greatest? What is the greatest profit?

EXERCISES 4.I B

If $A = \{-2, -4\}$ and $B = \{-1, 0, 1\}$, list the elements in each of Exercises 1–4.

1. $A \times B$ **2.** $B \times A$
3. $A \times A$ **4.** $B \times B$

For each of the relations in Exercises 5–8, state the domain and range.

5. $\{(2, 4), (3, 1), (3, 2), (4, 2)\}$
6. $\{(2, 0), (-3, 1), (3, -2), (-4, 2)\}$
7. $\{(2, -2), (3, -3), (5, -5)\}$ **8.** $\{(1, 1), (2, \frac{1}{2}), (-2, -\frac{1}{2})\}$

Which of 9–14 are solutions of $y = \frac{1}{2}x - 2$?

9. $(4, 0)$ **10.** $(2, -1)$
11. $(2, 0)$ **12.** $(10, 8)$
13. $(0, 2)$ **14.** $(1, 6)$

Specify each of the relations in Exercises 15–18 as (a) a table of values and (b) a graph.

15. $\{(x, y) \mid y = 2x, x$ is a natural number less than 5, y is a real number$\}$
16. $\{(x, y) \mid y = \sqrt{x + 1}, -2 \le x \le 2, x$ is an integer, y is a real number$\}$
17. $\left\{(m, n) \mid n = \dfrac{1}{m + 1}, 0 < m < 3, m$ is an integer, n is a real number$\right\}$
18. $\{x, y) \mid x + y = 0, x$ and y are integers$\}$

Specify each of the relations in Exercises 19–22 as (a) an equation and (b) a graph.

19. $\{(1, 1), (1, -1), (4, 2), (4, -2), (9, 3), (9, -3), (16, 4), (16, -4)\}$
20. $\{(1, 1), (2, 8), (3, 27), (4, 64)\}$
21. $\{(1, 1), (2, \frac{1}{2}), (3, \frac{1}{3}), (4, \frac{1}{4}), (5, \frac{1}{5}), (6, \frac{1}{6})\}$
22. $\{(1, 2), (1, 3), (1, 4), (1, 5), (1, 6), (1, 7)\}$

Find the domain for the relation defined by each of the equations in Exercises 23–26 if the domain is a subset of the set of real numbers.

23. $y = \dfrac{x + 1}{x - 1}$ **24.** $y = \dfrac{x - 2}{x(x + 2)}$

25. $y = \sqrt{x - 16}$ **26.** $y = \sqrt{4 - x}$

In Exercises 27–30, let $B = \{-4, -2, 0, 2, 4\}$. Find a relation that is a subset of $B \times B$ and that is specified by the given equation. List the ordered pairs in each relation and graph each relation.

27. $y \le x$ **28.** $y > x$

29. $x = 2y$ **30.** $y = \sqrt{x}$

31. Given the relation $\{(C, F) \mid F = \frac{9}{5}C + 32\}$, which describes the conversion of degrees centigrade C to degrees Fahrenheit F.

a. Make a table of values for this relation for domain values

$$\{0, 5, 10, 15, 20, 25, 30, 35, 40\}$$

b. Graph the relation for the specified values of C. (Let the horizontal axis be the C-axis.)

c. Draw a straight line through the points.

d. If normal body temperature is 98.6 degrees Fahrenheit, from the graph, determine the corresponding temperature in degrees centigrade.

32.

Speed v, in mph	20	30	40	50	60	70	80
Driver reaction distance x, in feet	22	33	44	55	66	77	88
Braking distance, in feet	20	45	80	125	180	245	320
Distance needed to stop s, in feet	42	78	124	180	246	322	408

a. List the ordered pairs in the relation $\{(v, x) \mid v$ is an entry in the table$\}$.

b. List the ordered pairs in the relation $\{(v, s) \mid v$ is an entry in the table$\}$.

c. Graph the relation in (a) and join the points with a smooth curve. Find an equation of the form $x = av + b$ to specify the relation.

d. On the same set of axes as used in (c), graph the relation in (b) and join the points with a smooth curve. Find an equation of the form $s = av + bv^2$ to specify the relation.

33. The graph in Figure 4.1.4 compares the average cost of education C per student with the school year y.

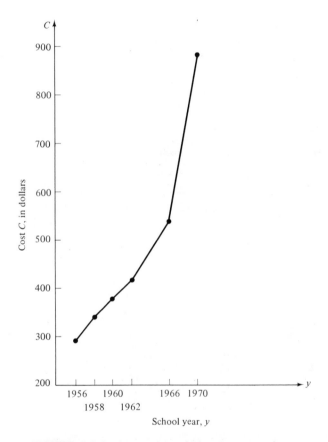

FIGURE 4.1.4 Average cost of education per student.

a. Using the graph, list the ordered pairs in the relation

$$\{(y, C) \mid y \in \{1956, 1958, 1960, 1962, 1966, 1970\}\}$$

b. If there are 180 days in a school year, find the cost per student per school day for the years 1966 and 1970.

c. Compare the rate of increase in the cost from 1966 to 1970 with that from 1962 to 1966.

4.2 FUNCTIONS

A relation was defined as *any* set of ordered pairs. A special relation which associates with each first component in its domain *exactly* one second component in its range is now defined.

DEFINITION

A **function** is a relation in which no two ordered pairs have the same first components and different second components.

If a relation consists of ordered pairs (x, y), then the relation is a function if and only if every x-value in the domain of the relation corresponds to one and only one y-value in the range of the relation.

Functions are often named by lowercase letters such as f, g, and h.

EXAMPLE 4.2.1 Which of the following relations are functions?

$$f = \{(2, 3), (4, 5), (1, 3), (6, 5)\}$$
$$g = \{(1, 1), (2, 2), (3, 3)\}$$
$$h = \{(3, 2), (5, 4), (3, 1), (5, 6)\}$$

Solution f is a function because each ordered pair has a different first component. g is a function because each ordered pair has a different first component. h is not a function because the pairs (3, 2) and (3, 1) have the first component, 3, associated with 2 and with 1. Also, 5 is paired with 4 and with 6.

EXAMPLE 4.2.2 State the domain and range of the functions f and g in Example 4.2.1.

Solution Domain of $f = \{2, 4, 1, 6\} = \{1, 2, 4, 6\}$
Range of $f = \{3, 5\}$
Domain of $g = \{1, 2, 3\}$
Range of $g = \{1, 2, 3\}$

The notation $f(x)$, read "f of x," is used to designate the unique value of y that is paired with a given value of x in the domain of the function f.

For example, if

$$f = \{(x, y) \mid y = 2x + 1\}$$

then $f(x) = 2x + 1$, since $y = f(x)$.

The function f could also be expressed by

$$f = \{(x, f(x)) \mid f(x) = 2x + 1\}$$

EXAMPLE 4.2.3 For $f(x) = 2x + 1$, find the ordered pairs in the function f for $-2 \leq x \leq 2$, where x is an integer and $f(x)$ is a real number.

Solution $y = f(x) = 2x + 1$

$$f(\ ) = 2(\ ) + 1$$

$$f(-2) = 2(-2) + 1 = -3, \text{ and } (-2, -3) \in f$$

$$f(-1) = 2(-1) + 1 = -1, \text{ and } (-1, -1) \in f$$

$$f(0) = 2(0) + 1 = 1, \text{ and } (0, 1) \in f$$

$$f(1) = 2(1) + 1 = 3, \text{ and } (1, 3) \in f$$

$$f(2) = 2(2) + 1 = 5, \text{ and } (2, 5) \in f$$

EXAMPLE 4.2.4 If $f = \{(x, y) \mid y = f(x) = x^2 + 2x + 1\}$, evaluate $f(1), f(-1), f(0), f(2), f(a), f(a + h), f(a + h) - f(a)$.

Solution $f(x) = x^2 + 2x + 1$

$$f(\ ) = (\ )^2 + 2(\ ) + 1$$

$$f(1) = (1)^2 + 2(1) + 1 = 4$$

$$f(-1) = (-1)^2 + 2(-1) + 1 = 0$$

$$f(0) = (0)^2 + 2(0) + 1 = 1$$

$$f(2) = (2)^2 + 2(2) + 1 = 9$$

$$f(a) = (a)^2 + 2(a) + 1 = a^2 + 2a + 1$$

$$f(a + h) = (a + h)^2 + 2(a + h) + 1$$

$$= a^2 + 2ah + h^2 + 2a + 2h + 1$$

$$f(a + h) - f(a) = (a^2 + 2ah + h^2 + 2a + 2h + 1) - (a^2 + 2a + 1)$$

$$= a^2 - a^2 + 2ah + h^2 + 2a - 2a + 2h + 1 - 1$$

$$= 2ah + h^2 + 2h$$

EXAMPLE 4.2.5   The equation $C = \frac{1}{2}x + 40$ represents the total cost C in dollars of manufacturing x items of a certain commodity. This equation defines a function f and $C = f(x)$.

a. Use a table of values to specify this function for the cost of manufacturing x items if $0 \leq x \leq 10$. (Since x is the *number* of items, x must be a natural number.)

b. Graph this function; draw a line through the points and continue the line, so that the cost of manufacturing up to 20 items can be read from the graph.

c. From the graph, what is the cost of manufacturing 0 items? (This is the *fixed* cost.) What is the cost of manufacturing 15 items? 20 items?

Solution

a. $C = \frac{1}{2}x + 40; C = f(x)$

x	0	1	2	3	4	5	6	7	8	9	10
$f(x)$	40	40.5	41	41.5	42	42.5	43	43.5	44	44.5	45

b.

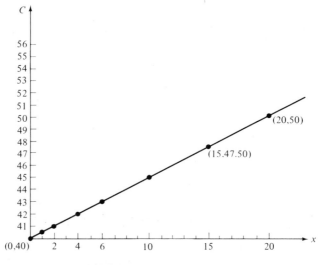

FIGURE 4.2.1 $C = \frac{1}{2}x + 40$.

Note that the intersection of the axes is at the point (0, 40).

c. The fixed cost is $40.

   The cost of manufacturing 15 items is $47.50.

   The cost of manufacturing 20 items is $50.00.

Suppose $y = f(x)$ is a rule for finding the ordered pairs of a relation. An easy way to check if this relation defines a function is to see that every substitution for x gives exactly one answer. For example, if $y = x^2$, then

$$f(x) = x^2$$

and for each x, there is exactly one y-value where $y = f(x)$.

$$f(1) = (1)^2 = 1$$
$$f(-1) = (-1)^2 = 1$$
$$f(2) = (2)^2 = 4$$

and so on.

However, if $y^2 = x$, then $y = \pm\sqrt{x}$. This time the rule yields *two* possible answers for every nonnegative number x.

For $x = 1$, $y = \pm\sqrt{1} = \pm 1$

For $x = 4$, $y = \pm\sqrt{4} = \pm 2$

Thus $y = \pm\sqrt{x}$ does *not* define a function.

EXERCISES 4.2 A

In Exercises 1–5, let r be a relation whose domain is $\{-1, 0, 1, 2, 3\}$.

a. List the ordered pairs corresponding to the rule stated in each of the following.
b. List the range for each relation.

1. $r = \{(x, y) \mid y = x - 3\}$ **2.** $r = \{(c, d) \mid d = 5 - c\}$

3. $r = \{(x, y) \mid y = \sqrt{x + 2}\}$ **4.** $r = \{(x, y) \mid y^2 = x + 1\}$

5. $r = \{(x, f(x)) \mid f(x) = 3x + 5\}$

6. Which of the relations in Exercises 1–5 are functions? Why?

If $f(x) = 3x^2 + 4x + 2$, find Exercises 7–12.

7. $f(2)$ **8.** $f(-1)$

9. $f(0)$ **10.** $f(a)$

11. $f(a + 1)$ **12.** $f\left(\dfrac{1}{a}\right)$

If $g(x) = x^3 + x - 3$, find Exercises 13–19.

13. $g(-3)$ **14.** $g(0)$

15. $2[g(1)]$ **16.** $g(2) + g(-2)$

17. $g(b)$ **18.** $g(b + h)$

19. $g(b) + g(h)$

20. If $f(x) = 3x + 4$, find:

    a. $f(x^2)$ b. $[f(x)]^2$

21. If the profit P from the sales of x refrigerators is related by the functional rule

$$P = \tfrac{1}{2}x^2 + 20x - 20 \qquad (0 \le x \le 1500)$$

    a. Make a table of values for the domain values $\{1, 2, 4, 5, 8, 10, 50, 75, 100\}$

    b. Find the profit when

$$5 \text{ refrigerators are sold}$$
$$50 \text{ refrigerators are sold}$$
$$100 \text{ refrigerators are sold}$$

22. If $f = \{(r, A) \mid A = \pi r^2\}$ defines the area of a circle as a function of its radius,

    a. List the ordered pairs in f when the domain of $f = \{1, 2, 3\}$.

    b. Is f a function? Why?

23. If $g = \{(t, d) \mid d = 50t\}$ is the functional relationship describing the distance in miles which a bus moving at 50 mph travels in t hours, find the elements of g when the domain of $g = \{1, 2, 5, 7\}$.

24. The volume V of a sphere is a function of its radius r. Thus $V = f(r) = \tfrac{4}{3}\pi r^3$. If $\pi = \tfrac{22}{7}$, find V when:

    a. $r = 1$ b. $r = 3$ c. $r = 5$

25. If f is a function relating the area A of a triangle to its base b and height 5 inches, write a formula for $A = f(b)$.

26. Evaluate the formula $f(b)$ in Exercise 25 for

    a. $f(3)$ b. $f(b + h)$ c. $f(b + h) - f(b)$

    d. $\dfrac{f(b + h) - f(b)}{h}$

EXERCISES 4.2 B

In Exercises 1–5, let r be a relation whose domain is $\{-1, 0, 1, 2, 3\}$.

a. List the ordered pairs corresponding to the rule stated in each of the following.

b. List the range for each relation.

1. $r = \{(x, y) \mid y = 2x + 3\}$ **2.** $r = \{(a, b) \mid b = 2 - a^2\}$

3. $r = \{(x, y) \mid y^2 = x + 1\}$

4. $r = \{(x, y) \mid y < x - 1 \text{ and } y \in \{-1, 0, 1, 2, 3\}\}$

5. $r = \{(x, f(x)) \mid f(x) = x^2 + 3\}$

6. Which of the relations in Exercises 1–5 are functions? Why?

If $f(x) = 2 + 3x - 4x^2$, *find Exercises 7–12.*

7. $f(2)$ **8.** $f(-1)$

9. $f(0)$ **10.** $f(a)$

11. $f(a + 1)$ **12.** $f\left(\dfrac{1}{a}\right)$

If $g(x) = 2x^2 + 3x + 7$, *find Exercises 13–19.*

13. $g(-3)$ **14.** $g(0)$

15. $2[g(1)]$ **16.** $g(2) + g(-2)$

17. $g(b)$ **18.** $g(b + h)$

19. $g(b) + g(h)$

20. If $g(x) = 2x - 1$, find:

     a. $g(x^2)$ b. $[g(x)]^2$

21. The income I from an investment of y dollars for one year at 5 percent is related by the functional rule

$$I = f(y) = 0.05y$$

Find the income when (a) \$1000 are invested and (b) \$10,000 are invested.

22. If $g = \{(x, g(x)) \mid g(x) = 1000(1.10)^x\}$ defines the amount $g(x)$ to which \$1000 accumulates at the end of x years when invested at a compound interest rate of 10 percent,

     a. List the ordered pairs in g if the domain of g is $\{1, 2, 3, 4, 5\}$.

     b. Graph the ordered pairs of (a).

     c. When will $g(x) = 2000$?

23. The intensity I of illumination on an object from a 120-candlepower source is a function of the distance x of the object from the source— that is,

$$I = I(x) = \frac{120}{x^2}$$

     a. Find $I(1)$, $I(-1)$, $I(2)$, $I(-2)$, $I(10)$, $I(20)$.

     b. Is $\{(x, I) \mid x$ and I are real numbers$\}$ a function? Why?

24. For a fixed distance of 600 miles, the uniform rate r at which a car travels is related to the time t by the equation

$$r = f(t) = \frac{600}{t}$$

     a. List the elements in $\{(t, r) \mid 0 < t \le 6$, where t is an integer$\}$.

     b. List the elements in $\{(r, t) \mid 0 < t \le 6\}$.

     c. Is r a function of t? Why?

     d. Is t a function of r? Why?

25. For $f(x) = \dfrac{1}{x}$, find:

a. $f(5)$ b. $f(x + 5)$ c. $f(x + 5) - f(5)$ d. $\dfrac{f(x + 5) - f(5)}{x}$

26. For $s(t) = 16t^2$, find:

a. $s(t + h)$ b. $s(t + h) - s(t)$ c. $\dfrac{s(t + h) - s(t)}{h}$

4.3   LINEAR FUNCTIONS

> **DEFINITION**
>
> A **linear function** is a function whose rule has the form
>
> $$f(x) = ax + b$$
>
> In set notation, a linear function f is the set
>
> $$f = \{(x, y) \mid y = ax + b\} = \{(x, f(x)) \mid f(x) = ax + b\}$$

The graph of every linear function is a straight line. The equation of every nonvertical line is a linear function. The reason for excluding vertical lines should be obvious. All points on a vertical line have the same x-value but different y-values. For example, draw a line through the points $(4, 2)$ and $(4, 4)$ (see Figure 4.3.1). Since the only value which x can assume is 4, the equation of this line is $x = 4$.

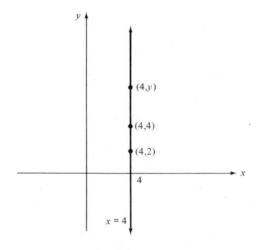

FIGURE 4.3.1   Graph of $x = 4$.

A linear function of the form

$$y = ax + b$$

yields several helpful bits of information to aid in graphing the line. For example, if $x = 0$, $y = a(0) + b = b$, so the point $(0, b)$ is on the graph, and the line crosses the y-axis where $y = b$. This y-value is called the **y-intercept** of the function.

EXAMPLE 4.3.1   If $y = 3x + 2$, find the y-intercept of the function.

Solution $b = 2$; therefore, the y-intercept is 2, or the graph passes through the point $(0, 2)$.

EXAMPLE 4.3.2   If $f(x) = \frac{1}{2}x - \frac{1}{5}$, name the point where the graph of $f(x)$ crosses the vertical axis.

Solution   The pattern still holds, and $b = -\frac{1}{5}$; therefore, the point where the graph of $f(x)$ crosses the vertical axis is $(0, -\frac{1}{5})$.

EXAMPLE 4.3.3   Consider the function $y = 3x + 2$ as discussed in Example 4.3.1, and graph this function.

Solution   It was shown that the line crosses the y-axis at 2. Now if $x = 1$, then $y = 5$, so the point $(1, 5)$ is also on the graph. To avoid error, select a third point, say $(-1, -1)$ (when $x = -1$, $y = 3(-1) + 2 = -1$). The graph is shown in Figure 4.3.2.

Another characteristic of a line is its slope. The pattern $f(x) = ax + b$ yields useful slope information, but before this can be discussed, slope must be defined.

Choose any two points on the graph shown in Figure 4.3.2, for example, point $P:(0, 2)$ and point $Q:(1, 5)$.

For the slope of the line,

$$\text{slope} = \frac{\text{change in } y\text{-value}}{\text{change in } x\text{-value}}$$

$$= \frac{5 - 2}{1 - 0} = \frac{3}{1} = 3$$

Therefore, the slope is $\frac{3}{1}$ or 3. The slope of a line is the same for *all* points on the line, not just for the two specific points selected in the above example.

A definition for the slope of a general line can be conveniently expressed by using subscripts to indicate two points on the line. For example, P_1 (read "P sub one") and P_2 (read "P sub two") name two points. The numeral 1 at the lower right of P_1 and the numeral 2 at the lower right of P_2 are subscripts used to indicate that P_1 is the first point and P_2 is the second point. Similarly, the coordinates of P_1 can be expressed as (x_1, y_1) and those of P_2 as (x_2, y_2). The letter m is the symbolic representation of the slope of a line, and a formal definition of slope m now follows.

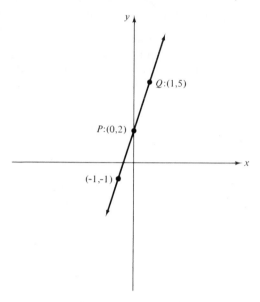

FIGURE 4.3.2 $y = 3x + 2$.

DEFINITION OF SLOPE m

If $P_1:(x_1, y_1)$ and $P_2:(x_2, y_2)$ are any two points on a line and if $x_1 \neq x_2$, then the slope m of the line joining P_1 and P_2 is given by

$$m = \frac{y_2 - y_1}{x_2 - x_1}$$

(See Figure 4.3.3.)

If a line is vertical (parallel to the y-axis), then $x_1 = x_2$ and $x_2 - x_1 = 0$. Since the denominator of the slope ratio is 0, the slope of a vertical line is undefined.

EXAMPLE 4.3.4   Find the slope of the line passing through the points whose coordinates are (3, 2) and (5, 1), respectively.

Solution   Using

$$m = \frac{y_2 - y_1}{x_2 - x_1} \quad \text{with} \quad (x_2, y_2) = (5, 1)$$

$$\text{and} \quad (x_1, y_1) = (3, 2)$$

$$m = \frac{1 - 2}{5 - 3} = -\frac{1}{2}$$

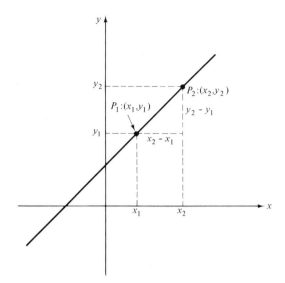

FIGURE 4.3.3 Slope of a line $m = \dfrac{y_2 - y_1}{x_2 - x_1}$.

Alternate Solution   Using

$$m = \frac{y_2 - y_1}{x_2 - x_1} \quad \text{with} \quad (x_2, y_2) = (3, 2)$$

$$\text{and} \quad (x_1, y_1) = (5, 1)$$

$$m = \frac{2 - 1}{3 - 5} = -\frac{1}{2}$$

Note in Example 4.3.4 that the value of the slope does not depend on which point is called the first point and which is the second.

Return now to the linear function $y = f(x) = ax + b$. It was established that b is the value of the y-intercept. It can now be stated that a is the slope of the line whose equation is $f(x) = ax + b$. Usually a is replaced by the letter m so that the linear function is stated

$$y = f(x) = mx + b$$

To verify that m is actually the slope, consider the following:
If $x = 0$, then $y = b$ and the ordered pair $(0, b)$ is a solution of $y = mx + b$. Selecting another value for x, let $x = 1$. Then $y = m + b$, and $(1, m + b)$ is a solution of $y = mx + b$. Since the slope of the line is determined by any two points on the line,

$$\text{slope} = \frac{y_2 - y_1}{x_2 - x_1} = \frac{(m + b) - b}{1 - 0} = m$$

DEFINITION

The equation $y = mx + b$ is called the **slope-intercept form** of the line whose slope is m and whose y-intercept is b.

EXAMPLE 4.3.5   Find the slope and the y-intercept of the line described by the linear function $y = f(x) = \frac{1}{2}x - 6$.

Solution   The slope $m = \frac{1}{2}$, and the y-intercept $b = -6$.

A linear function $f(x) = ax + b$ is called a constant function if $a = 0$. Thus the slope of a constant function is 0, and the general form of a constant function is $f(x) = b$.

DEFINITION

A function of the form $f(x) = b$, where b is a constant, is called a **constant function.**

The graph of a constant function is a horizontal line which crosses the y-axis at the point $(0, b)$.

EXAMPLE 4.3.6   Discuss and graph the function $\{(x, f(x)) \mid f(x) = 4\}$.

Solution $f(x) = 4$ is a constant function. The ordered pairs $(x, f(x))$ all have a second component, 4. All real numbers x are paired with 4. The graph is a line parallel to the x-axis and passing through the point $(0, 4)$ on the y-axis (see Figure 4.3.4).

Another linear function which is important enough to have a special name is the function

$$f(x) = x$$

DEFINITION

The linear function

$$f(x) = x$$

is called the **identity function.**

Some ordered pairs of the identity function

$$I = \{(x, f(x)) \mid f(x) = x\}$$

are $\{(-3, -3), (0, 0), (\frac{1}{2}, \frac{1}{2}), (1, 1), (\sqrt{2}, \sqrt{2}), \ldots\}$.

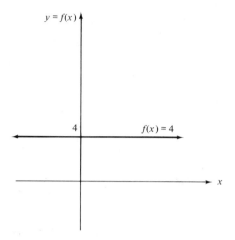

FIGURE 4.3.4 Graph of $f(x) = 4$, a constant function.

The first and second components of the identity function are identical. The graph of the identity function is shown in Figure 4.3.5.

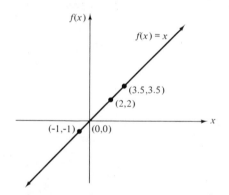

FIGURE 4.3.5 The identity function, $f(x) = x$.

EXAMPLE 4.3.7 If $3x + 2y + 6 = 0$, express y as a function of x and find the slope and y-intercept of the line described by this function.

Solution   To express y as a function of x, solve the equation for y:

$$3x + 2y + 6 = 0$$
$$2y = -3x - 6$$
$$y = -\tfrac{3}{2}x - 3$$

Now the function is in the form $y = mx + b$, so that the slope $m = -\tfrac{3}{2}$, and the y-intercept $b = -3$.

EXERCISES 4.3 A

Graph each of the functions in Exercises 1–10 and state the slope and y-intercept of each.

1. $f(x) = 2x - 1$ **2.** $f(x) = 3x + 2$

3. $f(x) = \frac{1}{2}x - 4$ **4.** $f(x) = -x - 1$

5. $f(x) = -2$ **6.** $f(x) = -x$

7. $f(x) = \frac{1}{2}x$ **8.** $f(x) = 0$

9. $f(x) = \frac{3}{5}x + \frac{1}{2}$ **10.** $f(x) = -\frac{1}{4}x - 1$

For the points in Exercises 11–15, find the slope of the line passing through the points and graph each line.

11. $P_1:(1, 2), P_2:(3, 4)$ **12.** $P_1:(-1, 2), P_2:(-4, 3)$

13. $P_1:(-5, -2), P_2:(1, 3)$ **14.** $P_1:(-4, -2), P_2:(-6, -1)$

15. $P_1:(1, 5), P_2:(0, 0)$

Express each of the equations in Exercises 16–20 in the form $y = f(x)$ and state the slope and y-intercept of the line whose equation is given.

16. $2x - 3y + 5 = 0$ **17.** $x + 4y - 1 = 0$

18. $x = 2y + 1$ **19.** $2x - \frac{1}{2}y = 3$

20. $x - y = 0$

21. Find the value of k so that the line through $(1, k)$ and $(5, 4k)$ has the stated slope:

     a. Slope: -6 b. Slope: 0

22. Given that $y = 3x + 6$

$$y = \frac{1}{3}x - 2$$

$$y = x$$

     a. Graph these three equations on the same set of axes.

     b. Discuss the apparent relationship between the graphs of the first two equations and the graph of $y = x$.

23. As an example of a slope, a highway grade expressed as a percentage means the number of feet the road changes in elevation for 100 feet measured horizontally.

     a. A certain highway has a $2\frac{1}{2}$ percent grade. How many feet does it rise in a 1-mile stretch (horizontal distance)? (*Note:* 1 mile = 5280 feet.)

     b. How many feet does a $-3\frac{1}{4}$ percent grade highway drop for a $\frac{1}{2}$-mile horizontal stretch?

24. A certain county specification requires that an inclined water pipe must have a slope greater than or equal to $\frac{1}{4}$. Which of the water pipes whose rises and runs are given below meets this specification?

   a. Rise = 20 feet, run = 64 feet

   b. Rise = 125 feet, run = 500 feet

   c. Rise = 60 feet, run = 250 feet

EXERCISES 4.3 B

Graph each of the functions in Exercises 1–10 and state the slope and y-intercept of each.

1. $f(x) = 3x + 4$ **2.** $f(x) = -2x + 1$

3. $f(x) = \frac{1}{4}x + 2$ **4.** $f(x) = -x + 1$

5. $f(x) = 5$ **6.** $f(x) = 2x$

7. $f(x) = -\frac{1}{4}x$ **8.** $f(x) = -2x + 2$

9. $f(x) = -x + 2$ **10.** $f(x) = -x + 3$

For the points in Exercises 11–15, find the slope of the line passing through the points and graph each line.

11. $P_1:(2, 3), P_2:(1, 6)$ **12.** $P_1:(-2, 3), P_2:(-1, 6)$

13. $P_1:(-3, -4), P_2:(2, 5)$ **14.** $P_1:(1, 1), P_2:(4, 4)$

15. $P_1:(-2, -1), P_2:(-5, -2)$

Express each of the equations in Exercises 16–20 in the form $y = f(x)$ and state the slope and y-intercept of the line whose equation is given.

16. $3x - 2y - 5 = 0$ **17.** $4y - x + 2 = 0$

18. $x = 3y - 2$ **19.** $3x - \frac{1}{4}y = 4$

20. $x + y = 0$

21. Find the value of k so that the line through $(2, k)$, $(3, 2k)$ has the stated slope:

   a. Slope: -3 b. Slope: 6

22. Given that $y = 2x - 4$

$$y = \tfrac{1}{2}x + 2$$

$$y = x$$

a. Graph these three equations on the same set of axes.

b. Discuss the apparent relationship between the graphs of the first two equations and the graph of $y = x$.

23. In construction, the pitch of a roof (a slope) is the ratio of the vertical rise to the horizontal half-span.

a. For a half-span of 18 feet, how high must the roof rise if the pitch of the roof is 4 to 12?

b. For a roof whose pitch is 3 to 12, find the half-span if the roof rises 7 feet.

24. Find the speed of a skier who skis down a hill 1320 feet high and having a slope of $\frac{5}{12}$ if he takes 26 seconds to reach the bottom. Express the answer in mph. (*Hint:* Use the theorem of Pythagoras.)

4.4 QUADRATIC FUNCTIONS

DEFINITION

A **quadratic function** is a function with the rule

$$f(x) = ax^2 + bx + c \quad (a \neq 0)$$

EXAMPLE 4.4.1 Draw the graph of the quadratic function $f(x) = 2x^2 - 4x + 1$ by making a table of values for some selected integral values for x. Plot the points from this table of values and draw a smooth curve through the points. From the graph, determine the range of the function.

Solution

x	y	$f(x) = 2x^2 - 4x + 1$
-3	31	$2(-3)^2 - 4(-3) + 1 = 18 + 12 + 1 = 31$
-2	17	$2(-2)^2 - 4(-2) + 1 = 8 + 8 + 1 = 17$
-1	7	$2(-1)^2 - 4(-1) + 1 = 2 + 4 + 1 = 7$
0	1	$2(0)^2 - 4(0) + 1 = 1$
1	-1	$2(1)^2 - 4(1) + 1 = 2 - 4 + 1 = -1$
2	1	$2(2)^2 - 4(2) + 1 = 8 - 8 + 1 = 1$
3	7	$2(3)^2 - 4(3) + 1 = 18 - 12 + 1 = 7$

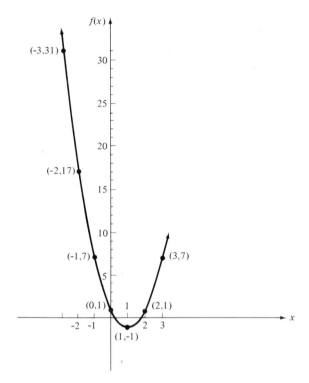

FIGURE 4.4.1   Graph of $f(x) = 2x^2 - 4x + 1$.

It is evident from the graph that $f(x) \geq -1$. It will shortly be shown this conjecture is correct, and a method for finding the range of a quadratic function will be developed.

Clearly the graph of $f(x) = 2x^2 - 4x + 1$ is not a straight line. As a matter of fact, the graph of a quadratic function is a *smooth* curve called a **parabola**.

> The parabola which describes the graph of the function
> $$y = ax^2 + bx + c$$
> where $a, b, c, x,$ and y are real numbers ($a \neq 0$), opens upward, $\cup$, if $a > 0$ or downward, $\cap$, if $a < 0$.

The turning point of the parabola is called its **vertex**, and the curve is symmetric with respect to a vertical line drawn through the vertex. This line is called the **axis of symmetry**. If the paper were to be folded along this axis, the part of the parabola on the right would coincide with the part on the left of the axis of symmetry.

The vertex always occurs on the axis of symmetry and is always either the low point or the high point of the graph of a quadratic function $y = ax^2 + bx + c$, depending on whether the parabola opens upward or downward.

Since the axis of symmetry is a vertical line, its equation has the form $x = k$, where k is constant.

To find k, the x-coordinate of the vertex, select two ordered pairs with the same y-value. The average of the x values is the x-coordinate of the vertex.

x	y
a	c
b	c

$$k = \frac{a + b}{2}$$

EXAMPLE 4.4.2   Find the coordinates of the vertex of the function $f(x) = 2x^2 - 4x + 1$.

Solution   This is the function which was discussed in Example 4.4.1. From the graph (Figure 4.4.1) and from the table of values, the ordered pairs $(0, 1)$ and $(2, 1)$ have the same y-value.

Let $a = 0$ and $b = 2; c = 1$.

Thus
$$k = \frac{a + b}{2} = \frac{0 + 2}{2} = 1$$

The coordinates of the vertex are $(k, f(k)) = (1, f(1)) = (1, -1)$.

Since the vertex is the lowest point on this parabola, -1 must be the smallest value for y. Therefore, $y \geq -1$, and the range of $f(x) = y = 2x^2 - 4x + 1$ is the set of real numbers greater than or equal to -1. The domain is the set of real numbers.

EXAMPLE 4.4.3

a. Graph $\{(x, y) \mid y = -x^2 + 3x + 4\}$.
b. State the coordinates of the vertex of the parabola.
c. State the domain and range of this function.

Solution

a. Making a table of values,

x	y	$f(x) = -x^2 + 3x + 4$
-2	-6	$f(-2) = -4 + 3(-2) + 4 = -6$
-1	0	$f(-1) = -1 + 3(-1) + 4 = 0$
0	4	$f(0) = 4$
1	6	$f(1) = -1 + 3 + 4 = 6$
2	6	$f(2) = -4 + 6 + 4 = 6$
3	4	$f(3) = -9 + 9 + 4 = 4$
4	0	$f(4) = -16 + 12 + 4 = 0$
5	-6	$f(5) = -25 + 15 + 4 = -6$

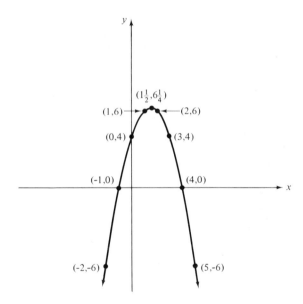

FIGURE 4.4.2 Graph of $y = -x^2 + 3x + 4$.

b. Using the pairs $(1, 6)$ and $(2, 6)$ to find the vertex,

$$x = \frac{1 + 2}{2} = \frac{3}{2}$$

$$y = f(\tfrac{3}{2}) = -\tfrac{9}{4} + \tfrac{9}{2} + 4 = \tfrac{25}{4}$$

Therefore, the vertex is $(\tfrac{3}{2}, \tfrac{25}{4})$ or $(1\tfrac{1}{2}, 6\tfrac{1}{4})$.

c. The range is $\{y \mid y \le 6\tfrac{1}{4}\}$. The domain is the set of real numbers.

EXERCISES 4.4 A

If $f(x) = 2x^2 - 3x + 4$, find Exercises 1–10.

1. $f(2)$	**2.** $f(-1)$
3. $f(0)$	**4.** $f(p)$
5. $f(p + h)$	**6.** $f(p) + h$
7. $f(-\frac{1}{2})$	**8.** $f(2x)$
9. $f\left(\dfrac{1}{x}\right)$ $\quad(x \neq 0)$	**10.** $f(x + 2)$

If $g(x) = 3 - x^2$, find Exercises 11–18.

11. $g(0)$	**12.** $g(3)$
13. $g(-3)$	**14.** $g(x + 1) - g(x)$
15. $g(x^2)$	**16.** $g(x - 1)$
17. $g(\frac{1}{3})$	**18.** $g\left(\dfrac{1}{x}\right)$ $\quad(x \neq 0)$

In Exercises 19–27,

a. Graph each for the stated interval in the domain of the function by plotting points and joining these points with a smooth curve.
b. Determine the coordinates of the vertex.
c. Determine the range of the function.

19. $\{(x, y) \mid y = 2x^2\}$
$-4 \leq x \leq 4$

20. $\{(x, f(x)) \mid f(x) = -\frac{1}{2}x^2\}$
$-4 \leq x \leq 4$

21. $\{(x, f(x)) \mid f(x) = x^2 + 2x + 1\}$
$-5 \leq x \leq 3$

22. $\{(x, y) \mid y = 3x^2 + 12x\}$
$-6 \leq x \leq 4$

23. $\{(x, y) \mid y = 5 - x^2\}$
$-3 \leq x \leq 3$

24. $\{(x, f(x)) \mid f(x) = 1 + 2x - x^2\}$
$-2 \leq x \leq 4$

25. $\{(x, f(x)) \mid f(x) = x^2 + 5x + 6\}$
$-6 \leq x \leq 1$

26. $\{(x, f(x)) \mid f(x) = 6x - x^2\}$
$-1 \leq x \leq 7$

27. $\{(x, f(x)) \mid f(x) = 4 - x^2\}$
$-3 \leq x \leq 3$

28. An arrow is shot vertically upward in the air. Its height h in feet after t seconds is given by the formula

$$h = 112t - 16t^2$$

a. Graph the function (let the t-axis be the horizontal axis).
b. From the graph, determine the greatest height the arrow reaches and the number of seconds it takes to reach this maximum height.

EXERCISES 4.4 B

If $f(x) = -2x^2 + 3x - 1$, find Exercises 1–10.

1. $f(-2)$ **2.** $f(2)$

3. $f(0)$ **4.** $f(-\frac{1}{4})$

5. $f(x + h)$ **6.** $f(x) + 2$

7. $f\left(\dfrac{1}{x}\right)$ $(x \neq 0)$ **8.** $f(x + 1)$

9. $f(2x)$ **10.** $f(x) + f(h)$

If $g(x) = 3x^2 + x - 7$, find Exercises 11–18.

11. $g(0)$ **12.** $g(x + 2)$

13. $g(x + 1) - g(x)$ **14.** $g(x + 3) - g(x)$

15. $g\left(\dfrac{1}{x}\right)$ $(x \neq 0)$ **16.** $g(x^2)$

17. $g(x) + 3$ **18.** $\dfrac{g(x + h) - g(x)}{h}$

In Exercises 19–27,

a. Graph each of the following for the stated interval in the domain of the function by plotting points and joining these points with a smooth curve.
b. Determine the coordinates of the vertex.
c. Determine the range of the function.

19. $\{(x, y) \mid y = 3x^2\}$ **20.** $\{(x, f(x)) \mid f(x) = -\frac{1}{2}x^2 + 1\}$
$\quad -4 \leq x \leq 4$ $\quad -4 \leq x \leq 4$

21. $\{(x, f(x)) \mid f(x) = x^2 + 4x + 1\}$ **22.** $\{(x, y) \mid y = 3 + 4x - x^2\}$
$\quad -6 \leq x \leq 3$ $\quad -1 \leq x \leq 5$

23. $\{(x, y) \mid y = 3 - x^2\}$ **24.** $\{(x, f(x)) \mid f(x) = 2x^2 + 4x\}$
$\quad -4 \leq x \leq 4$ $\quad -5 \leq x \leq 3$

25. $\{(x, f(x)) \mid f(x) = x^2 + 5x - 6\}$ **26.** $\{(x, f(x)) \mid f(x) = 10x - 2x^2\}$
$\quad -6 \leq x \leq 4$ $\quad -1 \leq x \leq 6$

27. $\{(x, f(x)) \mid f(x) = 9 - 4x^2\}$
$\quad -3 \leq x \leq 3$

28. After t seconds the height of a ball tossed into the air at 48 feet per second is given by the formula

$$h = 48t - 16t^2$$

a. Graph this function (let the t-axis be the horizontal axis).

b. From the graph, determine the maximum height the ball reaches, the number of seconds it takes to reach that height, and the number of seconds it takes the ball to come back to the ground again. (Assume the ball starts at $h = 0$.)

4.5 THE CIRCLE AND THE DISTANCE FORMULA

The relation which defines a circle algebraically is very important, not only for its geometric and algebraic significance but also in the study of trigonometry. As a matter of fact, trigonometry is also called the study of *circular functions*.

DEFINITION

A **circle** is the set of points in a plane that are equidistant from a fixed point called the **center** of the circle.

The distance between any point on the circle and the center of the circle is called the **radius** of the circle.

In order to obtain an equation for the circle, it is first necessary to express the geometric concept of distance between two points as an algebraic statement relating the coordinates of the two points.

THE DISTANCE FORMULA

$$|P_1P_2| = \sqrt{(x_2 - x_1)^2 + (y_2 - y_1)^2}$$

The distance formula is based on the theorem of Pythagoras. (In a right triangle, the square of the length c of the hypotenuse is equal to the sum of the squares of the lengths a and b of the other two sides. In symbols, $c^2 = a^2 + b^2$.)

The distance between any two points in a coordinate plane may be expressed in terms of the coordinates of these points by constructing a right triangle whose hypotenuse is the line segment joining the two points, as in Figure 4.5.1.

If the points are $P_1:(x_1, y_1)$ and $P_2:(x_2, y_2)$ and R is the point where the horizontal line through P_1 intersects the vertical line through P_2, then the coordinates of R are (x_2, y_1).

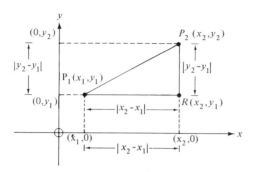

FIGURE 4.5.1

Now the distance $|P_1R|$ is $|x_2 - x_1|$ and the distance $|P_2R|$ is $|y_2 - y_1|$. By the theorem of Pythagoras,

$$(|P_1P_2|)^2 = (|P_1R|)^2 + (|P_2R|)^2$$

or

$$(|P_1P_2|)^2 = |x_2 - x_1|^2 + |y_2 - y_1|^2$$

By taking the square root of each side, the distance formula is obtained.

EXAMPLE 4.5.1   Find the distance between $P_1:(2, -1)$ and $P_2:(5, 3)$.

Solution

$$|P_1P_2| = \sqrt{(5 - 2)^2 + (3 - (-1))^2} = \sqrt{3^2 + 4^2} = \sqrt{25} = 5$$

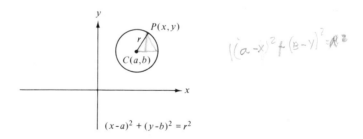

FIGURE 4.5.2

If $P:(x, y)$ designates any point on the circle and if $C:(a, b)$ is the center of the circle whose radius is r, then the equation of the circle is obtained by using the distance formula. Thus

$$\sqrt{(x - a)^2 + (y - b)^2} = r$$

By squaring both sides of this equation, the standard form of the equation of the circle is obtained.

EQUATION OF CIRCLE IN STANDARD FORM

$$(x - a)^2 + (y - b)^2 = r^2$$

Center is (a, b); radius $= \sqrt{r^2}$

The standard form for the equation of the circle is useful because the radius and the coordinates of the center can be read directly from the equation. The circle is symmetric to every line through its center; in particular, the circle is symmetric to the lines $x = a$ and $y = b$.

EXAMPLE 4.5.2   Write the equation of the circle with center $(3, -4)$ and radius 5.

Solution $(x - 3)^2 + (y + 4)^2 = 25$

EXAMPLE 4.5.3   Graph the circle whose equation is $x^2 + y^2 = 9$.

Solution   Since $x^2 + y^2 = 9$ can be written $(x - 0)^2 + (y - 0)^2 = 9$, the circle has its center at $(0, 0)$, the origin; the radius is 3. *Any equation of the form $x^2 + y^2 = r^2$ has as its graph a circle with center at the origin and radius r.*

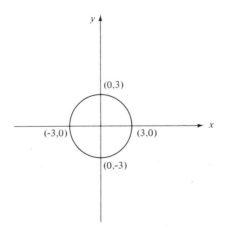

FIGURE 4.5.3

EXAMPLE 4.5.4 Draw the graph of the relation

$$\{(x, y) \mid (x + 2)^2 + (y - 3)^2 = 16\}, \text{ a circle}$$

Solution Draw the circle by using a compass with center at $(-2, 3)$ and with radius $= 4$ units (Figure 4.5.4).

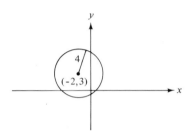

FIGURE 4.5.4

EXAMPLE 4.5.5 Determine an equation of the circle with center $C:(2, 3)$ and passing through the point $P:(5, 6)$.

Solution Since $|CP|$ determines the radius of the circle,

$$r = \sqrt{(5 - 2)^2 + (6 - 3)^2}$$
$$r = \sqrt{18}$$
$$r^2 = 18$$

An equation of the circle is: $(x - 2)^2 + (y - 3)^2 = 18$.

EXAMPLE 4.5.6 Show that $x^2 + y^2 + 8x - 10y + 5 = 0$ is the equation of a circle and then determine its center and radius.

Solution Completing the squares for x and for y,

$$(x^2 + 8x \quad) + (y^2 - 10y \quad) = -5$$
$$(x^2 + 8x + 16) + (y^2 - 10y + 25) = -5 + 16 + 25$$
$$(x + 4)^2 + (y - 5)^2 = 36$$

Thus the equation is that for a circle with center $(-4, 5)$ and with radius $= 6$.

EXERCISES 4.5 A

For Exercises 1–5, find the distance between the given points.

1. $(2, 7)$ and $(10, 1)$ **2.** $(18, -2)$ and $(-6, 5)$
3. $(-4, 3)$ and $(-7, 0)$ **4.** $(3, \frac{9}{2})$ and $(-3, 7)$
5. $(3, 1)$ and $(3, -4)$

For 6–10, write an equation of the circle with the given center and radius.

6. $(2, 3); r = 3$ **7.** $(-1, 2); r = 6$
8. $(3, -5); r = 5$ **9.** $(0, 0); r = 10$
10. $(-6, 2); r = \sqrt{3}$

For 11–15, write an equation of the circle with center C and passing through point P.

11. $C:(1, 5), P:(2, 2)$ **12.** $C:(-2, -3), P:(1, 5)$
13. $C:(-4, 1), P:(-3, -6)$ **14.** $C:(0, 0), P:(3, 4)$
15. $C:(3, 4), P:(0, 0)$

Show that each of Exercises 16–20 is the equation of a circle and determine the center and radius of each.

16. $x^2 + y^2 - 4x + 6y - 3 = 0$
17. $2x^2 + 2y^2 + 20x - 12y + 18 = 0$
18. $4x^2 + 4y^2 - 28x - 8y + 37 = 0$ **19.** $x^2 + y^2 - 14y = 0$
20. $x^2 + y^2 - 16x = 0$

Graph each of the relations in Exercises 21–25.

21. $\{(x, y) \mid (x - 2)^2 + (y - 3)^2 = 9\}$
22. $\{(x, y) \mid (x + 1)^2 + (y + 2)^2 = 4\}$
23. $\{(x, y) \mid x^2 + y^2 + 4x = 0\}$ **24.** $\{(x, y) \mid x^2 + y^2 - 2y = 0\}$
25. $\{(x, y) \mid 3x^2 + 3y^2 - 27 = 0\}$

EXERCISE 4.5 B

For Exercises 1–5, find the distance between the given points.

1. $(9, 5)$ and $(4, -7)$ **2.** $(\frac{1}{2}, -3)$ and $(2, -5)$
3. $(-6, 5)$ and $(-4, 1)$ **4.** $(9, 6)$ and $(\frac{11}{2}, -6)$
5. $(9, 6)$ and $(\frac{11}{2}, 6)$

For 6–10, write an equation of the circle with the given center and radius.

6. $(1, 4); r = 5$ **7.** $(-2, 3); r = 4$
8. $(4, -3); r = 7$ **9.** $(-5, -6); r = 12$
10. $(0, 0); r = \sqrt{7}$

For 11–15, write an equation of the circle with center C and passing through point P.

11. $C:(2, 4), P:(3, 1)$ **12.** $C:(-1, -5), P:(-4, -2)$
13. $C:(-3, 4), P:(1, 1)$ **14.** $C:(1, 1), P:(-4, -5)$
15. $C:(-4, -5), P:(1, 1)$

Show that each of Exercises 16–20 is the equation of a circle and determine the center and radius of each.

16. $x^2 + y^2 - 10x + 6y + 18 = 0$
17. $4x^2 + 4y^2 - 4x - 24y + 33 = 0$
18. $x^2 + y^2 - 4x + 2y - 6 = 0$ **19.** $x^2 + y^2 - 12y = 0$
20. $x^2 + y^2 + 20x = 0$

Graph each of the relations in Exercises 21–25.

21. $\{(x, y) \mid (x - 1)^2 + (y + 2)^2 = 16\}$
22. $\{(x, y) \mid (x + 2)^2 + (y - 4)^2 = 1\}$
23. $\{(x, y) \mid x^2 + y^2 + 6x = 0\}$ **24.** $\{(x, y) \mid x^2 + y^2 - 10y = 0\}$
25. $\{(x, y) \mid 4x^2 + 4y^2 - 64 = 0\}$

4.6 LINEAR INEQUALITIES: TWO VARIABLES

Equations and inequalities in one variable can be associated with points on a line, and the solution sets can be shown as line graphs.

Equations in two variables can be associated with points in a plane, using horizontal and vertical axes as references. Inequalities in two variables can also be graphed in a plane.

The graph of the linear equation $y = ax + b$ divides the plane into three mutually exclusive regions, or three disjoint sets of points:

$$1.\ y = ax + b$$

$$2.\ y > ax + b$$

$$3.\ y < ax + b$$

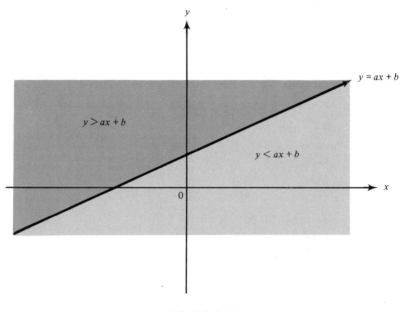

FIGURE 4.6.1

In Figure 4.6.1 the region above the line $y = ax + b$ represents all points satisfying the condition $\{(x, y) \mid y > ax + b\}$, whereas the region below the line $y = ax + b$ represents all points satisfying the condition $\{(x, y) \mid y < ax + b\}$. The line is the graph of the equation $y = ax + b$.

To graph an inequality of the form $y < ax + b$ or $y > ax + b$, first graph the equality, use a broken line to indicate that the graph of the equality is not part of the solution set, and then shade the desired region.

EXAMPLE 4.6.1   Graph the relation $\{(x, y) \mid y < 2x + 3\}$.

Solution   First graph the equation $y = 2x + 3$ and use a broken line to indicate that the line is *not* to be included in the solution set.
   To graph $y = 2x + 3$, use a table of values such as

x	0	1	2
y	3	5	7

to obtain the ordered pairs (0, 3), (1, 5), (2, 7). Plot these points and draw a dashed line through them.

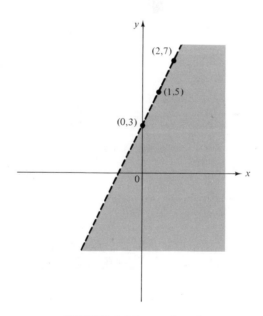

FIGURE 4.6.2 $y < 2x + 3$.

Since $y < 2x + 3$, shade the region *below* the broken line.

EXAMPLE 4.6.2   Graph $3x - 2y > x + y - 1$.

Solution   Solve the inequality for y:

$$3x - 2y > x + y - 1$$
$$- 3y > -2x - 1$$
$$3y < 2x + 1$$
$$y < \tfrac{2}{3}x + \tfrac{1}{3}$$

Graph the equality $y = \tfrac{2}{3}x + \tfrac{1}{3}$, using a broken line (see Figure 4.6.3). All points below the broken line are in the solution set

$$\{(x, y) \mid y < \tfrac{2}{3}x + \tfrac{1}{3}\}$$

Therefore, the shaded region is the graphical solution. (*Note:* The inequality does not define a function, but it does define a relation.)

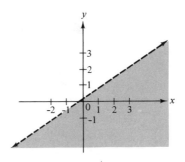

FIGURE 4.6.3

EXAMPLE 4.6.3   Find the graphical solution for the inequality $y \geq -2x + 4$.

Solution   Graph the equation $y = -2x + 4$. Use a solid line, since the line is part of the solution set. Shade the region above the line for the points

$$\{(x, y) \mid y > -2x + 4\}$$

The line and the shaded region in Figure 4.6.4 represent the solution set

$$\{(x, y) \mid y \geq -2x + 4\}$$

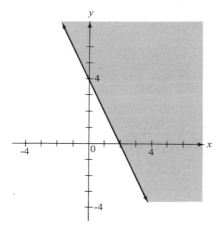

FIGURE 4.6.4

EXAMPLE 4.6.4 Graph the relation $\{(x, y) \mid x > 3\}$.

Solution The graph of the equation $x = 3$ is a vertical line through the point $(3, 0)$. All points (x, y) such that $x > 3$ lie to the right of this line (Figure 4.6.5).

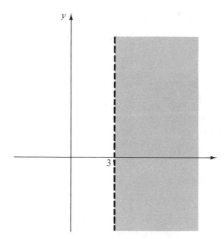

FIGURE 4.6.5 $\{(x, y) \mid x > 3\}$.

EXAMPLE 4.6.5 Graph the relation $\{(x, y) \mid y > 2\}$.

Solution The graph of the constant function $y = 2$ is a horizontal line through the point $(0, 2)$. Since $y > 2$, the shaded region in Figure 4.6.6 includes all points *above* this horizontal line.

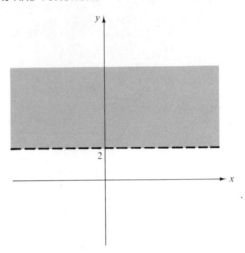

FIGURE 4.6.6 $\{(x, y) \mid y > 2\}$.

EXAMPLE 4.6.6   Find the graphical solution for the inequality $\{(x, y) \mid -2 \leq x < 3\}$.

Solution   First graph the vertical lines $x = -2$ and $x = 3$. Use a solid line to indicate the inclusion of $x = -2$ and a broken line to show that $x \neq 3$. Then shade the region *between* these lines, since $-2 \leq x < 3$ for *all* values of y. See Figure 4.6.7.

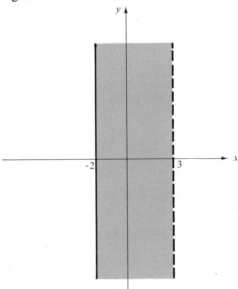

FIGURE 4.6.7 $\{(x, y) \mid -2 \leq x < 3\}$.

EXAMPLE 4.6.7   Find the graphical solution for

$$\{(x, y) \mid y > 2x + 1\} \cap \{(x, y) \mid y < 3 - x\}$$

Solution   Graph each relation separately.

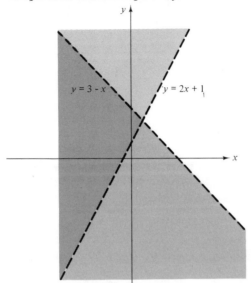

FIGURE 4.6.8

Since the solution is the intersection of the two sets, it will be the region where the shading overlaps.

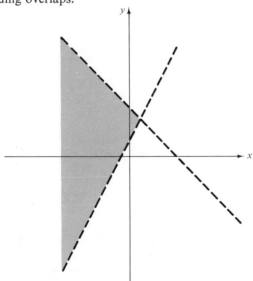

FIGURE 4.6.9 $\{(x, y) \mid y > 2x + 1\} \cap \{(x, y) \mid y < 3 - x\}$.

EXERCISES 4.6 A

Show a graphical solution for each of the relations in Exercises 1–20.

1. $\{(x, y) \mid x \geq 2\}$ **2.** $\{(x, y) \mid x < -1\}$

3. $\{(x, y) \mid y > -1\}$ **4.** $\{(x, y) \mid y \leq 4\}$

5. $\{(x, y) \mid -2 < x \leq 1\}$ **6.** $\{(x, y) \mid 1 \leq y < 3\}$

7. $\{(x, y) \mid y < 3x + 2\}$ **8.** $\{(x, y) \mid y > x - 5\}$

9. $\{(x, y) \mid y \leq 2x - 3\}$ **10.** $\{(x, y) \mid y > x\}$

11. $\{(x, y) \mid x + y \leq 2\}$ **12.** $\{(x, y) \mid y \geq x + 2\}$

13. $\{(x, y) \mid x \leq 2y - 4\}$ **14.** $\{(x, y) \mid x > 4\} \cup \{(x, y) \mid y \leq 2\}$

15. $\{(x, y) \mid y > 1\} \cap \{(x, y) \mid x \leq 3\}$ **16.** $\{(x, y) \mid x > 4\} \cap \{(x, y) \mid y \leq 2\}$

17. $\{(x, y) \mid y < x\} \cap \{(x, y) \mid y > -x\}$

18. $\{(x, y) \mid x > 0\} \cap \{(x, y) \mid y > 0\} \cap \{(x, y) \mid x + y \leq 5\}$

19. $\{(x, y) \mid y < x\} \cup \{(x, y) \mid x \geq 1\}$

20. $\{(x, y) \mid x + y \leq 2\} \cap \{(x, y) \mid y - x \leq 1\}$

EXERCISES 4.6 B

Show a graphical solution for each of the relations in Exercises 1–20.

1. $\{(x, y) \mid x > 4\}$ **2.** $\{(x, y) \mid x \leq -2\}$

3. $\{(x, y) \mid y \geq 4\}$ **4.** $\{(x, y) \mid y < 3\}$

5. $\{(x, y) \mid 1 \leq x \leq 3\}$ **6.** $\{(x, y) \mid 2 < y \leq 5\}$

7. $\{(x, y) \mid x < 0\}$ **8.** $\{(x, y) \mid y \geq 0\}$

9. $\{(x, y) \mid y \leq x\}$ **10.** $\{(x, y) \mid x + 2y < 2\}$

11. $\{(x, y) \mid 4 \geq 2y - x\}$ **12.** $\{(x, y) \mid 3x \leq y + 2\}$

13. $\{(x, y) \mid x - y < 2\}$ **14.** $\{(x, y) \mid y \leq x + 2\}$

15. $\{(x, y) \mid x \leq y + 2\}$ **16.** $\{(x, y) \mid x < 4\} \cap \{(x, y) \mid y > 2\}$

17. $\{(x, y) \mid y > x\} \cap \{(x, y) \mid y < -x\}$

18. $\{(x, y) \mid x < 0\} \cap \{(x, y) \mid y < 0\} \cap \{(x, y) \mid x + y \geq -3\}$

19. $\{(x, y) \mid y > 2\} \cup \{(x, y) \mid x > 3\}$

20. $\{(x, y) \mid x + 3y \geq 3\} \cap \{(x, y) \mid 2x + 2y \geq 5\}$

4.7   QUADRATIC INEQUALITIES: TWO VARIABLES

Just as a line separates the plane into three mutually exclusive regions, so does a parabola, and so does a circle.

Consider the case of a parabola first. Let $y = ax^2 + bx + c$ and assume $a > 0$. Then Figure 4.7.1 illustrates the three regions. The region

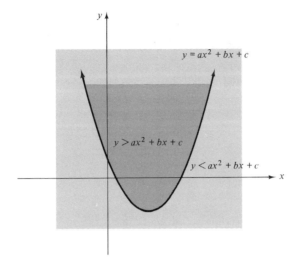

FIGURE 4.7.1

inside the parabola contains all points in the relation $y > ax^2 + bx + c$. The parabola corresponds to the equation $y = ax^2 + bx + c$, and the region *outside* the parabola contains all points in the relation $y < ax^2 + bx + c$.

If $a < 0$, then the situation is reversed, as illustrated in Figure 4.7.2. Now the region *inside* the parabola represents the solution $y < ax^2 + bx + c$, the parabola itself still corresponds to $y = ax^2 + bx + c$, and the region *outside* the parabola corresponds to $y > ax^2 + bx + c$.

It is also convenient to examine the graph of $f(x) = ax^2 + bx + c$ and determine for what values of x, $f(x)$ is positive or negative. The following example illustrates this concept.

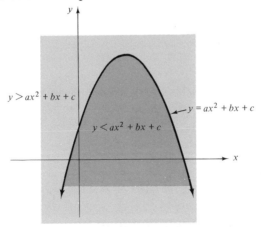

FIGURE 4.7.2

EXAMPLE 4.7.1   If $f(x) = x^2 + 2x - 8$, determine the values of x

a. For which $f(x) < 0$ by examination of the graph of $f(x)$
b. For which $f(x) > 0$ by examination of the graph

Solution   Since the coefficient of x^2 is 1 and $1 > 0$, the parabola opens upward.

Make a table of values by selecting some integral values for x.

x	$f(x) = x^2 + 2x - 8$
0	-8
1	-5
-1	-9
-2	-8
-4	0
-5	7
2	0
3	7

From the table, it is evident that when $x = -4$ or $x = 2$, the corresponding y-value is the same: $(-4, 0), (2, 0)$. The coordinates of the vertex are $(k, f(k))$, where

$$k = \frac{-4 + 2}{2} = -1$$

Therefore, $f(k) = f(-1) = -9$ and the coordinates of the vertex are $(-1, -9)$. Figure 4.7.3 is the graph of $f(x)$.

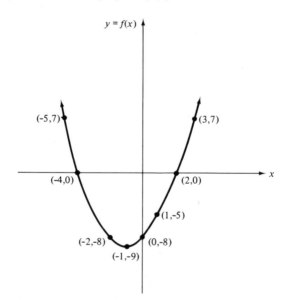

FIGURE 4.7.3 $f(x) = x^2 + 2x - 8$.

a. The graph clearly indicates that $f(x)$ is negative (its graph is below the x-axis) for

$$-4 < x < 2$$

b. $f(x) > 0$ when the graph of the function is *above* the x-axis. Therefore, from the graph it can be seen that $f(x) > 0$ when

$$x < -4 \quad \text{or} \quad x > 2$$

EXAMPLE 4.7.2   Draw a graph for $f(x) = -x^2 + 2x - 4$ and determine from the graph all x such that $f(x) < 0$.

Solution   Since the coefficient of x^2 is -1 and $-1 < 0$, the parabola opens downward.

x	$f(x) = -x^2 + 2x - 4$
0	-4
-1	-7
1	-3
-2	-12
2	-4
3	-7
4	-12

To find the coordinates of the vertex, select two x-values from the table, so that the corresponding y-values are equal: $(0, -4)$, $(2, -4)$.

$$k = \frac{2 + 0}{2} = 1$$

$$f(k) = -3$$

The coordinates of the vertex are $(1, -3)$ (see Figure 4.7.4). Since the entire graph is below the x-axis, $f(x) < 0$ for all real numbers x.

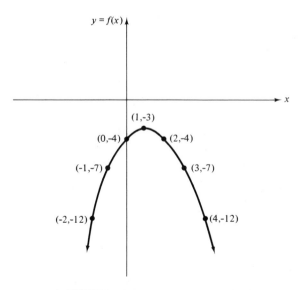

FIGURE 4.7.4 $y = -x^2 + 2x - 4$.

EXAMPLE 4.7.3 For $f(x)$ defined in Example 4.7.2, find all values of x such that $f(x) \geq 0$.

Solution It was shown that $f(x) < 0$ for all values x; therefore, there is no x such that $f(x) = -x^2 + 2x - 4 \geq 0$, and the solution set is the empty set, $\varnothing$.

EXAMPLE 4.7.4 Graph the solution set of the relation
$$y \geq 2x^2 - 3x - 2$$
Solution Graph the parabola whose equation is
$$y = 2x^2 - 3x - 2$$

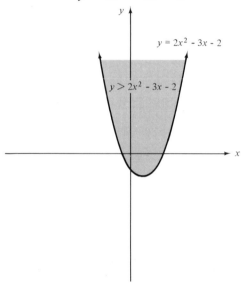

FIGURE 4.7.5 $y \geq 2x^2 - 3x - 2$.

Since the coefficient of x^2 is 2 and $2 > 0$, the parabola opens upward, and the region for which
$$y > 2x^2 - 3x - 2$$
is inside the parabola, as indicated by the shading in Figure 4.7.5. The solution set is the parabola *and* the shaded region.

A circle very clearly divides the plane into three regions: the region outside the circle, the circle, and the region inside the circle.

Recall that the equation of a circle with radius r and center (a, b) is
$$(x - a)^2 + (y - b)^2 = r^2$$

If $(x - a)^2 + (y - b)^2 < r^2$, then the region *inside* the circle is described.

If $(x - a)^2 + (y - b)^2 > r^2$, then the region *outside* the circle is described.

Clearly $(x - a)^2 + (y - b)^2 = r^2$ describes the circle itself.

EXAMPLE 4.7.5   Graph and discuss each of the following regions:

a. $x^2 + y^2 < 4$
b. $x^2 + y^2 > 4$

Solution $x^2 + y^2 = 4$ is the equation of a circle with center at the origin and radius 2.

    a. If $x^2 + y^2 < 4$, then all the circles with center at the origin and radius *less* than 2 must be considered. This yields the entire region *inside* the circle.

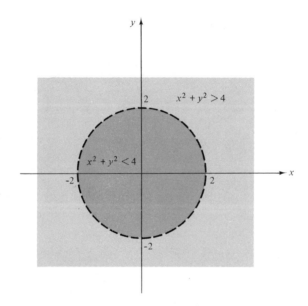

FIGURE 4.7.6

    b. If $x^2 + y^2 > 4$, then all the circles with center at the origin and radius *greater* than 2 must be considered. This yields the entire region *outside* the circle (see Figure 4.7.6).

EXERCISES 4.7 A

For Exercises 1–10,

a. Graph $y = f(x)$ for each of the following.
b. Using the graph, determine the solution set for each given inequality.

1. $f(x) = x^2 - 9$
    $x^2 - 9 > 0$

2. $f(x) = -x^2 + 4$
    $-x^2 + 4 \leq 0$

3. $f(x) = x^2 + 3$

$x^2 + 3 \geq 0$

5. $f(x) = 6x - x^2$

$6x - x^2 > 0$

7. $f(x) = 2x^2 + 5x - 3$

$f(x) \leq 0$

9. $f(x) = x^2 + 4x + 1$

$f(x) < 0$

4. $f(x) = 2x^2 - 4x$

$2x^2 - 4x < 0$

6. $f(x) = 2x^2 + 5x - 3$

$f(x) > 0$

8. $f(x) = x^2 + 2x + 1$

$f(x) \geq 0$

10. $f(x) = 4 - 4x - x^2$

$f(x) \leq 0$

Show a graphical solution for each of the relations in Exercises 11–20.

11. $y > x^2 - 5$

13. $y > 2 - x^2$

15. $y > 2x^2 - 4x$

17. $y > 2x - 4x^2$

19. $y \leq x^2 - 4x + 4$

12. $y \leq x^2 + 2$

14. $y < -1 - x^2$

16. $y \leq 4x^2 - 2x$

18. $y > x^2 + 2x + 1$

20. $y > -x^2 + 6x - 9$

Graph each of the relations in Exercises 21–25.

21. $x^2 + y^2 < 9$

23. $(x - 1)^2 + (y - 2)^2 < 1$

25. $(x - 3)^2 + (y + 4)^2 \leq 16$

22. $x^2 + y^2 \geq 4$

24. $(x + 2)^2 + (y - 1)^2 \geq 9$

EXERCISES 4.7 B

For Exercises 1–10,

a. Graph $y = f(x)$ for each of the following.

b. Using the graph, determine the solution set for each given inequality.

1. $f(x) = x^2 + 1$

$x^2 + 1 > 0$

3. $f(x) = x^2 - 1$

$x^2 - 1 > 0$

5. $f(x) = 2x - x^2$

$2x - x^2 < 0$

7. $f(x) = x^2 - 4x + 4$

$f(x) \leq 0$

9. $f(x) = -2x^2 + 4x - 1$

$f(x) > 0$

2. $f(x) = -x^2 - 4$

$-x^2 - 4 \leq 0$

4. $f(x) = 3x^2 + 12x$

$3x^2 + 12x < 0$

6. $f(x) = x^2 + 6x + 9$

$f(x) \geq 0$

8. $f(x) = -2x^2 + 4x - 1$

$f(x) \leq 0$

10. $f(x) = x^2 - 8x + 6$

$f(x) \leq 0$

Show a graphical solution for each of the relations in Exercises 11–20.

11. $y > x^2 + 1$ **12.** $y \leq x^2 - 1$

13. $y < 4 - x^2$ **14.** $y \geq -2 - x^2$

15. $y > 3x^2 - 6x$ **16.** $y \leq x^2 - 3x$

17. $y > 3x - x^2$ **18.** $y > x^2 - 2x + 1$

19. $y \leq x^2 + 4x + 4$ **20.** $y < -x^2 + 4x - 4$

Graph each of the relations in Exercises 21–25.

21. $x^2 + y^2 < 4$ **22.** $x^2 + y^2 \geq 1$

23. $(x + 1)^2 + (y + 2)^2 > 9$ **24.** $(x - 2)^2 + (y - 1)^2 \leq 4$

25. $(x + 3)^2 + (y - 2)^2 > 16$

4.8 ABSOLUTE VALUE FUNCTIONS AND RELATIONS

Several different functions have been discussed in this chapter: linear functions, quadratic functions, constant functions, the identity function. Another important function is the absolute value function $y = |x|$.

In order to graph the function

$$y = |x|$$

it is necessary to consider the definition of $|x|$:

If $x \geq 0$, $y = x$, and if $x < 0$, $-y = x$ and $y = -x$

The graph of Figure 4.8.1 is a composite of two graphs on the same set of axes. The graph $\{(x, y) \mid x \geq 0\} \cap \{(x, y) \mid y = x\}$ is the graph in the first quadrant, including the origin, whereas the graph

$$\{(x, y) \mid x < 0\} \cap \{(x, y) \mid y = -x\}$$

is the graph in the second quadrant.

Another example of the graph of an absolute value function is the following.

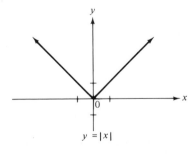

$y = |x|$

FIGURE 4.8.1

EXAMPLE 4.8.1 $y = |x| + 2$

Solution   Again two cases must be considered:

Case 1: $x \geq 0$; then $|x| = x$, and $y = x + 2$

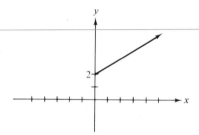

FIGURE 4.8.2

Case 2: $x < 0$; then $|x| = -x$, and $y = -x + 2$

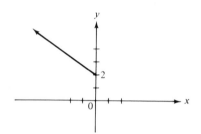

FIGURE 4.8.3

Combining both graphs on one set of axes yields the complete graph of the solution:

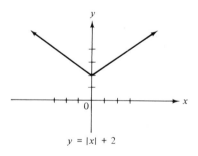

$y = |x| + 2$

FIGURE 4.8.4

Absolute value inequalities are graphed in a manner similar to the inequality relations previously discussed.

EXAMPLE 4.8.2 Graph the relation $y < |x + 2|$.

Solution Use a broken line to graph the absolute value function,

$$y = |x + 2|$$

Case 1: $\quad x + 2 \geq 0 \quad$ and $\quad y = x + 2$

$\quad\quad\quad\quad x \geq -2 \quad$ and $\quad y = x + 2$

Case 2: $\quad x + 2 < 0 \quad$ and $\quad y = -(x + 2)$

$\quad\quad\quad\quad x < -2 \quad$ and $\quad y = -x - 2$

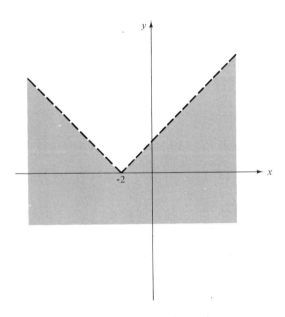

FIGURE 4.8.5 $y < |x + 2|$.

Since $y < |x + 2|$, the solution set is the shaded region *below* the broken line graph of $y = |x + 2|$, as illustrated in Figure 4.8.5.

EXAMPLE 4.8.3   Graph the relation $y \geq 2 + |x + 1|$.

Solution   This time the graph $y = 2 + |x + 1|$ is included, so graph this function first:

$$Case\ 1: \quad x + 1 \geq 0 \quad \text{and} \quad y = 2 + (x + 1)$$
$$x \geq -1 \quad \text{and} \quad y = x + 3$$
$$Case\ 2: \quad x + 1 < 0 \quad \text{and} \quad y = 2 - (x + 1)$$
$$x < -1 \quad \text{and} \quad y = -x + 1$$

The relation $y > 2 + |x + 1|$ is the shaded region *above* the graph of $y = 2 + |x + 1|$, as illustrated in Figure 4.8.6.

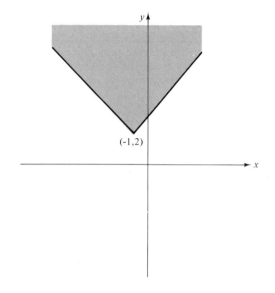

(-1,2)

FIGURE 4.8.6 $y \geq 2 + |x + 1|$.

EXERCISES 4.8 A

Graph Exercises 1–15.

1. $y = |x - 5|$ **2.** $y \geq |x - 5|$

3. $y = |x + 2|$ **4.** $y < |x + 2|$

5. $y \leq |3x - 12|$ **6.** $y > |2x + 6|$

7. $y = 3 + |x - 2|$ **8.** $y = -4 + |1 - x|$

9. $y = 2 - |x + 4|$

10. $y \geq 2 - |x + 4|$

11. $y = 3 + 2|3 - x|$

12. $y \leq 3 + 2|3 - x|$

13. $y \geq 4 + |2x - 8|$

14. $y \leq 5 - |4x + 12|$

15. $y > |x| + x + 1$

Graph Exercises 16–20.

16. $\{(x, y) \mid y = -|x|\}$

17. $\{(x, y) \mid |x| = -x\}$

18. $|x - y| = x - y$

19. $|x| + |y| = 1$

20. $|x| + |y| < 1$

EXERCISES 4.8 B

Graph Exercises 1–15.

1. $y = |2x|$

2. $y \geq |2x|$

3. $y = |x - 3|$

4. $y < |x - 3|$

5. $y \leq |2x - 1|$

6. $y > |3x + 2|$

7. $y = 2 + |x + 2|$

8. $y = -1 + |2 - x|$

9. $y = 3 - |x + 1|$

10. $y > 2 + |x + 4|$

11. $y = 4 + 2|1 - x|$

12. $y \leq 4 + 2|1 - x|$

13. $y \geq 3 + |2x - 5|$

14. $y < -2 - |3x + 6|$

15. $y > |x| + 2 - x$

Graph Exercises 16–20.

16. $\{(x, y) \mid |x| = x\}$

17. $\{(x, y) \mid |x| > 1\}$

18. $|x - y| = 1$

19. $|x - y| = y - x$

20. $\{(x, y) \mid |y| \geq 3\}$

4.9 APPLICATIONS (Optional)

Inequalities have many applications, not only in other branches of mathematics such as geometry, trigonometry, and calculus but also in science, technology, social sciences, and especially in business, economics, and military tactics.

Linear programming is one application which has become increasingly important ever since its introduction in the 1940s. In general, linear programming is concerned with finding a solution of a system of inequalities in two or more variables so that the solution produces a greatest value (maximum) or a least value (minimum) of a specified function of the variables.

The geometric technique for the case of two variables consists in graphing the intersection of the solution sets of the inequalities of the system. If a convex polygonal region is obtained, then the greatest or least value of the function will occur at one of the vertices.

EXAMPLE 4.9.1   If $x \geq 0$, $y \geq 0$, $3x + 2y \geq 15$, $4x + y \geq 10$, find the least possible value for $C = 2x + y$.

Solution (*vertex method*)   Graph $x = 0$, $y = 0$, $3x + 2y = 15$, $4x + y = 10$ and shade the region indicated by the system of inequalities.

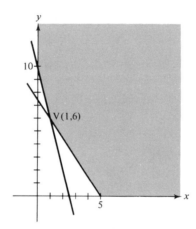

FIGURE 4.9.1

Solve the system $3x + 2y = 15$, $4x + y = 10$. The solution is $(1, 6)$, the coordinates of vertex V. In Figure 4.9.2, the same region is sketched with selected parallel lines from the set $2x + y = C$—namely, $2x + y = 6$, $2x + y = 8$, $2x + y = 10$.

To verify that $(1, 6)$ is actually the solution, check the value at each vertex.

    a. At $(0, 10)$, $C = 2(0) + 10$; $C = 10$.
    b. At $(5, 0)$, $C = 2(5) + 0$; $C = 10$.
    c. At $(1, 6)$, $C = 2(1) + 6$; $C = 8$.

Therefore, $(1, 6)$ yields the smallest possible value for C.

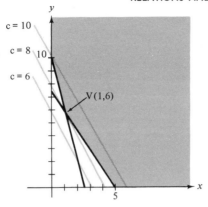

FIGURE 4.9.2

*Alternate Solution (**algebraic method**)* It can be shown algebraically that the value for C determined in this way is the minimum value desired.

1. Solve C for y: $\quad y = C - 2x$
2. Substitute in the inequalities containing both x and y:

$$3x + 2(C - 2x) \geq 15 \quad \text{or} \quad -x + 2C \geq 15$$

$$4x + (C - 2x) \geq 10 \quad \text{or} \quad 2x + C \geq 10$$

3. Eliminate the variable x and solve for C:

$$2(-x + 2C \geq 15) \rightarrow -2x + 4C \geq 30$$

$$\underline{2x + \quad C \geq 10}$$

$$5C \geq 40$$

$$C \geq 8$$

Since $C \geq 8$, the smallest value for $C = 8$.

4. In the two inequalities in Step 2, replace C by 8 and solve for x and then y:

$$-x + 16 \geq 15 \qquad 2x + 8 \geq 10$$

$$-x \geq -1 \qquad 2x \geq 2$$

$$x \leq 1 \qquad x \geq 1$$

The common solution of $x \leq 1$ and $x \geq 1$ is $x = 1$. If $x = 1$, then

$$y = C - 2x = 8 - 2(1) = 6$$

Thus $(1, 6)$ produces the minimum value $C = 8$.

EXAMPLE 4.9.2 A manufacturer must use each of two different machines for the production of each of two different products. The number of hours required on each machine to produce the two products is given in the table, along with the maximum number of hours each machine can be run per week.

Machine	Product A	B	Maximum Hours for Machine
I	3	2	48
II	1	4	46

If the profit on product A is \$20 per item and the profit on product B is \$30 per item, how many of each product should be manufactured for maximum profit? What is the maximum profit?

Solution Let x = number of items of A, y = number of items of B. Then $x \geq 0$, $y \geq 0$, $3x + 2y \leq 48$, $x + 4y \leq 46$,

$$\text{and the profit } P = 20x + 30y$$

1. Graph the intersection of the inequalities.

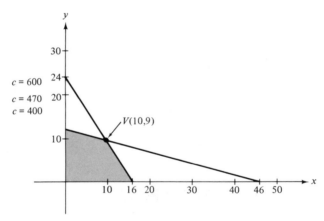

FIGURE 4.9.3

2. Solve the two equations to obtain the coordinates of the vertex V:

$$-2(3x + 2y) = 48 \rightarrow -6x - 4y = -96$$
$$x + 4y = 46$$
$$\overline{-5x = -50}$$
$$x = 10$$

$$4y = 46 - 10 = 36 \quad \text{and} \quad y = 9$$

The vertex V is $(10, 9)$.

8. A man wants to buy some 6-cent and 10-cent postage stamps. He does not want to spend over $9. He wants to buy at least 3 times as many 6-cent stamps as 10-cent stamps. What is the largest number of 10-cent stamps he can buy under these conditions?

9. A grower can ship three kinds of citrus fruits from California to Chicago. He intends to ship 800 boxes of fruit on each truck. He must ship at least 200 boxes of oranges, at least 100 boxes of limes, and at least 200 boxes of lemons. If the profit on oranges is $2 a box, on lemons $3 a box, and on limes $1 a box, how should his truck be loaded for maximum profit? (*Hint:* Let x = number of boxes of oranges, y = number of boxes of lemons, $800 - x - y$ = number of boxes of limes.)

10. A student publication containing 80 pages is to contain at least 40 pages of literary articles and at most 20 pages of advertising. It also contains art work limited to one-fourth the number of pages for the literary articles. If the cost per page is $50 for literary pages, $100 for art pages, and $-$50 for advertisement pages (the advertiser pays for the page), how should the publication be composed for minimum cost? (*Hint:* Let x = number of literary pages, y = number of art pages, and $80 - x - y$ = number of pages of advertisements.) What is the minimum cost?

EXERCISES 4.9 B

1. A 6-foot by 8-foot rectangle is to be constructed so that the measurement of the perimeter is correct within 0.1 foot. What is the largest error possible for the measurement of the sides? What is the error in the measurement of the area?

2. A merchant wants to make a profit of at least 25 percent on the sale of an electric frying pan. Similar pans are selling for $25. If his cost is $15 each, what range of prices could he charge and still sell for less than his competitors?

3. In a triangle, the sum of any two sides must be greater than the third side. If a, b, and c are the lengths of the sides, then $a + b > c$ and $a + c > b$ and $b + c > a$.

a. Show that if $a < b$, then $b - a < c < b + a$.
b. Between what two values does b lie if $c < a$?

4. If an object is shot directly upward with an initial velocity of v feet per second, then the maximum height it reaches is $h = v^2/64$. With what velocity should the object be shot so that the height is at least 100 feet but no more than 225 feet?

In Exercises 5–7, graph the solution set of each of the systems of inequalities and find the coordinates of the vertices.

5. $x \geq 0$ **6.** $3y \leq 4x$ **7.** $0 \leq x \leq 12$

    $y \geq 0$ $x \geq 0$ $y \leq 2x$

    $x + 2y \leq 16$ $4 \leq y \leq 12$ $x + y \leq 15$

    $2x + y \leq 20$

8. A grocer wants to combine domestic rice worth 20 cents per pound with wild rice worth \$1.60 per pound to obtain a mixture whose value will not exceed 60 cents per pound. He has only 40 pounds of wild rice available to him, and he must use at least one fourth as much wild rice as domestic rice so that the resulting product will have good sales. State the inequalities that describe this situation and graph the solution set of the system. What combination uses the most domestic rice?

9. Two foods A and B are to be combined in such a way that the combination supplies at least 120 units of vitamins and at least 80 units of minerals. Food A supplies 2 units of vitamins and 2 units of minerals per ounce. Food B supplies 3 units of vitamins and 1 unit of minerals per ounce. If the cost of Food A is 10 cents per ounce and the cost of Food B is 6 cents per ounce, how should the foods be combined for minimum cost?

10. A promoter wants to combine two entertainment groups for a 60-minute special. Group I provides 1 minute of music for every 3 minutes of comedy. Group II provides 2 minutes of music for each minute of comedy. At least 20 minutes of music and at least 30 minutes of comedy are wanted for maximum viewer appeal. How many minutes of each group should he have if he wants to minimize the cost?

SUMMARY

☐ A set whose elements are ordered pairs is called a **relation.**

☐ The set of all first components of the ordered pairs in a relation is called the **domain** of the relation.

☐ The set of all second components of the ordered pairs in a relation is called the **range** of the relation.

☐ A **function** is a relation in which no two ordered pairs have the same first components and different second components.

☐ A **linear function** is a function whose rule has the form $f(x) = ax + b$, where a and b are constants.

☐ A **constant function** is a function whose rule has the form $f(x) = b$, where b is a constant.

32. An object is thrown vertically upward according to the equation $h = 80t - 16t^2$, where h is the height at any given time and t is the time in seconds.

a. Graph the function.

b. From the graph, determine the maximum height the object reaches.

c. From the graph, determine how long it takes the object to return to earth if the initial height $h = 0$.

33. Find the distance between $A:(-3, 6)$ and $B:(-9, 4)$.

34. Find the center and the radius of the circle whose equation is $x^2 + y^2 - 6x + 10y + 9 = 0$. Graph this circle.

Show a graphical solution for each of Exercises 35–46.

35. $y < 2x + 3$ **36.** $y \geq x - 1$

37. $\{(x, y) \mid -2 \leq x < 2\}$ **38.** $\{(x, y) \mid -2 \leq y < 2\}$

39. $y < x^2 + 3x + 2$

40. $f(x) \geq 0$ and $f(x) = x^2 - 6x - 7$

41. $y > 2x - 4x^2$ **42.** $y = |2x - 8|$

43. $y < 2 + |2x|$ **44.** $y \geq |3 - 2x|$

45. $x^2 + y^2 \leq 25$ **46.** $(x - 1)^2 + (y + 2)^2 > 4$

CHAPTER **5**

LOGARITHMIC AND EXPONENTIAL FUNCTIONS

5.1 FUNCTIONS AND INVERSES

A *relation* has been defined as a set of ordered pairs having a *domain*, the set of all first components of the ordered pairs in the relation, and a *range*, the set of all second components of the ordered pairs of the relation.

A *function* was defined as a special relation such that each first component in its domain is paired with exactly one second component in its range.

If the components of the ordered pairs of a relation r are interchanged, then another set of ordered pairs is obtained, and thus this set is also a relation. It is called the *inverse* of the relation r and is designated symbolically as r^{-1} (read "r inverse").

DEFINITION

The **inverse r^{-1} of a relation r** is the set of ordered pairs obtained by interchanging the components of r.

EXAMPLE 5.1.1 Given $r = \{(2, 3), (4, 5), (6, 7)\}$.

a. Find the domain and range of r.
b. List the elements in r^{-1}.
c. Find the domain and range of r^{-1}.

However, if a function has the property that there is a one-to-one correspondence between the elements in its domain and the elements in its range, then the function is said to be one-to-one, and its inverse is a function.

DEFINITION

The **function f is one-to-one** if and only if each element in its domain is paired with exactly one element in its range and each element in its range is paired with exactly one element in its domain.

THE INVERSE FUNCTION THEOREM

If the function f is one-to-one, then its inverse is a function, designated as f^{-1} (read "f inverse").

EXAMPLE 5.1.5   Given the function $f = \{(x, y) \mid y = x^2$ where $x \geq 0\}$.

a. Find $f(x)$ and the domain and range of f.
b. Find f^{-1} and its domain and range.
c. Find $f^{-1}(x)$.
d. Graph f and f^{-1} on the same set of axes.

Solution

a. $f(x) = x^2$; domain $= \{x \mid x \geq 0\}$; range $= \{y \mid y \geq 0\}$
b. $f^{-1} = \{(x, y) \mid x = y^2$ where $y \geq 0\}$; domain $= \{x \mid x \geq 0\}$; range $= \{y \mid y \geq 0\}$
c. Solving $x = y^2$ for y, $y = \sqrt{x}$ since $y \geq 0$. Thus $f^{-1}(x) = \sqrt{x}$.
d.

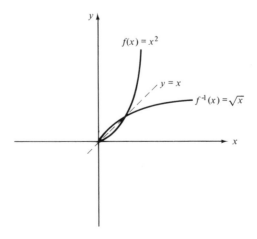

FIGURE 5.1.2   Graph of a one-to-one function and its inverse function.

The preceding example shows that it is possible to obtain a one-to-one function by restricting the domain of a function that is not one-to-one.

There is an important relationship between a one-to-one function and its inverse function. Suppose (a, b) is an ordered pair of the one-to-one function f. Then (b, a) is an ordered pair of the inverse function f^{-1}. This

means $f(a) = b$ and $f^{-1}(b) = a$

Then $f^{-1}(f(a)) = f^{-1}(b) = a$

and $f(f^{-1}(b)) = f(a) = b$

This result is stated as the following theorem:

THE FUNCTION-INVERSE FUNCTION THEOREM

Let f be a one-to-one function. For each x in the domain of f and for each x in the domain of f^{-1},

$$f^{-1}(f(x)) = x \quad \text{and} \quad f(f^{-1}(x)) = x$$

Geometrically, the graphs of a function and its inverse are symmetrical to the $y = x$ line, as may be observed in Figures 5.1.1 and 5.1.2. This means that the graphs would coincide if the coordinate plane were folded along the $y = x$ line.

EXAMPLE 5.1.6   Given $f(x) = 5x - 2$.

a. Find $f^{-1}(x)$, using $f(f^{-1}(x)) = x$.
b. Show that $f^{-1}(f(x)) = x$.

Solution

a. $f(\quad x \quad) = 5x - 2$
   $f(\qquad) = 5(\qquad) - 2$
   $f(f^{-1}(x)) = 5(f^{-1}(x)) - 2 = x$
   $\qquad 5\, f^{-1}(x) \qquad\quad = x + 2$
   $\qquad\qquad f^{-1}(x) \qquad = \dfrac{x + 2}{5}$

b. $f^{-1}(\quad x \quad) = \dfrac{x + 2}{5}$

   $f^{-1}(\qquad) = \dfrac{(\qquad) + 2}{5}$

   $f^{-1}(5x - 2) = \dfrac{(5x - 2) + 2}{5} = \dfrac{5x}{5} = x$

31. Use the results of Exercises 23 and 24 to graph f and f^{-1} on the same set of axes. (Plot the ordered pairs in the tables and join the points with a smooth curve.)

Solve Exercises 32–40 if $f(x) = 10^x$.

32. $f^{-1}(x) = 2$ **33.** $f^{-1}(x) = 0$

34. $f^{-1}(x) = -1$ **35.** $f^{-1}(x) = 0.5$

36. $f^{-1}(x) = -1.5$ **37.** $f^{-1}(0.01) = y$

38. $f^{-1}(1000) = y$ **39.** $f^{-1}(\sqrt[3]{10}) = y$

40. $f^{-1}(1) = y$

EXERCISES 5.1 B

For each of the functions in Exercises 1–15, determine (a) $f(x)$, (b) the inverse of f, and (c) $f^{-1}(x)$ if the inverse of f is a function.

1. $f = \{(x, y) \mid y = 3x\}$ **2.** $f = \{(x, y) \mid y = x + 7\}$

3. $f = \{(x, y) \mid y = x - 4\}$ **4.** $f = \left\{(x, y) \mid y = \dfrac{x}{10}\right\}$

5. $f = \{(x, y) \mid y = 3x + 12\}$ **6.** $f = \{(x, y) \mid 4x - y = 8\}$

7. $f = \{(x, y) \mid 5x + 2y = 10\}$ **8.** $f = \{(x, y) \mid y = x\}$

9. $f = \{(x, y) \mid y + 5 = 0\}$ **10.** $f = \{(x, y) \mid xy + 6 = 0\}$

11. $f = \{(x, y) \mid x^2 - y = 16\}$ **12.** $f = \{(x, y) \mid y + \sqrt{4 - x^2} = 0\}$

13. $f = \{(x, y) \mid x^2 + y^3 = 0\}$ **14.** $f = \{(x, y) \mid xy = |x|\} \quad (x \neq 0)$

15. $f = \{(x, y) \mid \sqrt{x} + \sqrt{y} = 4\}$

For Exercises 16–22,

a. Find $f^{-1}(x)$ using $f(f^{-1}(x)) = x$.
b. Show that $f^{-1}(f(x)) = x$.
c. Graph f and f^{-1} on the same set of axes.

16. $f(x) = 5x + 10$ **17.** $f(x) = -\sqrt{x - 2}$

18. $f(x) = \dfrac{-12}{x}$ **19.** $f(x) = \sqrt[3]{x + 1}$

20. $f(x) = \sqrt{9 - x^2} \quad (x \geq 0)$ **21.** $f(x) = 4 - x^2 \quad (x \geq 0)$

22. $f(x) = (1 - \sqrt{x})^2 \quad (0 \leq x \leq 1)$

23. Complete the following table of values for $f(x) = 4^{-x}$.

x	-2	$-\frac{3}{2}$	-1	$-\frac{1}{2}$	0	$\frac{1}{2}$	1	$\frac{3}{2}$	2
$f(x)$									

24. Using the results of Exercise 23, make a table of values for $(x, f^{-1}(x))$ by interchanging the ordered pairs in the table for $(x, f(x))$.

Use the table in Exercise 24 to solve Exercises 25–30.

25. $f^{-1}(2)$ **26.** $f^{-1}(16)$
27. $f^{-1}(8)$ **28.** $f^{-1}(1)$
29. $f^{-1}(\frac{1}{8})$ **30.** $f^{-1}(\frac{1}{2})$

31. Use the results of Exercises 23 and 24 to graph f and f^{-1} on the same set of axes. (Plot the ordered pairs in the tables and join the points with a smooth curve.)

Solve Exercises 32–40 if $f(x) = 10^{-x}$.

32. $f^{-1}(x) = 3$ **33.** $f^{-1}(x) = -2$
34. $f^{-1}(x) = 0$ **35.** $f^{-1}(x) = 0.5$
36. $f^{-1}(x) = -2.5$ **37.** $f^{-1}(10,000) = y$

38. $f^{-1}(0.001) = y$ **39.** $f^{-1}\left(\dfrac{\sqrt{10}}{10}\right) = y$

40. $f^{-1}(10) = y$

5.2 EXPONENTIAL FUNCTIONS

The expression b^x has been defined for any nonzero real number b and for any rational number x. For b^x to be a real number, it was shown that the base b must be restricted to the set of nonnegative real numbers whenever a rational exponent indicates the extraction of an even root. The definitions and theorems for rational exponents previously developed are now restated.

Figure 5.2.1 shows that the value of 2^x increases as x increases. This is true in general. If $b > 0$ and $b \neq 1$, then $b^{x_2} > b^{x_1}$ if $x_2 > x_1$.

What value should be assigned to $2^{\sqrt{3}}$ can now be determined by approximating $\sqrt{3}$ by rational numbers and by assuming that $2^r < 2^{\sqrt{3}} < 2^s$ if $r < \sqrt{3} < s$ (Figure 5.2.2).

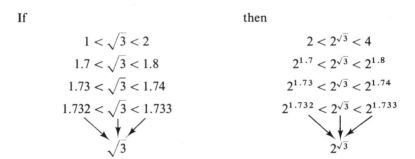

If	then
$1 < \sqrt{3} < 2$	$2 < 2^{\sqrt{3}} < 4$
$1.7 < \sqrt{3} < 1.8$	$2^{1.7} < 2^{\sqrt{3}} < 2^{1.8}$
$1.73 < \sqrt{3} < 1.74$	$2^{1.73} < 2^{\sqrt{3}} < 2^{1.74}$
$1.732 < \sqrt{3} < 1.733$	$2^{1.732} < 2^{\sqrt{3}} < 2^{1.733}$

$$\sqrt{3}$$

$$2^{\sqrt{3}}$$

FIGURE 5.2.2

As the approximations to $\sqrt{3}$ get better and better, the approximations to $2^{\sqrt{3}}$ get closer and closer to exactly one real number. This real number, called the limiting value, is assigned as the definition of $2^{\sqrt{3}}$.

In general, if x is an irrational real number and b is a positive real number, then b^x **is defined as the limiting value of the approximations to b^x**, as illustrated in the special case above.

By using the methods of higher mathematics, it can be shown that the theorems for rational exponents are still valid when the exponent is an irrational real number.

Accepting these results, the definition of an exponential function can now be stated.

DEFINITION OF EXPONENTIAL FUNCTION

The exponential function to the base $b =$

$\{(x, y) \mid y = b^x$ where $b > 0, b \neq 1$, and x is any real number$\}$

EXAMPLE 5.2.4   Simplify $2^{1-\sqrt{5}}2^{\sqrt{5}-3}$.

Solution $2^{1-\sqrt{5}}2^{\sqrt{5}-3} = 2^{1-\sqrt{5}+\sqrt{5}-3}$ (Using $b^x b^y = b^{x+y}$)

$= 2^{-2} = \frac{1}{4}$ (Using $b^{-n} = 1/b^n$)

EXAMPLE 5.2.5 Simplify $(5^{\sqrt{2}})^{1/\sqrt{8}}$.

Solution　$(5^{\sqrt{2}})^{1/\sqrt{8}} = 5^{\sqrt{2}/\sqrt{8}}$ 　　　　　(Using $(b^x)^y = b^{xy}$)

　　　　　　　　　　$= 5^{1/2} = \sqrt{5}$ 　　　　　(Using $b^{1/n} = \sqrt[n]{b}$)

EXERCISES 5.2 A

Solve Exercises 1–10, using the theorem $b^x = b^y$ if and only if $x = y$.

1. $5^x = 125$ 　　　　　　　　　　**2.** $5^x = \frac{1}{25}$

3. $5^x = 5\sqrt{5}$ 　　　　　　　　　**4.** $2^{1-x} = \frac{1}{8}$

5. $3^{x-2} = \sqrt{27}$ 　　　　　　　**6.** $10^x = 0.001$

7. $10^x = 100,000$ 　　　　　　　**8.** $10^x = 1$

9. $8^{x+1} = \frac{1}{4}$ 　　　　　　　　**10.** $(6^x)^2 = \sqrt{18}\sqrt{12}$

Graph Exercises 11–20 by plotting points for $x \in \{-2, -1, -\frac{1}{2}, 0, \frac{1}{2}, 1, 2\}$ and joining these points with a smooth curve.

11. $y = 4^x$ 　　　　　　　　　　**12.** $y = 3^x$

13. $y = 2^{-x}$ 　　　　　　　　　**14.** $y = 4^{-x}$

15. $y = 5^x$ 　　　　　　　　　　**16.** $y = 5^{-x}$

17. $y = 2^{|x|}$ 　　　　　　　　　**18.** $y = 2^{x^2}$

19. $y = 2^{-x^2}$ 　　　　　　　　**20.** $y = (\frac{1}{3})^x$

★ **21.** Graph $y = f^{-1}(x)$ for $f(x) = 4^x$ (use the results of Exercise 11).

★ **22.** Graph $y = f^{-1}(x)$ for $f(x) = 2^{-x}$ (use the results of Exercise 13).

EXERCISES 5.2 B

Solve Exercises 1–10, using the theorem $b^x = b^y$ if and only if $x = y$.

1. $4^x = \frac{1}{64}$ 　　　　　　　　**2.** $4^{2-x} = \frac{1}{8}$

3. $10^x = \sqrt{1000}$ 　　　　　　**4.** $10^{2x-1} = 0.01$

5. $9^{4x-12} = 1$ 　　　　　　　　**6.** $3^{x-1} = \sqrt{\frac{1}{3}}$

7. $2^x 3^x = \frac{1}{36}$ 　　　　　　　**8.** $\dfrac{10^x}{2^x} = \dfrac{3}{\sqrt{45}}$

9. $7^{-x} = 7\sqrt[3]{7}$ 　　　　　　**10.** $8^{4-2x} = 4^{x+4}$

EXAMPLE 5.3.2   Write in logarithmic form:

a. $10^3 = 1000$
b. $2^{-4} = \frac{1}{16}$
c. $125^{-2/3} = \frac{1}{25}$

Solution   Since

$$b^y = x \text{ if and only if } \log_b x = y$$

a. $10^3 = 1000 \quad \rightarrow \quad \log_{10} 1000 = 3$
b. $2^{-4} = \frac{1}{16} \quad \rightarrow \quad \log_2 \left(\frac{1}{16}\right) = -4$
c. $125^{-2/3} = \frac{1}{25} \quad \rightarrow \quad \log_{125} \left(\frac{1}{25}\right) = -\frac{2}{3}$

EXAMPLE 5.3.3   Write in exponential form:

a. $\log_{10} 100 = 2$
b. $\log_2 \left(\frac{1}{8}\right) = -3$
c. $\log_{25} 5 = \frac{1}{2}$

Solution

$$\text{Log}_b x = y \text{ if and only if } b^y = x$$

a. $\log_{10} 100 = 2 \quad \rightarrow \quad 10^2 = 100$
b. $\log_2 \left(\frac{1}{8}\right) = -3 \quad \rightarrow \quad 2^{-3} = \frac{1}{8}$
c. $\log_{25} 5 = \frac{1}{2} \quad \rightarrow \quad 25^{1/2} = 5$

EXAMPLE 5.3.4   Evaluate $\log_5 0.04$.

Solution   Let $x = \log_5 0.04$. Then

$$5^x = 0.04 = \frac{4}{100} = \frac{1}{25} = 5^{-2}$$

$$5^x = 5^{-2}$$

Thus $x = -2$, since $b^x = b^y$ if and only if $x = y$. Therefore, $\log_5 0.04 = -2$.

EXAMPLE 5.3.5   Solve for x: $\log_3 x = -4$.

Solution $\text{Log}_3 x = -4$ if and only if $3^{-4} = x$. Thus $x = \frac{1}{81}$.

EXAMPLE 5.3.6   Solve for x: $\log_x 8 = \frac{3}{2}$.

Solution $\text{Log}_x 8 = \frac{3}{2}$ if and only if

$$x^{3/2} = 8$$
$$(x^{3/2})^{2/3} = 8^{2/3}$$
$$x = (\sqrt[3]{8})^2 = 4$$

EXERCISES 5.3 A

Write Exercises 1–8 in logarithmic form.

1. $10^4 = 10{,}000$ **2.** $5^0 = 1$
3. $9^{1/2} = 3$ **4.** $4^{-3} = \frac{1}{64}$
5. $3^1 = 3$ **6.** $64^{2/3} = 16$
7. $9^{-3/2} = \frac{1}{27}$ **8.** $(\frac{1}{6})^{-2} = 36$

Write Exercises 9–16 in exponential form.

9. $\log_3 9 = 2$ **10.** $\log_{10} 1 = 0$
11. $\log_{10} 10 = 1$ **12.** $\log_2 0.125 = -3$
13. $\log_{36} 6 = \frac{1}{2}$ **14.** $\log_{10} 100{,}000 = 5$
15. $\log_{10} 0.1 = -1$ **16.** $\log_8 0.25 = -\frac{2}{3}$

Evaluate Exercises 17–24.

17. $\log_{10} 1000$ **18.** $\log_{10} 0.01$
19. $\log_5 0.0016$ **20.** $\log_4 32$
21. $\log_9 \frac{1}{27}$ **22.** $\log_2 0.0625$
23. $\log_{10} 10^{-2.5}$ **24.** $10^{\log_{10} 3}$

Solve Exercises 25–40.

25. $\log_2 64 = x$ **26.** $\log_3 x = -4$
27. $\log_{10} x = 0$ **28.** $\log_2 0.25 = x$
29. $\log_x 0.04 = -2$ **30.** $\log_x 16 = \frac{2}{2}$
31. $\log_5 x = 1$ **32.** $\log_{10} x = -3$
33. $\log_{10} x = 4$ **34.** $\log_{10} x = 1.5$
35. $\log_5 125 = x$ **36.** $\log_x 8 = -\frac{3}{2}$
37. $\log_x 6\sqrt{6} = \frac{3}{2}$ **38.** $\log_{27} x = -\frac{4}{3}$
39. $\log_x 9 = -2$ **40.** $\log_x 0.125 = 3$

EXAMPLE 5.4.2   Express $\frac{1}{3}(\log_b 5 + \log_b 7 - \log_b 2)$ as the logarithm of a single number.

Solution $\frac{1}{3}(\log_b 5 + \log_b 7 - \log_b 2) = \frac{1}{3}(\log_b 35 - \log_b 2)$

(Theorem A)

$$= \frac{1}{3} \log_b \frac{35}{2} \qquad \text{(Theorem B)}$$

$$= \log_b \sqrt[3]{\frac{35}{2}} \qquad \text{(Theorem D)}$$

EXERCISES 5.4 A

By using logarithmic operations theorems, express each of Exercises 1–10 in terms of the logarithms of prime integers.

1. $\log_b 15$

2. $\log_b \frac{3}{5}$

3. $\log_b 3^5$

4. $\log_b \sqrt{3}$

5. $\log_b \sqrt{\frac{5}{3}}$

6. $\log_b \dfrac{1}{\sqrt[3]{5}}$

7. $\log_b \frac{81}{25}$

8. $\log_b 25\sqrt[4]{3}$

9. $\log_b \dfrac{5\sqrt{5}}{7}$

10. $\log_b \sqrt[3]{\frac{625}{9}}$

By using logarithmic operations theorems, express each of Exercises 11–20 as the logarithm of a single number.

11. $\log_b 8 + \log_b 9$

12. $\log_b 21 - \log_b 8$

13. $5 \log_b 3$

14. $\frac{1}{2} \log_b 17$

15. $2 \log_b 5 + 4 \log_b 3$ 2025

16. $\frac{1}{3} \log_b 7 - 2 \log_b 6$

17. $\frac{1}{5}(\log_b 3 + 2 \log_b 5)$

18. $3(\log_b 4 - 3 \log_b 6)$

19. $2 \log_b 3 + \frac{1}{4} \log_b 5 - 3 \log_b 7$

20. $\frac{1}{3}(2 \log_b 9 + \log_b 4 - \log_b 85)$

$\log b 9 \sqrt[4]{5}/7^3$

EXERCISES 5.4 B

By using logarithmic operations theorems, express each of Exercises 1–10 in terms of the logarithms of prime integers.

1. $\log_b 6$

2. $\log_b \frac{3}{2}$

3. $\log_b 2^6$

4. $\log_b \sqrt[3]{2}$

5. $\log_b \frac{1}{6}$

6. $\log_b 12$

7. $\log_b \frac{27}{64}$

8. $\log_b \sqrt{1.5}$

9. $\log_b 4\sqrt{3}$

10. $\log_b 6\sqrt{6}$

By using logarithmic operations theorems, express each of Exercises 11–20 as the logarithm of a single number.

11. $\log_b 25 + \log_b 18$

12. $\log_b 25 - \log_b 18$

13. $\frac{1}{2} \log_b 5 + \frac{1}{3} \log_b 17$

14. $\log_b 7 + 5 \log_b 3$

15. $\frac{1}{3}(\log_b 56 - \log_b 45)$

16. $\log_b 1 - \log_b 7 - \log_b 10$

17. $3 \log_b b - \frac{1}{2} \log_b 25$

18. $\log_b \frac{5}{2} + \log_b \frac{1}{3} - \log_b 6$

19. $\frac{1}{5} \log_b 64 - \log_b \sqrt[5]{2}$

20. $\frac{1}{2} \log_b 5 + \frac{3}{4} \log_b 7 - \frac{3}{2} \log_b 6$

5.5 COMMON LOGARITHMS

Since our numeral system is a base 10 positional system, the logarithmic function whose base is 10 is most useful for computations. The values of $\log_{10} x$ are called common logarithms or logarithms to the base 10. To reduce the amount of writing involved in a calculation using common logarithms, the numeral 10 designating the base is usually omitted.

CONVENTION

$$\log x = \log_{10} x$$

Finding the common logarithm of a number depends on the principle of scientific notation, restated below.

DEFINITION

Let r be a positive real number.
Let x be a real number between 1 and 10.
Let k be an integer.
Then the **scientific notation** for r is $x \times 10^k$;

$$r = x \times 10^k = x(10^k)$$

EXAMPLE 5.5.1 Express each of the following numbers in scientific notation:

    a. 285,000   b. 0.032   c. 7.2

    Solution

    a. $285{,}000 = 2.85 \times 10^5$

    b. $0.032 \quad = 3.2 \times 10^{-2}$

    c. $7.2 \quad\;\; = 7.2 \times 10^0$

EXAMPLE 5.5.5   Find antilog $(0.4518 - 1)$.

Solution $x = $ antilog $(0.4518 - 1)$

$\log x = 0.4518 - 1$

$\log x = \log 2.83 - 1$ (Using the tables)

$\log x = \log (2.83 \times 10^{-1})$

$x = 2.83 \times 10^{-1}$ (Scientific notation)

$x = 0.283$ (Ordinary notation)

EXERCISES 5.5 A

In Exercises 1–16, express each logarithm in (a) computational form and (b) formula form.

1. $\log 4.76$
2. $\log 890$
3. $\log 0.123$
4. $\log 405{,}000$
5. $\log 0.0948$
6. $\log 5$
7. $\log 0.007$
8. $\log 6520$
9. $\log 0.564$
10. $\log 2810$
11. $\log 36{,}700$
12. $\log 84.9$
13. $\log 0.000151$
14. $\log 0.35$
15. $\log 3.14$
16. $\log 0.08$

In Exercises 17–32, express each antilogarithm in (a) scientific notation and (b) ordinary notation.

17. antilog 1.8407
18. antilog 4.6232
19. antilog $(0.0682 - 1)$
20. antilog 0.9750
21. antilog $(0.8344 - 3)$
22. antilog 6.9031
23. antilog $(0.2900 - 2)$
24. antilog 0.5740
25. antilog 2.5911
26. antilog 3.7832
27. antilog $(0.2430 - 1)$
28. antilog $(0.5717 - 4)$
29. antilog $(0.4232 - 2)$
30. antilog (1.8)
31. antilog (0.0374)
32. antilog (-1.2)

EXERCISES 5.5 B

In Exercises 1–16, express each logarithm in (a) *computational form and* (b) *formula form.*

1. log 0.345
2. log 75,000
3. log 6.32
4. log 0.00419
5. log 1.5
6. log 0.0842
7. log 929
8. log 0.00053
9. log 2.72
10. log 68.1
11. log 0.266
12. log 2,000
13. log 0.0752
14. log 123,000,000
15. log 0.000 000 456
16. log 32.2

In Exercises 17–32, express each antilogarithm in (a) *scientific notation and* (b) *ordinary notation.*

17. antilog $(0.2330 - 2)$
18. antilog 0.9991
19. antilog 5.7694
20. antilog 1.5065
21. antilog $(0.3201 - 1)$
22. antilog 3.6021
23. antilog $(0.8500 - 3)$
24. antilog 0.4346
25. antilog $(0.1875 - 4)$
26. antilog (2.85)
27. antilog (0.48)
28. antilog (-0.52)
29. antilog (-1.15)
30. antilog (-2.71)
31. antilog (-3.8962)
32. antilog (-4.2)

5.6 LINEAR INTERPOLATION

The table of common logarithms inside the cover of this book lists the mantissas of the logarithms of numbers having at most three significant digits. For example, the values $\log 2.340 = 0.3692$ and $\log 2.350 = 0.3711$ may be read directly from the table. By using a process called *linear interpolation*, the logarithms of numbers having *four* significant digits may be **approximated**. The graph of the logarithmic function $\{(x, y) \mid y = \log x\}$ is approximated by the straight line joining two points on the curve whose ordinates are successive entries in the table. Figure 5.6.1 illustrates the procedure for determining log 2.346 by linear interpolation.

The value of log 2.346 will be approximated by $\log 2.340 + d = 0.3692 + d$. To determine d, an equation is found by using the fact from plane geometry that corresponding sides of similar triangles are proportional.

EXAMPLE 5.6.2   Find antilog 3.1888.

Solution

$$
\begin{array}{ccc}
 & x & \log x \\
 & 1540 & 3.1875 \\
10 \;\; d\!\downarrow & & \quad\;\downarrow 13 \\
 & 1550 & 3.1888 \;\;\; 28 \\
 & & 3.1903
\end{array}
$$

$$\frac{d}{10} = \frac{13}{28} \quad \text{or} \quad d = \frac{130}{28} = 5 \text{ approximately}$$

Thus antilog 3.1888 = 1545.

EXERCISES 5.6 A

Find each logarithm in Exercises 1–12.

1. log 3.142   **2.** log 78.25
3. log 0.4356   **4.** log 0.09213
5. log 223.8   **6.** log 1234
7. log 5277   **8.** log 0.006 489
9. log 0.8461   **10.** log 6.125
11. log 29.56   **12.** log 0.01206

Find each antilogarithm in Exercises 13–24.

13. antilog 0.5258   **14.** antilog 2.8714
15. antilog (0.6960 − 1)   **16.** antilog (0.4000 − 3)
17. antilog 1.7770   **18.** antilog 3.1780
19. antilog (0.2795 − 2)   **20.** antilog 0.3696
21. antilog (−0.9200)   **22.** antilog (−1.2847)
23. antilog (−2.8190)   **24.** antilog (−3.2956)

EXERCISES 5.6 B

Find each logarithm in Exercises 1–12.

1. log 2.718   **2.** log 0.8354
3. log 31.47   **4.** log 0.06421
5. log 5989   **6.** log 179.6
7. log 0.004563   **8.** log 70,420
9. log 900,900   **10.** log 0.1478
11. log 0.09346   **12.** log 0.004904

Find each antilogarithm in Exercises 13–24.

13. antilog 0.3612

14. antilog (0.8435 − 1)

15. antilog 4.5805

16. antilog (0.9517 − 2)

17. antilog 1.2180

18. antilog 1.9999

19. antilog (0.2590 − 1)

20. antilog (0.6205 − 3)

21. antilog 2.8480

22. antilog 3.9308

23. antilog (−1.6194)

24. antilog (−2.1495)

5.7 COMPUTATIONS

By using the theorems for logarithms, computations involving products, quotients, powers, and roots can be replaced by the simpler calculations involving sums, differences, products, and quotients.

EXAMPLE 5.7.1 Compute $(0.725)^4(34.7)$.

Solution Let $N = (0.725)^4(34.7)$.

$$\log N = 4 \log 0.725 + \log 34.7$$

$$\log 0.725 = 0.8603 - 1$$

$$4 \log 0.725 = 3.4412 - 4 = 0.4412 - 1$$

$$\log 34.7 = 0.5403 + 1$$

$$\text{sum} = \quad \log N \quad = 0.9815$$

$$N = 9.58 \text{ to three significant digits}$$

EXAMPLE 5.7.2 Compute $\sqrt[3]{\dfrac{1.380}{24.2}}$.

Solution Let $N = \sqrt[3]{\dfrac{1.380}{24.2}}$.

$$\log N = \tfrac{1}{3}(\log 1.380 - \log 24.2)$$

$\log 1.380 = 0.1399 = 2.1399 - 2$ (2 is added and subtracted so

$\log 24.2 = 1.3838 = 1.3838$ that the subtraction will

yield a positive decimal)

Difference =   0.7561 − 2   (1 is added and subtracted so

=   1.7561 − 3   that the division by 3 will

$\log N = \tfrac{1}{3}$ difference =   0.5854 − 1   yield an integer for the

$N = 0.3849$ characteristic)

$N = 0.385$ approximated to
three significant digits

EXAMPLE 5.7.3 Compute $\dfrac{\sqrt[4]{0.566}}{68.5}$.

Solution   Let $N = \dfrac{\sqrt[4]{0.566}}{68.5}$.

$$\log N = \tfrac{1}{4} \log 0.566 - \log 68.5$$

$$\log 0.566 = 0.7528 - 1$$

$$+3 \qquad -3 \qquad \text{(Adding and subtracting 3}$$
$$\log 0.566 = 3.7528 - 4 \qquad \begin{array}{l}\text{to yield } -4, \text{ exactly divis-}\\ \text{ible by 4)}\end{array}$$

$$\tfrac{1}{4} \log 0.566 = 0.9382 - 1$$

$$\log 68.5 = 1.8357$$

Now 1 must be added and subtracted from $\tfrac{1}{4} \log 0.566$ so the subsequent subtraction will yield a logarithm having a positive mantissa.

$$\tfrac{1}{4} \log 0.566 = 0.9382 - 1$$

$$+1 \qquad -1$$

$$\tfrac{1}{4} \log 0.566 = 1.9382 - 2$$

$$\log 68.5 = 1.8357$$

$$\overline{\rule{6cm}{0.4pt}}$$

$$\log N = \text{difference} = 0.1025 - 2$$

$$N = 1.266(10^{-2}) = 0.01266$$

$$N = 0.0127 \text{ approximated to three}$$
$$\text{significant digits}$$

EXERCISES 5.7 A

Compute Exercises 1–20 by using logarithms.

1. $(32.6)(0.854)$

2. $\dfrac{642}{79.1}$

3. $(2.43)^4$

4. $(0.589)^{-3}$

5. $\sqrt{608}$

6. $\sqrt{0.922}$

7. $\sqrt[3]{47.5}$

8. $\sqrt[3]{0.137}$

9. $\dfrac{(4.92)(0.0658)}{786}$

10. $(0.00729)(2.06)^8$

11. $\sqrt[4]{\dfrac{936}{288}}$ 12. $\dfrac{\sqrt[3]{46.6}}{98.8}$

13. $\sqrt[5]{38.3}$ 14. $\sqrt[5]{0.0383}$

15. $0.956\sqrt{0.645}$ 16. $\dfrac{450\sqrt[3]{75.2}}{82.4}$

17. $3.142(24.65)^2$ 18. $\sqrt{(6574)(21.25)}$

19. $\sqrt[3]{\dfrac{(72.8)^2}{-2610}}$ 20. $\dfrac{(5.96)(7.82)(0.937)}{0.0568}$

EXERCISES 5.7 B

Compute Exercises 1–20 by using logarithms.

1. $(0.547)(0.0829)$ 2. $\dfrac{5.96}{0.95}$

3. $(64.3)^{-2}$ 4. $(0.412)^5$

5. $\sqrt{3.142}$ 6. $\sqrt[3]{2.718}$

7. $\sqrt{0.0894}$ 8. $\sqrt[3]{0.909}$

9. $3.14(0.956)^3$ 10. $\dfrac{\sqrt{0.843}}{54.9}$

11. $\sqrt[3]{\dfrac{72.4}{8.79}}$ 12. $\sqrt{\dfrac{12}{32.2}}$

13. $\sqrt[6]{100}$ 14. $\sqrt[6]{0.7846}$

15. $(1.025)^{20}(3995)$ 16. $\sqrt[3]{\dfrac{(4.32)^2}{(0.402)^4}}$

17. $(-35.7)(-2.93)(-0.648)$ 18. $\dfrac{\sqrt{0.0534}}{(0.841)(0.346)}$

19. $\dfrac{(865)(12.2)(0.0555)}{0.988}$ 20. $\dfrac{1}{(0.246)(0.575)(0.888)}$

5.8   LOGARITHMIC AND EXPONENTIAL EQUATIONS

An *exponential equation* is an equation in which the variable appears in an exponent. Exponential equations can be solved by using the property that the exponential function is one to one for $b > 0$ and $b \neq 1$—that is,

$$b^x = b^y \text{ if and only if } x = y$$

and $$\log_b x = \log_b y \text{ if and only if } x = y$$

EXAMPLE 5.8.1   Solve $2^x = 0.125$.

Solution $0.125 = (0.5)^3 = (\frac{1}{2})^3 = 2^{-3}$

Thus $2^x = 2^{-3}$ and $x = -3$.

EXAMPLE 5.8.2   Solve $2^x = 3$.

Solution   Since 3 is not a rational power of 2, the logarithms of the two numbers are equated:

$$\log_{10} 2^x = \log_{10} 3$$

$$x \log 2 = \log 3$$

$$x = \frac{\log 3}{\log 2} \quad \text{exact answer}$$

$$x = \frac{0.4771}{0.3010} = 1.585 \quad \text{approx.}$$

A *logarithmic equation* is an equation that contains logarithms. It is also solved by using the one-to-one property of the logarithmic and exponential functions.

EXAMPLE 5.8.3   Solve $2 \log x - \log (x + 3) + \log 5 = \log 4$.

Solution $(\log x^2 + \log 5) - \log (x + 3) = \log 4$

$$\log \frac{5x^2}{x + 3} = \log 4$$

$$\frac{5x^2}{x + 3} = 4$$

$$5x^2 - 4x - 12 = 0$$

$$(x - 2)(5x + 6) = 0$$

$$x = 2 \quad \text{or} \quad x = -\tfrac{6}{5}$$

Check: $x = 2$: $2 \log 2 - \log 5 + \log 5 = 2 \log 2 = \log 2^2 = \log 4$. Thus 2 is a solution.

$x = -\tfrac{6}{5}$: $2 \log x = 2 \log (-\tfrac{6}{5})$, which is undefined. Thus $-\tfrac{6}{5}$ is *not* a solution.

Since the domain of the logarithmic function is the set of positive real numbers, the original equation requires the restriction that $x > 0$ and $x + 3 > 0$. It is important, then, to check all proposed solutions of a logarithmic equation.

EXAMPLE 5.8.4   Solve $x = \log_5 12$.

Solution   Rewriting this equation in exponential form,

$$5^x = 12$$

$$x \log_{10} 5 = \log_{10} 12$$

$$x = \frac{\log 12}{\log 5} \quad \text{exact answer}$$

$$x = \frac{1.0792}{0.6990} = 1.54 \text{ correct to three}$$
$$\text{significant digits}$$

EXERCISES 5.8 A

Solve for x in Exercises 1–15, using tables only when necessary.

1. $3^{x-2} = 243$

2. $4^x = \frac{1}{32}$

3. $5^x = 2$

4. $4^{1-x} = 64$

5. $6^{x-3} = 4.5$

6. $(2.5)^{3x} = 6.25$

7. $(3^x)^2 = \frac{1}{27}$

8. $7^{3x+5} = 1$

9. $2 \log(x - 1) - \log(x - 4) = 4 \log 2$

10. $2 \log(x + 5) - 3 \log 2 = 0$

11. $\log 64 - 3 \log x = 3 + \log 8$

12. $3^{2\log_3 5} = x$

13. $10^{-2\log x} = 3$

14. $x - 2 = \log_3 10$

15. $x = \log_2 7$

EXERCISES 5.8 B

Solve for x in Exercises 1–15.

1. $9^{x+2} = 1$

2. $(125)^{2-x} = 0.04$

3. $2^{x-1} = \sqrt[3]{16}$

4. $8^{2x+1} = 15$

5. $2(5^{3x}) = 5$

6. $(\frac{1}{3})^x = 81$

7. $2y = y(1.05)^x$

8. $(\frac{1}{8})^{x-1} = 4^{1-2x}$

9. $2 \log(x + 3) - \log(x + 7) + 1 = \log 2$

10. $10^{-3\log 2} = x$

11. $5^{2\log_5 x} = 9$

12. $x = \log_2 26$

13. $\log(x + 4) + \log(x - 4) - \log 9 = 0$

14. $0.5 \log 3 + \log x = 2 \log 5 - \log 2$

15. $3 \log x + \log 3 - 2 \log 5 = -3 + \log 15$

5.9 NATURAL LOGARITHMS

Common logarithms, or logarithms to the base 10 (also called Briggsian after their inventor, Henry Briggs), are the most convenient for numerical computation. However, for more advanced mathematics, especially that involving calculus, *natural logarithms* (also called Naperian after their originator, John Napier), or logarithms to the base e, are more appropriate. The number e is irrational, with an approximate value of 2.71828.

The logarithm of a number to the base b can be obtained from the logarithm of the number to the base a by using the following theorem:

THE CONVERSION OF BASES THEOREM

$$\log_b x = \frac{\log_a x}{\log_a b}$$

Proof: $\log_b x = \log_b a^{\log_a x}$ since $x = a^{\log_a x}$

$$= (\log_a x) \log_b a \qquad \text{(By logarithmic operations theorem C)}$$

Now, letting $x = b$,

$$\log_b b = (\log_a b) \log_b a$$

$$1 = (\log_a b) \log_b a$$

Thus

$$\log_b a = \frac{1}{\log_a b}$$

and

$$\log_b x = \frac{\log_a x}{\log_a b}$$

For the special case that $b = e$ and $a = 10$, then

$$\log_e x = \frac{\log_{10} x}{\log_{10} 2.71828} = \frac{\log_{10} x}{0.4343} = 2.303 \log_{10} x$$

Therefore, the natural logarithm of a number is obtained by multiplying its common logarithm by 2.303.

EXAMPLE 5.9.1   Find $\log_e 54.6$.

Solution $\log_e 54.6 = 2.303 \log_{10} 54.6$

$$= 2.303(1.7372)$$

$$= 4.001 \text{ to four significant digits}$$

EXAMPLE 5.9.2   Find $\log_e 0.932$.

Solution $\log_e 0.932 = 2.303 \log_{10} 0.932$

$$= 2.303(0.9694 - 1)$$

$$= 2.303(-0.0306)$$

$$= -0.0705 \text{ to three significant digits}$$

EXAMPLE 5.9.3   Find $\log_5 28$.

Solution   Using $\log_b x = \dfrac{\log_a x}{\log_a b}$,

$$\log_5 28 = \frac{\log_{10} 28}{\log_{10} 5}$$

$$= \frac{1.4472}{0.6990}$$

$$= 2.070 \text{ to four significant digits}$$

EXERCISES 5.9 A

Find each logarithm in Exercises 1–10.

1. $\log_e 10$

2. $\log_e 100$

3. $\log_e 1.56$

4. $\log_e 3.142$

5. $\log_e 25.2$

6. $\log_{10} \dfrac{1}{e^2}$

7. $\log_e 0.0825$

8. $\log_5 39$

9. $\log_{\sqrt{3}} 0.316$

10. $\log_{12} 7800$

EXERCISES 5.9 B

Find each logarithm in Exercises 1–10.

1. $\log_e 0.2$ **2.** $\log_e 97$

3. $\log_e \sqrt{e^3}$ **4.** $\log_e 0.012$

5. $\log_e 2(3)(4)(5)(6)$ **6.** $\log_7 0.83$

7. $\log_{21} 236$ **8.** $\log_8 4.3$

9. $\log_{1/2} 750$ **10.** $\log_e 10^{\log_{10} 6}$

5.10   APPLICATIONS

Some of the many practical applications of logarithms are presented in this section.

5.10.1   pH

In chemistry, the pH of a solution is a measure of the acidity or alkalinity of the solution. If $[H^+]$ designates the hydrogen ion concentration measured in moles per liter, then the pH is defined as follows:

DEFINITION

$$pH = -\log_{10} [H^+]$$

EXAMPLE 5.10.1   Find the pH of a solution whose hydrogen ion concentration is 2.0×10^{-4}.

    Solution $pH = -\log_{10} (2.0 \times 10^{-4})$

$$= -(0.3010 - 4) = -0.3010 + 4 = 3.6990$$

$= 3.7$ correct to the nearest tenth (pH values are usually stated correct to the nearest tenth)

EXAMPLE 5.10.2   Find the hydrogen ion concentration of a solution whose pH is 4.7.

    Solution $4.7 = -\log_{10} [H^+]$

$$\log_{10} [H^+] = -4.7 = -5 + 0.3$$

$[H^+] = 2.0 \times 10^{-5}$ approximated (usually calculated to two significant digits)

5.10.2   Exponential Growth and Decay

Radioactive decay, population growth, and other phenomena which change at a rate directly proportional to the amount present at a given time are described by the exponential function $y = ae^{bt}$, where the natural logarithm base $e = 2.71828 \ldots$, the variable t measures the time, and a and b are constants. If $t = 0$, then $y = ae^{bt(0)} = ae^0 = a(1) = a$. Thus a measures y at the time when the measurements are begun—that is, when $t = 0$. For this reason, y_0 is often used instead of the letter a, or $y = y_0e^{bt}$.

EXAMPLE 5.10.3   The number of bacteria in a certain culture is determined from the relation $y = 500\, e^{0.38t}$, where t is measured in hours. How many bacteria are present at the end of 15 hours?

Solution $y = 500\, e^{0.38(15)} = 500\, e^{5.70}$

$$\log y = \log 500 + 5.70 \log e$$

$$= 2.6990 + 5.7(0.4343) = 5.1745$$

$$y = 149{,}400 \text{ approx.}$$

EXAMPLE 5.10.4   What is the half-life of a radioactive substance that decays according to the rule $y = y_0e^{-0.035t}$ if t is measured in years?

Solution   To find the half-life means to find t when $y = \tfrac{1}{2}y_0$:

$$\tfrac{1}{2}y_0 = y_0e^{-0.035t} \text{ or } \tfrac{1}{2} = e^{-0.035t}$$

$$\log \tfrac{1}{2} = \log e^{-0.035t} = -\,0.035t \log e$$

$$-0.035t \log e = -\log 2$$

$$t = \frac{\log 2}{0.035 \log e} = \frac{0.3010}{0.035(.4343)}$$

$$= 20 \text{ years approx.}$$

5.10.3   Compound Interest

If P designates the sum of money invested, i is the compound interest rate per conversion period, and n is the number of conversion periods, then the amount A that the money is worth at the end of n conversion periods is given by the formula

$$A = P(1 + i)^n$$

Four-place logarithm tables are not accurate enough for most applications of this formula, although they will yield useful approximations. For results accurate to the nearest cent, tables of values of $(1 + i)^n$ are available, and they provide the most practical means of computation.

EXAMPLE 5.10.5   What will be the amount of $100 invested at 5 percent converted quarterly at the end of 5 years?

Solution

$$A = P(1 + i)^n, \text{ where } P = 100, i = \frac{0.05}{4} = 0.0125, n = 5 \times 4 = 20$$

$$A = 100(1.0125)^{20}$$

$$\log A = \log 100 + 20 \log 1.0125$$

$$= 2 + 0.1080$$

$$A = 10^2 \times 1.28 = \$128 \text{ approx.}$$

Using the tables for $(1 + i)^n$ with $i = 1\frac{1}{4}$ percent and $n = 20$,

$$A = 100(1.2820) = \$128.20 \quad \text{correct to nearest cent}$$

EXAMPLE 5.10.6   How much money must be invested at 7 percent converted monthly to amount to $5000 at the end of 3 years?

Solution $A = P(1 + i)^n$ where $A = 5000, i = \dfrac{0.07}{12}, n = 12(3) = 36$

$$P = A(1 + i)^{-n} = 5000\left(1 + \frac{0.07}{12}\right)^{-36}$$

$$\log P = \log 5000 - 36 \log 1.0058$$

$$= 3.6990 - 36(0.0025) = 3.6990 - 0.0900$$

$$= 3.6090$$

$$P = \$4065 \text{ approx.}$$

Using the tables for $(1 + i)^{-n}$ with $i = \frac{7}{12}$ percent and $n = 36$,

$$P = 5000(0.811079) = \$4055.40 \quad \text{correct to the}$$
$$\text{nearest cent}$$

EXAMPLE 5.10.7   How long will it take a sum of money to double itself if it is invested at 6 percent converted monthly?

Solution $A = P(1 + i)^n$, $A = 2P$, $i = \dfrac{0.06}{12} = 0.005$

$$2P = P(1.005)^n \quad \text{or} \quad 2 = (1.005)^n$$

$$n \log 1.005 = \log 2$$

$$n = \frac{\log 2}{\log 1.005}$$

$$= \frac{0.3010}{0.00215}$$

$$= 140 \text{ months or } 11 \text{ years } 8 \text{ months approx.}$$

Using the tables, $n = 139$ months approx.

5.10.4   Electrical Circuits

In an RL electrical circuit consisting of a battery of E volts, a resistance of R ohms, an inductance of L henrys, and a switch S (Figure 5.10.1), then after the switch is closed, the current i in amperes is given by the formula

$$i = \frac{E}{R}(1 - e^{-Rt/L})$$

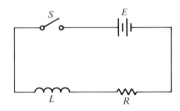

FIGURE 5.10.1

where the time t is measured in seconds. E/R is called the steady-state current. If a steady current is flowing in the circuit and the battery is short-circuited, then the decay of the current is given by the equation

$$i = \frac{E}{R}e^{-Rt/L}$$

EXAMPLE 5.10.8 If $E = 14$ volts, $R = 5$ ohms, $L = 2.5$ henrys, find the current in an RL circuit 0.25 second after the switch is closed.

Solution $i = \frac{14}{5}(1 - e^{-5(0.25)/2.5}) = 2.8(1 - e^{-0.5}) = 2.8 - 2.8e^{-0.5}$

$\log 2.8e^{-0.5} = \log 2.8 - 0.5 \log 2.718$

$\qquad\qquad\qquad = 0.4472 - 0.2172 = 0.2300$

$2.8e^{-0.5} = 1.70$ amperes   and $i = 2.8 - 1.7 = 1.1$ amperes

EXAMPLE 5.10.9 If the battery of an RL circuit is short-circuited after a steady current is flowing, find E/R for $i = 20$ amperes, $R = 6$ ohms, $L = 2$ henrys, $t = 0.4$ second.

Solution $20 = \dfrac{E}{R} e^{-6(0.4)/2} = \dfrac{E}{R} e^{-1.2}$

$\dfrac{E}{R} = 20e^{1.2}$ and $\log \dfrac{E}{R} = \log 20 + 1.2 \log e = 1.8222$

$\dfrac{E}{R} = 66.4$ amperes

EXERCISES 5.10 A

1. Find the pH of a solution whose hydrogen ion concentration is given below:

   a. 4.2×10^{-3} b. 2.9×10^{-7} c. 7.5×10^{-5}

2. Find the hydrogen ion concentration for each of the solutions whose pH is given below:

   a. pH $= 7.0$ b. pH $= 2.6$ c. pH $= 8.7$

3. a. Determine the half-life of the decay of polonium 218 into lead 214 if its decay relation is $y = y_0 10^{-0.0987t}$, where t is measured in minutes.

   b. If 1000 grams are present at $t = 0$, how much polonium is left at the end of $\frac{1}{2}$ hour?

4. If the half-life of radium 226 is 1590 years, find the decay constant b so that $y = y_0 e^{-bt}$.

5. Bacteria in a certain culture grew according to the rule $y = 100e^{bt}$. If y increased to 354 at the end of 1 hour, how many bacteria were in the culture at the end of $1\frac{1}{2}$ hours?

6. Find approximately to the nearest dollar the amount of money accumulated at the end of 5 years if $4000 is invested at 6 percent converted yearly.

7. Approximately how long will it take a sum of money to double itself at 8 percent converted quarterly?

8. An *RL* circuit has an electromotive force *E* of 16.5 volts, a resistance *R* of 0.1 ohm, and an inductance *L* of 0.05 henry.

a. Find the current i $\frac{1}{4}$ second after the switch is closed.

b. If the circuit is short-circuited after a steady current is flowing, find the current at the end of 1 second.

9. The build-up of the current i in a circuit with a condenser of capacitance *C* farads, a resistance of *R* ohms, and a source of electromotive force of *E* volts is expressed by the equation

$$i = \frac{E}{R} e^{-t/CR}$$

Find *E* if $i = 0.03$ ampere, $R = 250$ ohms, $C = 2 \times 10^{-6}$ farad, and $t = 0.001$ second.

10. The depreciation of a certain machine is calculated by the constant percentage method, $S = C(1 - r)^n$, where *C* is the original cost, *S* is the scrap value after a useful life of *n* years, and *r* is the constant percentage of depreciation. Find the scrap value of a machine costing $20,000 if its useful life is 10 years and it depreciates by a constant percentage of 18 percent.

11. If a projectile fired vertically upward with initial velocity v_0 feet per second is subject only to the force of gravity *g* and to the air resistance kv, then the maximum height *H* it reaches is given by

$$H = \frac{1}{k}\left(v_0 - \frac{g}{k}\log_e\frac{g + v_0 k}{g}\right)$$

Find *H* if $v_0 = 680$ feet/second, $k = 2.40$, and $g = 32$. ($e = 2.71828\ldots$)

12. By Newton's law of cooling, the temperature *T* of a body at time *t* is expressed by the equation

$$t = \frac{1}{k}\log_e\frac{T_0 - a}{T - a}$$

where *a* is the constant temperature of the surrounding air, T_0 is the temperature of the body at time $t = 0$, and *k* is the constant rate of cooling.

a. If $a = 20$ degrees centigrade, find *k* if it takes 25 minutes for a substance to cool from 100 degrees centigrade to 50 degrees centigrade.

b. How long will it take the substance to cool to 25 degrees centigrade? to 20 degrees centigrade?

EXERCISES 5.10 B

1. Find the pH of a solution whose hydrogen ion concentration is given below:

   a. 5.1×10^{-6} b. 3.4×10^{-2} c. 6.3×10^{-8}

2. Find the hydrogen ion concentration for each of the solutions whose pH is given below:

   a. pH = 5.5   b. pH = 1.4   c. pH = 3.9

3. Determine the half-life of the decay of thorium 234 into protactinium 234 if its decay equation is $y = y_0 10^{-0.0123t}$, where t is measured in years. If 100 grams are present at $t = 0$, how much thorium is left at the end of 2 years?

4. If the half-life of uranium 238 is 4.5×10^9 years, find the decay constant b so that $y = y_0 e^{-bt}$.

5. In a unimolecular chemical reaction, the number of molecules N present at time t is given by the equation $N = N_0 e^{-kt}$, where k is a positive constant called the *reaction rate*.

   a. If the concentration of a dilute sugar solution is given by the equation $y = 0.01 e^{-kt}$, find the reaction rate k if the concentration is $\frac{1}{250}$ gram per cubic centimeter at the end of 5 hours.

   b. Find the concentration at the end of 10 hours.

6. What was the approximate value of the original investment that amounted to $3000 at the end of 6 years if the interest rate was 6 percent converted semiannually?

7. At what rate of compound interest will a sum of money double itself at the end of 10 years?

8. Find the constant percentage of depreciation of a car initially costing $5000 if its scrap value is $250 at the end of a useful life of 15 years (see Exercise 10, 5.10A).

9. The gain G in decibels (a measure of loudness) due to an increase in power from w_1 watts to w_2 watts may be calculated from the equation $G = 10 \log_{10} w_2/w_1$.

   a. What was the decibel gain of a television station that increased its power from 35 kilowatts to 150 kilowatts?

   b. If power (watts) = (volts)^2/ohms, what is the decibel loss across a long radio transmission line if an input of 20 volts across a resistance of 600 ohms is reduced to a voltage of 4 volts measured across a resistance of 5000 ohms?

10. The population of a certain city increased from 100,000 in 1940 to 125,000 in 1960. Assume $P = P_0 10^{kt}$.

    a. What will be the population in 1980?

    b. In what year will the population be 200,000? (Let $t = 0$ for 1940.)

11. If

$$\frac{1}{\sqrt{5}}\left(\frac{1 + \sqrt{5}}{2}\right)^n$$

where n is a positive integer, is approximated to the nearest integer, then this nearest integer is the nth Fibonacci number. A Fibonacci number is the sum of the two preceding Fibonacci numbers, where 1, 1, 2, 3, 5, 8, 13, and 21 are the first eight Fibonacci numbers.

a. Find the fifteenth Fibonacci number in two different ways.

b. Find the twentieth Fibonacci number.

(One of the interesting features of Fibonacci numbers is the role they play in describing various kinds of biological growth.)

SUMMARY

☐ A **function** is a set of ordered pairs with exactly one second component assigned to each first component.

☐ A **one-to-one** function is a function with each element in its range paired with exactly one element in its domain.

☐ The **inverse** r^{-1} **of a relation** r is the set of ordered pairs obtained by interchanging the components of r.

☐ **Theorem.** If a function f is one to one, then its inverse is a function, designated by f^{-1}.

☐ **Definition.** The **logarithmic function with base** b = $\{(x, y) \mid x = b^y$ where $b > 0$, $b \neq 1$, and x is a positive real number$\}$.

☐ **Convention.** If $f(x) = b^x$, then $f^{-1}(x) = \log_b x$.

☐ **Theorem.** $b^{\log_b x} = x$ and $\log_b b^x = x$ for $b > 0$, $b \neq 1$; $\log_b b = 1$ and $\log_b 1 = 0$.

☐ **Theorem.** $b^x = b^y$ if and only if $x = y$. $\log_b x = \log_b y$ if and only if $x = y$.

☐ **The logarithmic operations theorems.**

A. $\log_b xy = \log_b x + \log_b y$

B. $\log_b \dfrac{x}{y} = \log_b x - \log_b y$

C. $\log_b x^y = y \log_b x$

D. $\log_b \sqrt[n]{x} = \dfrac{1}{n} \log_b x$

☐ A **common logarithm** has base 10. A **natural logarithm** has base $e = 2.71828\ldots$; $\log_e x = 2.303 \log_{10} x$.

☐ **The common logarithm theorem.** If $0 \le x < 10$, then $\log_{10} x(10^k) = k + \log_{10} x$, where the integer k is called the **characteristic** and the positive decimal $\log_{10} x$ is called the **mantissa**.

REVIEW EXERCISES

Evaluate each of Exercises 1–6, if possible.

1. $\log_{10} 10$ **2.** $\log_{10} 1$

3. $\log_{10} 0$ **4.** $\log_{10} (-10)$

5. $\log_{10} 10^{-3}$ **6.** $10^{\log_{10} 2}$

Evaluate each of Exercises 7–12, given that $\log_b 2 = 0.4$.

7. $\log_b 8$ **8.** $\log_b \frac{1}{2}$

9. $\log_b \sqrt{2}$ **10.** $\log_b \sqrt[3]{0.25}$

11. $\log_b (\log_b \sqrt{b})$ **12.** $\dfrac{\log_b 2}{\log_b b^{\sqrt{3}}}$

Complete Exercises 13–20. Do not use tables.

13. If $N = 5^{-4.6}$, then $\log_5 N =$ **14.** If $\log_5 N = -\frac{1}{2}$, then $N =$

15. If $5^x = 3$, then $x =$

16. If $\log_{10} N = -1 + 0.37$, then $\log_{10} \sqrt[3]{N} =$

17. If $\log_{10} x = 0.4$ and $\log_{10} y = -2 + 0.7$, then $\log_{10} \dfrac{x}{y} =$

18. If $\log x = \log 15 + 3 \log 2 - \frac{1}{4} \log 81$, then $x =$

19. $(10^{\sqrt{5}+3} 10^{\sqrt{5}-3})^{-1/\sqrt{20}} =$ **20.** If $f(x) = 3^x$, then $f^{-1}(\frac{1}{9}) =$

21. a. Write the logarithmic equation that would be used to compute

$$\sqrt[3]{\frac{(42.21)(23.50)}{5650}}$$

     b. Using tables, compute the number designated in (a).

22. By using logarithms, find the radius r of a sphere whose volume V is 52 cubic inches. Use $V = \frac{4}{3}\pi r^3$ and $\pi = 3.142$.

23. Solve for x: $(16)^{x-3} = (0.25)^{4x+3}$. (Do not use tables.)

24. Solve for x correct to the nearest hundredth: $2^{3x} = 40$.

HISTORICAL NOTE

Logarithms were devised as a method for rapid and accurate computations related to problems in astronomy, engineering, surveying, navigation, and other areas. One of the earliest contributions was the work of the Swabian Michael Stifel. In his *Arithmetica Integra*, published in Nuremburg in 1544, he stated the four laws of exponents for rational numbers and referred to the "upper numbers" as "exponents." He also presented the following table which could be considered as a primitive table of logarithms.

which was essentially based on adding areas under the hyperbolic curve, $y = 1/x$.

About the same time but independently of Napier, the Swiss instrument maker Jobst Bürgi (1552–1632) calculated a logarithm table. His *Arithmetische und geometrische Progresstabuln* was published in Prague in 1620. This was actually a list of antilogarithms, with the logarithms written in red and the antilogarithms in black. Thus Bürgi referred to the logarithm as "Die Rothe Zahl" ("the red number"). Whereas Napier

x	-3	-2	-1	0	1	2	3	4	5	6
2^x	$\frac{1}{8}$	$\frac{1}{4}$	$\frac{1}{2}$	1	2	4	8	16	32	64

Since decimal fractions were not developed until after 1600, it would have been impossible for Stifel to make a table of logarithms suitable for practical calculations.

The Scotsman John Napier (1550–1617), who worked for about twenty years on the theory, is generally acknowledged as the founder of logarithms. Although he first used the term *artificial number*, he finally adopted the term *logarithm*, which in Greek literally means "ratio number."

Napier's work, *Mirifici logarithorum canonis descriptio* (*A Description of the Marvelous Law of Logarithms*), published in Edinburgh in 1614, contained the first real table of logarithms. The response was very enthusiastic. In a later work Napier also explained how to calculate a table of logarithms by a method

selected the base $b = 0.9999999 = 1 - 10^{-7}$ and used geometric methods, Bürgi selected $b = 1.0001$ and used algebraic methods. Both selected a base close to 1 so that the powers of b would be close together, and thus the antilogarithms could be listed in intervals of 0.0000001 or 0.0001, respectively.

Henry Briggs (1561–1631), a professor of geometry at Gresham College, London, as a result of a mutual agreement resulting from a conversation with Napier, developed a table of logarithms using the base 10. His *Arithmetica logarithmica*, published in 1624, contained fourteen-place tables for the integers from 1 to 20,000 and from 90,000 to 100,000. The interval from 20,000 to 90,000 was completed by the Dutch bookseller and publisher Adriaen Vlacq (1600–66), who published a complete

HISTORICAL NOTE

fourteen-place table of logarithms in 1628.

It was Briggs who introduced the word *mantissa* (originally meaning an "addition" and later an "appendix") and who also suggested the term *characteristic*. In the early tables, the characteristic was printed and was not dropped until about the middle of the eighteenth century.

Tables more accurate than those of Briggs and Vlacq were not calculated until the years 1924–1949, when twenty-place tables were made.

Later developments were concerned with establishing the theory of logarithms on a logically sound foundation. The Swiss mathe-

matician Leonhard Euler (1707–1783) made important contributions in his *Introductio*. It was Euler who introduced the letter *e* to represent the base of the Naperian or natural logarithms. Euler wrote, "Ponamus autem brevitatis gratia pro numero hoc 2.71828 . . . constanter litteram *e*" which is the Latin for "For the sake of brevity we shall let the literal constant *e* represent this number 2.71828 . . . ."

The theory of logarithms was finally established through the works of the French mathematician Augustin Cauchy (1789–1857), particularly in his *Cours d'Analyse*, published in Paris in 1821.

CHAPTER **6**

LINEAR AND QUADRATIC SYSTEMS

6.1 BASIC CONCEPTS

It was previously seen that the solution of an equation in two variables x and y is an ordered pair (a, b) such that the equation becomes true when x is replaced by a and y by b. It was also noted that in general there are infinitely many ordered pairs in the solution set of an equation in two variables.

However, when the graphs of *two* equations in two variables are graphed on the same set of coordinate axes, there are three possibilities:

1. The graphs may have no point in common (they do not intersect).
2. The graphs may have a finite number of points in common.
3. The graphs may have infinitely many points in common.

Algebraically, a set of two or more equations in two or more variables is called a system of equations, and the solution set of a system corresponds to the geometric points of intersection of the graphs of the equations.

DEFINITION

A **system of equations** is a set of equations in two or more variables.

A system of equations is also called a set of *simultaneous equations*.

DEFINITION

The **solution set of a system of equations in two variables** is the set of all ordered pairs that are common solutions to all the equations in the system.

Using the language of set theory, the solution set of a system of two equations in two variables is the *intersection* of the solution set of one of the equations with the solution set of the other.

EXAMPLE 6.1.1   Show that $(9, -2)$ is a solution of the system

$$2x + 3y = 12 \quad \text{and} \quad x - 5y = 19$$

Solution   For $x = 9$ and $y = -2$,

$$2x + 3y = 2(9) + 3(-2) = 18 - 6 = 12$$

$$x - 5y = 9 - 5(-2) = 9 + 10 = 19$$

Since $(9, -2)$ is a solution of each equation in the system, it is a solution of the system.

EXAMPLE 6.1.2   Show that $(-4, 3)$ is a solution of the system

$$x^2 + y^2 = 25 \quad \text{and} \quad y = 2x^2 + 5x - 9$$

Solution   For $x = -4$ and $y = 3$,

$$x^2 + y^2 = (-4)^2 + (3)^2 = 16 + 9 = 25$$

$$2x^2 + 5x - 9 = 2(-4)^2 + 5(-4) - 9$$

$$= 32 - 20 - 9 = 3 = y$$

The *real solutions* of a system of equations in two variables can be determined graphically by graphing the equations on the same set of axes. An ordered pair corresponding to a point of intersection is a solution of the system.

EXAMPLE 6.1.3   Graphically solve the system $x + y = 5$ and $y = x + 2$.

Solution

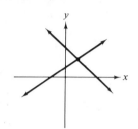

$x + y = 5$			$y = x + 2$	
x	y		x	y
0	5		0	2
5	0		1	3
1	4		2	4

FIGURE 6.1.1 $A = \{(x, y) \mid x + y = 5\}$
$B = \{(x, y) \mid y = x + 2\}$
$A \cap B = \{(\frac{3}{2}, \frac{7}{2})\}$

From the graph, the solution is read as $(\frac{3}{2}, \frac{7}{2})$.

Check: For $x = \frac{3}{2}$ and $y = \frac{7}{2}$,

$$x + y = 5 \text{ becomes } \frac{3}{2} + \frac{7}{2} = 5, \frac{10}{2} = 5, 5 = 5, \text{ true}$$

$$y = x + 2 \text{ becomes } \frac{7}{2} = \frac{3}{2} + 2, \frac{7}{2} = \frac{3}{2} + \frac{4}{2}, \frac{7}{2} = \frac{7}{2}, \text{ true}$$

EXAMPLE 6.1.4   Graphically solve the system $x + y = 5$ and $x + y = 2$.

Solution

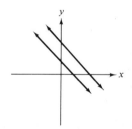

$x + y = 5$			$x + y = 2$	
x	y		x	y
0	5		0	2
5	0		2	0

FIGURE 6.1.2 $A = \{(x, y) \mid x + y = 5\}$
$B = \{(x, y) \mid x + y = 2\}$
$A \cap B = \varnothing$

The lines are parallel and the solution set is empty.

Check: Finding the slopes of each equation,

For $x + y = 5$; $y = -x + 5$, and the slope is -1.

For $x + y = 2$, $y = -x + 2$, and the slope is -1.

Nonvertical parallel lines always have the same slope, so the lines are parallel.

EXAMPLE 6.1.5 Graphically solve the system $x + y = 5$ and $3x + 3y = 15$.

Solution

$$x + y = 5 \qquad 3x + 3y = 15$$

x	y
0	5
5	0

x	y
0	5
5	0

FIGURE 6.1.3 $A = \{(x, y) \mid x + y = 5\}$
$B = \{(x, y) \mid 3x + 3y = 15\}$
$A \cap B = \{(x, y) \mid x + y = 5\}$

The lines coincide, and the solution set of the system is the solution set of either equation of the system.

EXAMPLE 6.1.6 Graphically solve the system $x^2 + y^2 = 25$ and $y = 3x - 5$.

Solution

1. The graph of $x^2 + y^2 = 25$ is a circle with center at the origin and with radius 5.
2. $y = 3x - 5$

x	y
0	−5
2	1

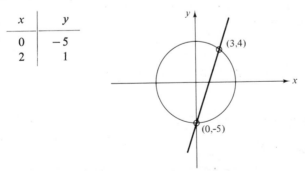

FIGURE 6.1.4 $\{(x, y) \mid x^2 + y^2 = 25\} \cap \{(x, y) \mid y = 3x - 5\} = \{(3, 4), (0, -5)\}$.

Referring to the graph, the solution set is $\{(3, 4), (0, -5)\}$.

Check:

For (3, 4),
$$x^2 + y^2 = 25 \qquad y = 3x - 5$$
$$3^2 + 4^2 = 25 \qquad 4 = 3(3) - 5$$
$$9 + 16 = 25 \qquad 4 = 9 - 5$$
$$25 = 25 \qquad 4 = 4$$

For (0, −5),
$$x^2 + y^2 = 25 \qquad y = 3x - 5$$
$$0 + (-5)^2 = 25 \qquad -5 = 0 - 5$$
$$25 = 25 \qquad -5 = -5$$

DEFINITION

A **linear system** of equations in the variables x and y is a system each of whose equations has the form

$$Ax + By + C = 0$$

where A, B, and C are constants with A and B not both 0.

The **general quadratic equation in two variables** x and y has the form
$$Ax^2 + Bxy + Cy^2 + Dx + Ey + F = 0$$
where A, B, C, D, E, and F are constants and at least one of the constants A, B, and C is different from 0.

The second degree terms are Ax^2, Bxy, and Cy^2.

The first degree, or linear, terms are Dx and Ey.

The constant term F is said to have degree 0 because F can be considered as the coefficient of x^0—that is, $Fx^0 = F \cdot 1 = F$.

Since the degree of a polynomial equation is defined as the greatest of the degrees of its terms, the general quadratic equation is an equation of second degree.

DEFINITION

A **quadratic system** of two equations in the variables x and y is a system one of whose equations is quadratic in x and y and the other equation is linear or quadratic.

The systems in Examples 6.1.3, 6.1.4, and 6.1.5 are linear systems. The system in Example 6.1.6 is a quadratic system.

Another example of a quadratic system is shown in Example 6.1.7.

EXAMPLE 6.1.7   Graphically solve the system $4x^2 + 4y^2 = 169$ and

$$y = \frac{x^2 - 26}{4}$$

Solution

1. $4x^2 + 4y^2 = 169$ is equivalent to $x^2 + y^2 = \frac{169}{4}$. This is a circle with center at the origin and radius $= \frac{13}{2}$.

2.

x	$y = \frac{1}{4}(x^2 - 26)$. This is a parabola.
0	-6.5
± 2	-5.5
± 4	-2.5
± 6	2.5
± 8	9.5

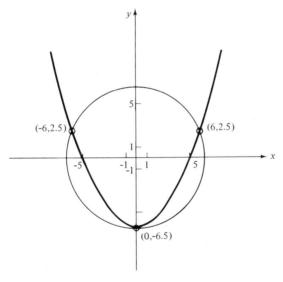

FIGURE 6.1.5 $\left\{ (x, y) \mid 4x^2 + 4y^2 = 169 \text{ and } y = \dfrac{x^2 - 26}{4} \right\}.$

From the graph, the solution set is read as the ordered pairs that correspond to the points of intersection of the circle and the parabola—that is, $\{(0, -6.5), (6, 2.5), (-6, 2.5)\}$.

EXERCISES 6.1 A

Show that each ordered pair stated in Exercises 1–5 is a solution of the given system of equations.

1. $x + y = 8$ and $x - y = 2$; $(5, 3)$
2. $3x + y = 1$ and $y = 2x + 6$; $(-1, 4)$
3. $x^2 + y^2 = 9$ and $y = x - 3$; $(0, -3)$, $(3, 0)$
4. $y = x^2 - 4x - 1$ and $4x + y = 3$; $(2, -5)$, $(-2, 11)$
5. $x^2 + y^2 = 25$ and $y = 5 - x^2$; $(0, 5)$, $(3, -4)$, $(-3, -4)$

Graphically solve each of the systems in Exercises 6–20.

6. $x + y = 7$ and $x - y = 1$ 7. $4x - 3y = 0$ and $2x - 3y = 6$
8. $2x + y = 7$ and $y = 5 - 2x$ 9. $2x + y = 2$ and $4x - y = 13$
10. $y = 3x - 2$ and $6x - 2y = 4$ 11. $y = x^2 - 4$ and $y = 4x - 7$
12. $y = 3x - x^2$ and $x + y + 5 = 0$
13. $x^2 + y^2 = 16$ and $x - y + 4 = 0$
14. $x^2 + y^2 = 25$ and $x - y = 1$
15. $y = x^2 - 4x + 7$ and $y = 2x + 2$
16. $y = 8 + 2x - x^2$ and $3x + y = 8$
17. $x^2 + y^2 = 25$ and $y = x^2 - 5$
18. $4x^2 + 4y^2 = 25$ and $2y = 13 - 4x^2$
19. $x^2 + y^2 + 2x - 6y = 15$ and $x + y = 1$
20. $x^2 + y^2 = 36$ and $x + y = 10$

EXERCISES 6.1 B

Show that each ordered pair stated in Exercises 1–5 is a solution of the given system of equations.

1. $2x + y = 10$ and $y = x - 8$; $(6, -2)$
2. $5x - 2y = 2$ and $3x + y + 12 = 0$; $(-2, -6)$
3. $x^2 + y^2 = 25$ and $x = 3y - 15$; $(-3, 4)$, $(0, 5)$
4. $y = 6x - x^2$ and $5x - y = 0$; $(0, 0)$, $(1, 5)$
5. $x^2 + y^2 = 25$ and $y = x^2 - 13$; $(4, 3)$, $(-4, 3)$, $(3, -4)$, $(-3, -4)$

Graphically solve each of the systems in Exercises 6–20.

6. $x + 2y = 4$ and $2x - y = 3$ 7. $x - y = 0$ and $3x + 2y = 10$
8. $4x + y = 4$ and $4(x - 1) + y = 0$

9. $x - 2y = 5$ and $2y - x = 1$

10. $4x = y - 11$ and $2x + 3y = 12$ **11.** $y = 9 - x^2$ and $y = x + 3$

12. $y = x^2 - 4x$ and $2x + y = 3$ **13.** $x^2 + y^2 = 100$ and $x + 3y = 10$

14. $x^2 + y^2 = 36$ and $x - y = 6$

15. $y = x^2 - 6x + 5$ and $y = 2x - 7$

16. $y = 1 - x - x^2$ and $2x + y + 1 = 0$

17. $x^2 + y^2 = 25$ and $y = 13 - x^2$ **18.** $x^2 + y^2 = 100$ and $y = \dfrac{x^2}{4} - 10$

19. $y = x^2 - 2x - 1$ and $x + y = 5$

20. $x^2 + y^2 - x + 4y = 38$ and $12x + 5y + 4 = 0$

6.2 PARABOLAS AND SYSTEMS

It has been seen that the graph of $y = ax^2 + bx + c$ is a parabola that opens upward if $a > 0$ and downward if $a < 0$. The graph of $x = ay^2 + by + c$ is also a parabola that opens toward the right if $a > 0$ and toward the left if $a < 0$.

A parabola can easily be graphed by locating the vertex and several points on either side of the vertex.

It was shown earlier that if (r, t) and (s, t) are two points on the graph of $y = ax^2 + bx + c$, then the x-coordinate of the vertex is $(r + s)/2$. For the special case that $t = 0$, then $y = 0$, and r and s are the roots of $ax^2 + bx + c = 0$. However, the sum of the roots can be obtained from the quadratic equation—that is, $r + s = -b/a$. Therefore, the x-coordinate of the vertex is $(r + s)/2 = -b/2a$. This same value can be obtained upon completing the square with respect to the x-terms:

$$\left(x + \frac{b}{2a}\right)^2 = \frac{1}{a}\left(y + \frac{b^2 - 4ac}{4a}\right)$$

Equating each side to 0 and solving each resulting equation yields the **coordinates of the vertex**:

$$x = \frac{-b}{2a} \quad \text{and} \quad y = \frac{4ac - b^2}{4a}$$

Similarly for $x = ay^2 + by + c$,

$$\left(y + \frac{b}{2a}\right)^2 = \frac{1}{a}\left(x + \frac{b^2 - 4ac}{4a}\right)$$

and the **coordinates of the vertex** are

$$x = \frac{4ac - b^2}{4a} \quad \text{and} \quad y = \frac{-b}{2a}$$

EXAMPLE 6.2.1   Graph $x = 2y^2 - 12y + 11$.

Solution

1. First find the vertex by completing the square with respect to the y-terms:

$$x - 11 = 2y^2 - 12y$$

$$x - 11 + 18 = 2(y^2 - 6y + 9)$$

$$x + 7 = 2(y - 3)^2$$

Equating the left side and the right side to 0,

$$x + 7 = 0 \quad \text{and} \quad 2(y - 3)^2 = 0$$

$$x = -7 \quad \text{and} \quad y = 3$$

the vertex is at $(-7, 3)$.

2. Make a table of values for points near the vertex, plot the points, and join them with a smooth curve. Select y so that $0 \le y \le 6$.

x	y	$x = 2y^2 - 12y + 11$
11	0	
1	1	$x = 2(1) - 12(1) + 11 = 1$
-5	2	$x = 2(4) - 12(2) + 11 = -5$
-7	3	$x = 2(9) - 12(3) + 11 = -7$
-5	4	$x = 2(16) - 12(4) + 11 = -5$
1	5	$x = 2(25) - 12(5) + 11 = 1$
11	6	$x = 2(36) - 12(6) + 11 = 11$

3.

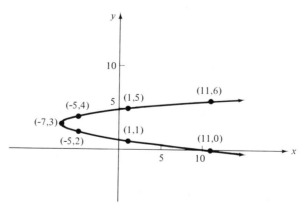

FIGURE 6.2.1   Graph of $x = 2y^2 - 12y + 11$.

EXAMPLE 6.2.2   Graph $x = 4y - y^2$.

Solution

1. Completing the square to find the vertex,

$$x = -y^2 + 4y$$
$$x - 4 = -(y^2 - 4y + 4)$$
$$x - 4 = -(y - 2)^2$$
$$x - 4 = 0 \quad \text{and} \quad y - 2 = 0$$
$$x = 4 \quad \text{and} \quad y = 2$$

The vertex is at $(4, 2)$.

2. Selecting values of y near $y = 2$ and making a table,

x	y	$x = 4y - y^2$
-5	-1	$x = 4(-1) - 1 = -5$
0	0	
3	1	$x = 4 - 1 = 3$
4	2	$x = 4(2) - 4 = 4$
3	3	$x = 4(3) - 9 = 3$
0	4	$x = 4(4) - 16 = 0$
-5	5	$x = 4(5) - 25 = -5$

3.

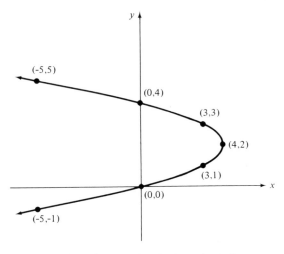

FIGURE 6.2.2   Graph of $x = 4y - y^2$.

In general, the vertex of the graph of $y = ax^2 + bx + c$ is the point $(-b/2a, \ c - (b^2/4a))$, and the vertex of $x = ay^2 + by + c$ is the point $(c - (b^2/4a), \ -b/2a)$.

Whether the parabola opens upward, downward, to the right, or to the left can be determined by learning the general graphs shown in Figure 6.2.3.

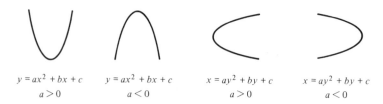

| $y = ax^2 + bx + c$ | $y = ax^2 + bx + c$ | $x = ay^2 + by + c$ | $x = ay^2 + by + c$ |
| $a > 0$ | $a < 0$ | $a > 0$ | $a < 0$ |

FIGURE 6.2.3

A quadratic system involving one or more parabolas is solved graphically by graphing the equations on the same set of coordinate axes. The real solutions of the system are the ordered pairs corresponding to the points of intersection of the graphs.

EXAMPLE 6.2.3   Graphically solve the system

$$y^2 = x + 3 \quad \text{and} \quad 6y = -x^2 + x + 12$$

Solution

1. For $y^2 = x + 3$, the vertex is $(-3, 0)$.

For $6y = -x^2 + x + 12$,

$$6y - 12 = -(x^2 - x)$$

$$6y - 12 - \tfrac{1}{4} = -(x^2 - x + \tfrac{1}{4})$$

$$6(y - \tfrac{49}{24}) = -(x - \tfrac{1}{2})^2$$

The vertex is $(\tfrac{1}{2}, 2\tfrac{1}{24})$.

2.

x	y	$x = y^2 - 3$	x	y	$y = \tfrac{1}{6}(-x^2 + x + 12)$
6	-3		-5	-3	
1	-2		-3	0	
-2	-1		-2	1	
-3	0		0	2	
-2	1		$\tfrac{1}{2}$	$2\tfrac{1}{4}$	
1	2		1	2	
6	3		3	1	
			4	0	
			6	-3	

3.

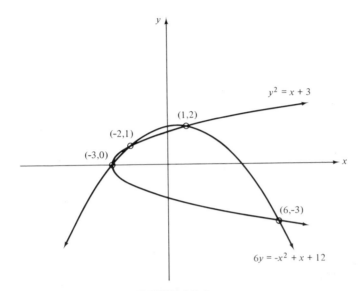

FIGURE 6.2.4

4. From the graph, the real solutions are read as $(-3, 0)$, $(-2, 1)$, $(1, 2)$, $(6, -3)$.

EXERCISES 6.2 A

Solve each of the systems in Exercises 1–10 graphically.

1. $x = y^2 - 4$ and $x = 4y - 7$ 2. $x = 5y - y^2$ and $y = x$
3. $y^2 = x + 6$ and $x = y - 4$ 4. $x - y^2 = 0$ and $y = 2x$
5. $y^2 + 8y + 21 = x$ and $x = 6$
6. $y^2 + x - 8y + 12 = 0$ and $x = 2y - 7$
7. $y = x^2$ and $x = -y^2$
8. $y = x^2 + 4x + 4$ and $y^2 = 8x + 16$
9. $y^2 = 64 - 8x$ and $2y = x^2 - 10x + 16$
10. $4x = y^2$ and $3y = 7x - x^2$

EXERCISES 6.2 B

Solve each of the systems in Exercises 1–10 graphically.

1. $x = 9 - y^2$ and $x + 2y = 1$ 2. $x = y^2 + 3y$ and $y = x - 3$
3. $x + y^2 = 0$ and $y = x$

4. $2y^2 + 2y + x + 4 = 0$ and $x + 2y + 6 = 0$
5. $2x = y^2 + 6y + 17$ and $x = 6$
6. $y^2 + 2x + 6y + 7 = 0$ and $y = 2x + 15$
7. $x^2 + y = 0$ and $x - y^2 = 0$ 8. $x^2 = 21 - 3y$ and $y^2 = 3x + 7$
9. $x^2 - 4x + 9y = 18$ and $x = y^2$
10. $y^2 + 4x + 2y = 7$ and $x^2 + 3x - 3y = 13$

6.3 ELLIPSES AND HYPERBOLAS

There are two other types of quadratic graphs that are of interest because of their wide area of applications. These are the ellipse and the hyperbola.

DEFINITION

A standard ellipse is the graph of

$$\frac{x^2}{a^2} + \frac{y^2}{b^2} = 1 \quad (a > 0 \text{ and } b > 0)$$

The graphs of the two types of standard ellipses are shown in Figure 6.3.1.

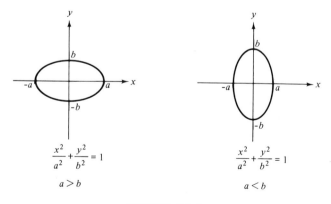

FIGURE 6.3.1

The domain and range of an ellipse can be readily observed from the graphs shown in Figure 6.3.1.

The domain of a standard ellipse $= \{x \mid -a \leq x \leq a\}$.

The range of a standard ellipse $= \{y \mid -b \leq y \leq b\}$.

EXAMPLE 6.3.1   Graph $4x^2 + 9y^2 = 36$.

Solution

1. Identify the curve as an ellipse. Dividing each side by 36,

$$\frac{x^2}{9} + \frac{y^2}{4} = 1$$

   This is a standard ellipse with $a = 3$ and $b = 2$.
2. Since $a = 3$, the domain is $\{x \mid -3 \leq x \leq 3\}$.
   Since $b = 2$, the range is $\{y \mid -2 \leq y \leq 2\}$.
   Make a table of values for $-3 \leq x \leq 3$, approximating irrational numbers to the nearest tenth.
   Solving for y,

$$y = \pm\tfrac{2}{3}\sqrt{9 - x^2}$$

x	-3	-2	-1	0	1	2	3
y	0	±1.5	±1.9	±2	±1.9	±1.5	0

3. Graph: Plot the tabulated points. Join with a smooth curve.

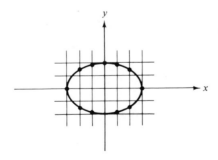

FIGURE 6.3.2

EXAMPLE 6.3.2   Graph $4x^2 + y^2 - 100 = 0$.

Solution

1. Standard form:

$$\frac{x^2}{25} + \frac{y^2}{100} = 1$$

Thus the curve is an ellipse with $a = 5$ and $b = 10$.
Domain: $-5 \le x \le 5$.
Range: $-10 \le y \le 10$.

2. Table: $y = \pm 2\sqrt{25 - x^2}$

x	0	± 1	± 2	± 3	± 4	± 5
y	± 10	± 9.8	± 9.2	± 8	± 6	0

3. Graph: Plot the tabulated points. Join with a smooth curve.

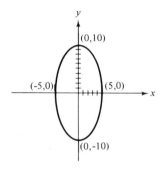

FIGURE 6.3.3

DEFINITION

A standard hyperbola is the graph of

1. $\dfrac{x^2}{a^2} - \dfrac{y^2}{b^2} = 1$ $(a > 0, b > 0)$

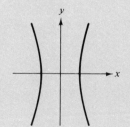

FIGURE 6.3.4

or

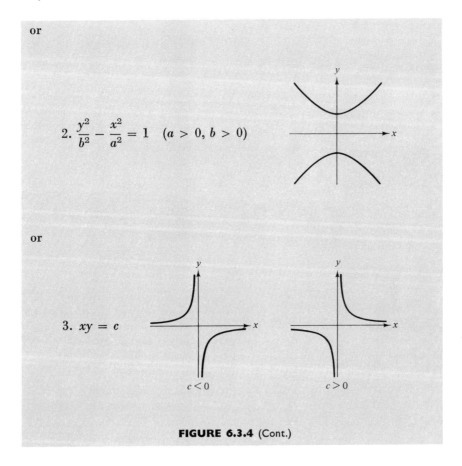

2. $\dfrac{y^2}{b^2} - \dfrac{x^2}{a^2} = 1 \quad (a > 0, b > 0)$

or

3. $xy = c$

$c < 0$ $c > 0$

FIGURE 6.3.4 (Cont.)

Once a hyperbola is recognized from the equation of a standard hyperbola, the graph of the hyperbola is obtained by finding the domain for the hyperbola and plotting points using values for x in the domain.

Domain and Range of Standard Hyperbolas

Equation (1), solving for y, yields $y = \pm(b/a)\sqrt{x^2 - a^2}$; if y is to be a real number, $x^2 \geq a^2$, $|x| \geq a$ implies that $x \geq a$ or $x \leq -a$.

Equation (2), solving for y, yields $y = \pm(b/a)\sqrt{a^2 + x^2}$. In this equation y is a real number for all real values of x. Therefore, the domain of (2) is all x.

Equation (3) is defined for all $x \neq 0$.

The range of standard hyperbolas is determined in a similar way.

Therefore,

For $\dfrac{x^2}{a^2} - \dfrac{y^2}{b^2} = 1$, the domain is $\{x \mid x \leq -a \text{ or } x \geq a\}$ and

the range is $\{y \mid y \text{ is a real number}\}$

For $\dfrac{y^2}{b^2} - \dfrac{x^2}{a^2} = 1$, the domain is $\{x \mid x \text{ is a real number}\}$ and

the range is $\{y \mid y \leq -b \text{ or } y \geq b\}$

For $xy = c$, the domain is $\{x \mid x \text{ is a real number and } x \neq 0\}$ and
the range is $\{y \mid y \text{ is a real number and } y \neq 0\}$

Asymptotes

The lines obtained by replacing the constant term by 0 are called the **asymptotes** of the hyperbola. These lines are useful as an aid in graphing.

For example, if $(x^2/a^2) - (y^2/b^2) = 1$, then the asymptotes of this hyperbola are the lines $y = bx/a$ and $y = -bx/a$, obtained from $(x^2/a^2) - (y^2/b^2) = 0$.

From the figures accompanying the definition, it is seen that a hyperbola has two branches. These branches of the hyperbola lie entirely within two of the regions formed by the asymptotes as shown in Figures 6.3.5 and 6.3.6.

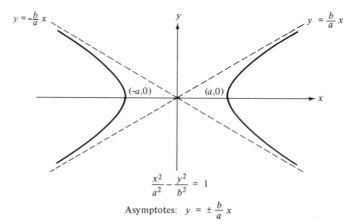

$$\frac{x^2}{a^2} - \frac{y^2}{b^2} = 1$$

Asymptotes: $y = \pm\dfrac{b}{a}x$

FIGURE 6.3.5

For larger and larger values of x, the distance between a branch of the hyperbola and its asymptote becomes smaller and smaller. Thus in graphing the hyperbola, the asymptotes can be drawn to serve as useful guidelines.

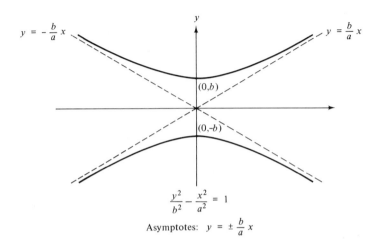

$y = -\dfrac{b}{a}x$

$y = \dfrac{b}{a}x$

$(0,b)$

$(0,-b)$

$$\frac{y^2}{b^2} - \frac{x^2}{a^2} = 1$$

Asymptotes: $y = \pm\dfrac{b}{a}x$

FIGURE 6.3.6

EXAMPLE 6.3.3   Graph $25x^2 - 16y^2 = 400$.

Solution

1. Identify the curve by obtaining the equation of a standard hyperbola. Dividing both sides by 400,

$$\frac{x^2}{16} - \frac{y^2}{25} = 1 \quad (a = 4, \, b = 5)$$

2. Domain: $y = \pm\frac{5}{4}\sqrt{x^2 - 16}$. Thus $|x| \geq 4$; that is, $x \leq -4$ or $x \geq 4$.

3. Asymptotes: $y = \frac{5}{4}x$ and $y = -\frac{5}{4}x$.

4. Table:

x	± 4	± 5	± 6	± 8
y	0	± 3.8	± 5.6	± 8.7

5. Graph: Draw the asymptotes. Plot the tabulated points. Join the points with a smooth curve.

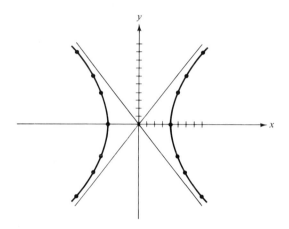

FIGURE 6.3.7

EXAMPLE 6.3.4   Graph $25x^2 - 16y^2 + 400 = 0$.

Solution

1. $\dfrac{y^2}{25} - \dfrac{x^2}{16} = 1$. Thus the curve is a hyperbola with $b = 5$ and $a = 4$.

2. Domain: all x since $y = \pm\frac{5}{4}\sqrt{x^2 + 16}$.

   Range: $|y| \geq 5$ since $x = \pm\frac{4}{5}\sqrt{y^2 - 25}$.

3. Asymptotes: $y = \frac{5}{4}x$ and $y = -\frac{5}{4}x$.

4. Table:

x	0	± 2	± 3	± 4	± 9.6
y	± 5	± 5.6	± 6.3	± 7.1	± 12

5. Graph: Draw the asymptotes. Plot the tabulated points. Join the points with a smooth curve.

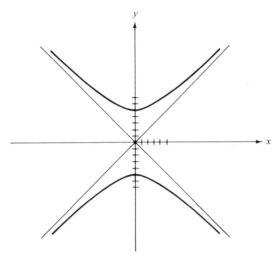

FIGURE 6.3.8

EXERCISES 6.3 A

Discuss and graph each of the equations in Exercises 1–10.

1. $x^2 + 25y^2 = 100$ 2. $16x^2 = 1600 - 25y^2$
3. $5x^2 + 2y^2 - 50 = 0$ 4. $x^2 - y^2 = 9$
5. $2x^2 - 2y^2 + 18 = 0$ 6. $9x^2 = 16y^2 - 576$
7. $xy = 12$ 8. $xy = -6$
9. $y^2 - 4x^2 = 64$ 10. $y^2 - 4x^2 = 0$

EXERCISES 6.3 B

Discuss and graph each of the equations in Exercises 1–10.

1. $9x^2 + y^2 = 225$ 2. $y^2 = 52 - 4x^2$
3. $x^2 + 4y^2 - 169 = 0$ 4. $4x^2 = y^2 + 16$
5. $4x^2 = y^2 - 16$ 6. $xy = 8$
7. $xy = -4$ 8. $4x^2 - 25y^2 = 0$
9. $x^2 + y^2 = 0$ 10. $4x^2 - 9y^2 + 36 = 0$

HISTORICAL NOTE

The Greek mathematician Apollonius (c. 262 B.C.–200 B.C.) is noted for his work *Conic Sections*, in which he made a very thorough study of these curves using geometrical methods. The names *parabola*, *ellipse*, and *hyperbola* were established by Apollonius.

The French mathematician René Descartes (1596–1650) is credited with establishing a correspondence between algebraic equations and geometric curves by means of a rectangular (Cartesian) coordinate system. His famous work on this subject, *La Géométrie*, was published in 1637.

6.4 GRAPHIC SOLUTION OF QUADRATIC SYSTEMS

6.4.1 Conic Sections (Optional)

The graphs of quadratic equations in two variables are called *conic sections*, or *conics*, because the methods of analytic geometry can be used to show that the graph of each quadratic equation can be obtained as the geometric intersection of a plane and a right circular cone having two nappes, illustrated in Figure 6.4.1. The cone has a line of symmetry called its *axis*. A line that lies on the surface of the cone is called a *generator*, and all generators of a cone pass through a special point called the *vertex*.

If the cutting plane does not pass through the vertex, then the curve of intersection is a circle, an ellipse, a parabola, or a hyperbola.

If the cutting plane does pass through the vertex, then the intersection will be a point, a line, or two intersecting lines. These possibilities are called the *degenerate cases*.

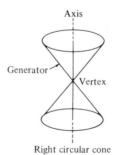

Right circular cone

FIGURE 6.4.1

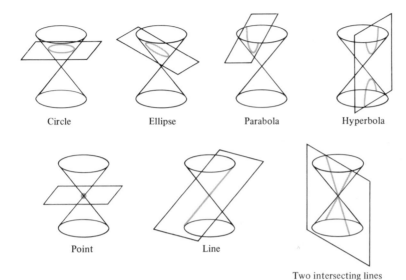

Circle   Ellipse   Parabola   Hyperbola

Point   Line

Two intersecting lines

FIGURE 6.4.2

The different kinds of intersections are illustrated in Figure 6.4.2. The definition of a conic (curve) is as follows:

DEFINITION

A *conic* is the set of points, in a plane, such that the distance from a point on the conic to a fixed point, called the focus, is equal to a constant e multiplied by the distance from the point on the conic to a fixed line, called the directrix. The constant e is called the *eccentricity* of the conic.

If $0 < e < 1$, then the conic is called an *ellipse*.
If $e = 1$, then the conic is called a *parabola*.
If $e > 1$, then the conic is called a *hyperbola*.

The circle is considered as a limiting position of a conic, with $e = 0$.

By using the methods of analytic geometry, the following theorem can be established:

THEOREM

The graph of $Ax^2 + Bxy + Cy^2 + Dx + Ey + F = 0$ is a conic, a limiting form of a conic, or a degenerate conic.

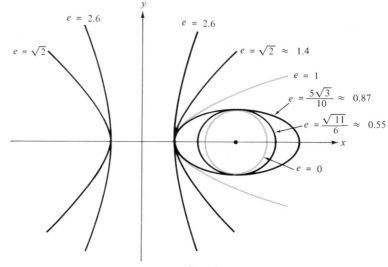

1. $e = 0$, circle: $(x - 15)^2 + y^2 = 25$

2. $e \approx 0.55$, ellipse: $25(x - 15)^2 + 36y^2 = 900$

3. $e \approx 0.87$, ellipse: $25(x - 15)^2 + 100y^2 = 2500$

4. $e = 1$, parabola: $y^2 = 5x - 25$

5. $e = \sqrt{2} \approx 1.4$, hyperbola: $x^2 - y^2 = 25$

6. $e = 2.6$, hyperbola: $144x^2 - 25y^2 = 3600$

FIGURE 6.4.3

Eccentricity

The eccentricity e may be considered as a measure of the deviation of the shape of a curve from the shape of a circle. The effect of the eccentricity on the shape of the curve is illustrated in Figure 6.4.3.

6.4.2   Solution Techniques

The solution set of a system of equations is the intersection of the solution sets of the equations in the system. Recall that a real solution (a, b) of a system of two equations in two variables x and y is also the name of a geometric point of intersection of the graphs of the two equations. Thus the graphs can be used to determine or to approximate the *real* solutions of the quadratic system.

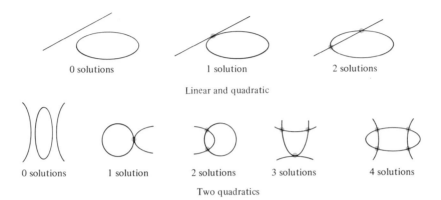

0 solutions　　　　　　　1 solution　　　　　　　2 solutions

Linear and quadratic

0 solutions　　　1 solution　　　2 solutions　　　3 solutions　　　4 solutions

Two quadratics

FIGURE 6.4.4

If the quadratic system consists of one linear equation and one quadratic equation, then the system will have 0, 1, or 2 real solutions.

If the quadratic system consists of two quadratic equations, then the system will have 0, 1, 2, 3, or 4 real solutions.

The different possibilities are illustrated in Figure 6.4.4.

The ideas developed in the preceding sections are summarized below as an aid to graphing a quadratic equation. It is useful to develop the ability to identify the shape of the curve by recognizing the pattern of its equation, as in the following summary (Figures 6.4.5 A–E).

1. *Parabola.*

   $y = Ax^2 + Bx + C$

   $x = Ay^2 + By + C$

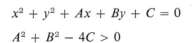

$A > 0$　　　　$A < 0$

$A > 0$　　　　$A < 0$

FIGURE 6.4.5A

2. *Circle.*

   $x^2 + y^2 + Ax + By + C = 0$

   $A^2 + B^2 - 4C > 0$

FIGURE 6.4.5B

3. *Ellipse.*

$Ax^2 + By^2 = C$

$A > 0, B > 0, C > 0$

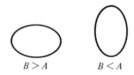

$B > A$ $\qquad$ $B < A$

FIGURE 6.4.5C

4. *Hyperbola.*

$Ax^2 - By^2 = C$
$A > 0, B > 0, C \neq 0$
$xy = C$

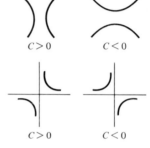

$C > 0$ $\qquad$ $C < 0$

$C > 0$ $\qquad$ $C < 0$

FIGURE 6.4.5D

5. *Degenerate cases.*

a. No point.
$(x - a)^2 + (y - b)^2 = -k^2$ $\qquad$ $x^2 + y^2 = -4$

b. One point.
$(x - a)^2 + (y - b)^2 = 0$ $\qquad$ $x^2 + y^2 = 0$

c. One line
$(ax + by + c)^2 = 0$ $\qquad$ $x^2 - 2xy + y^2 = 0$

d. Two lines.
$(ax + by + c)(dx + ey + f) = 0$ $\quad$ $x^2 - y^2 = 0$

FIGURE 6.4.5E

The graphic solution of a quadratic system yields the real solutions *only*. Moreover, due to the limitations imposed by eyesight and drawing techniques, these real solutions, as a general rule, can only be approximated. Thus it is essential to check a proposed solution obtained by the method of graphs.

EXAMPLE 6.4.1   Estimate graphically the real solutions of the quadratic system $x^2 + 16y^2 = 169$, $x^2 - y^2 = 16$.

Solution

1. Graph both equations on the same axes (Figure 6.4.6).

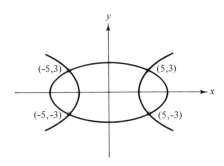

FIGURE 6.4.6

2. Determine the coordinates of the points of intersection: $(5, 3)$, $(-5, 3)$, $(-5, -3)$, $(5, -3)$.
3. Check each of these proposed solutions. The check is illustrated for $(5, 3)$. The other checks are made similarly.

     If $x = 5$ and $y = 3$, then

$$x^2 + 16y^2 = 5^2 + 16(3^2) = 25 + 144 = 169$$

and

$$x^2 - y^2 = 5^2 - 3^2 = 25 - 9 = 16$$

Thus $(5, 3)$ is a solution.
4. List the solution set: $\{(5, 3), (-5, 3), (-5, -3), (5, -3)\}$.

EXERCISES 6.4 A

Estimate graphically, to the nearest tenth, the real solutions of each of the quadratic systems in Exercises 1–6. Check each solution.

1. $x^2 - 2x - y = 5$
   $x + y = 1$

2. $9x^2 + y^2 = 9$
   $y = 4x + 5$

3. $x^2 - y^2 = 16$
   $x - 3y = 0$

4. $y^2 - 2x^2 = 4$
   $y^2 = 9x$

5. $xy = 6$
   $9x^2 + 16y^2 = 144$

6. $4x + y^2 = 36$
   $x^2 + y^2 = 81$

EXERCISES 6.4 B

Estimate graphically, to the nearest tenth, the real solutions of each of the quadratic systems in Exercises 1–6. Check each solution.

1. $x^2 + y^2 - 4x - 10y + 4 = 0$ **2.** $xy + 5 = 0$
    $3x + y = 6$ $4y - 5x = 20$
3. $y = 9 - x^2$ **4.** $x^2 + 3y^2 = 12$
    $y = x + 10$ $x^2 + y + 2 = 0$
5. $x^2 - y^2 = 5$ **6.** $x^2 - 2x - y + 1 = 0$
    $3y^2 - x^2 = 3$ $x^2 + y^2 - 2x - 4y + 1 = 0$

6.5 ALGEBRAIC SOLUTION: SUBSTITUTION METHOD

    If one of the equations in a quadratic system can be solved for one variable, expressed explicitly as a function of the other variable, then the expression obtained may be *substituted* for the variable in the other equation. This replacement yields an equation in one variable. The solution is completed by solving this equation for the one variable, and then by using the function to obtain the corresponding values of the other variable.

    EXAMPLE 6.5.1 (One linear and one quadratic) Solve and check $x^2 + 4y^2 = 13$, $x + y = 2$.

Solution

1. Solve the linear equation for y: $y = 2 - x$.
2. Replace y by $2 - x$ in the quadratic equation:

$$x^2 + 4(2 - x)^2 = 13$$

3. Solve for x:

$$x^2 + 4(4 - 4x + x^2) = 13$$
$$5x^2 - 16x + 3 = 0$$
$$(5x - 1)(x - 3) = 0$$
$$5x - 1 = 0 \quad \text{or} \quad x - 3 = 0$$
$$x = \tfrac{1}{5} \quad \text{or} \quad x = 3$$

4. Solve for y, using the linear equation $y = 2 - x$:

$$\text{If } x = \tfrac{1}{5}, \text{ then } y = 2 - \tfrac{1}{5} = \tfrac{9}{5}.$$

$$\text{If } x = 3, \text{ then } y = 2 - 3 = -1.$$

5. Check each solution in both equations:

$(\tfrac{1}{5}, \tfrac{9}{5}):$ $x^2 + 4y^2 = \tfrac{1}{25} + 4(\tfrac{81}{25}) = \tfrac{325}{25} = 13$

$$x + y = \tfrac{1}{5} + \tfrac{9}{5} = \tfrac{10}{5} = 2$$

$(3, -1):$ $x^2 + 4y^2 = 9 + 4(1) = 13$

$$x + y = 3 + (-1) = 2$$

6. State the solution set: $\{(\tfrac{1}{5}, \tfrac{9}{5}), (3, -1)\}.$

EXAMPLE 6.5.2 (Two quadratic equations) Solve and check

$$4x^2 + y^2 = 16, \; x^2 - y = 4$$

Solution

1. Solve $x^2 - y = 4$ for x^2: $x^2 = 4 + y$.
2. Replace x^2 by $4 + y$ in the other equation:

$$4(4 + y) + y^2 = 16$$

3. Solve for y: $y^2 + 4y = 0$

$$y(y + 4) = 0$$

Thus $y = 0$ or $y = -4$.

4. Solve for x in the first equation used ($x^2 = 4 + y$):

If $y = 0$, then $x^2 = 4 + 0 = 4$ and $x = 2$ or $x = -2$.
If $y = -4$, then $x^2 = 4 - 4 = 0$. Thus $x = 0$.

5. Check each solution in both equations:

$(0, -4)$ $4x^2 + y^2 = 4(0) + (-4)^2 = 0 + 16 = 16$

$$x^2 - y = 0^2 - (-4) = 4$$

$(2, 0)$ $4x^2 + y^2 = 4(2^2) + 0^2 = 4(4) = 16$

$$x^2 - y = 2^2 - 0 = 4$$

$(-2, 0)$ $4x^2 + y^2 = 4(-2)^2 + 0^2 = 4(4) = 16$

$$x^2 - y = (-2)^2 - 0 = 4$$

6. State the solution set: $\{(0, -4), (2, 0), (-2, 0)\}.$

EXAMPLE 6.5.3 (Two quadratic equations) Solve and check

$$4x^2 - y^2 = 5, \; xy = 3$$

Solution

1. If $xy = 3$, then $y = \dfrac{3}{x}$ (for $x \neq 0$).

2. $4x^2 - \left(\dfrac{3}{x}\right)^2 = 5$

   $4x^2 - \dfrac{9}{x^2} = 5$

   $4x^4 - 9 = 5x^2$

   $4x^4 - 5x^2 - 9 = 0$

   $(4x^2 - 9)(x^2 + 1) = 0$

   $4x^2 - 9 = 0$ or $x^2 + 1 = 0$

   $x^2 = \frac{9}{4}$ or $x^2 = -1$

   Thus $x = \frac{3}{2}$, $x = -\frac{3}{2}$, $x = i$, or $x = -i$.

3. Using $y = \dfrac{3}{x}$ to find y,

   If $x = \dfrac{3}{2}$, then $y = \dfrac{3}{3/2} = 2$.

   If $x = -\frac{3}{2}$, then $y = -2$.

   If $x = i$, then $y = \dfrac{3}{i} = \dfrac{3i}{ii} = -3i$.

   If $x = -i$, then $y = 3i$.

4. Check each solution in both equations. (The check is left to the student.)

5. The solution set is $\{(\frac{3}{2}, 2), (-\frac{3}{2}, -2), (i, -3i), (-i, 3i)\}$.

EXERCISES 6.5 A

Solve each of the systems in Exercises 1–10 by the substitution method. Check. Leave irrational answers in simplified radical form. Express imaginary answers in the $a + bi$ form.

1. $x^2 + y^2 - 2x = 9$
   $x + 2y + 4 = 0$

2. $x^2 + 2x + 4 = y$
   $x = y - 16$

3. $x^2 - y^2 - 16 = 0$
   $3x + y = 8$

4. $4x^2 + y^2 = 25$
   $2x - y = 1$

5. $xy + 12 = 0$
   $2x + 3y = 6$

6. $y^2 = 4x$
   $x^2 + 3 = y^2$

7. $4a^2 + b^2 = 16$
$b + 4 = 2a^2$

8. $c^2 - 5d^2 = 1$
$cd = 2$

9. $4x^2 + 4y^2 = 25$
$2y = 13 - 4x^2$

★ **10.** $4x = y^2$
$3y = 7x - x^2$

EXERCISES 6.5 B

Solve each of the systems in Exercises 1–10 by the substitution method. Check. Leave irrational answers in simplified radical form. Express imaginary answers in the a + bi form.

1. $3x - y = 6$
$y^2 + 3x - 8y = 0$

2. $xy = 4$
$x + 3y = 10$

3. $2x^2 + y^2 = 12$
$x - 2y = 2$

4. $2x^2 - 2xy + y^2 = 10$
$2x = y - 2$

5. $9x^2 - 4y^2 = 0$
$y^2 = 4x + 1$

6. $xy + y = 1$
$xy - x = 4$

7. $y^2 - x = 9$
$y^2 + x - 2y + 5 = 0$

8. $y^2 - x^2 = 16$
$2y + 2x^2 + 32 = 0$

9. $x^2 + y^2 = 100$
$x^2 = 4y + 40$

★ **10.** $x^2 - 4x + 9y = 18$
$x = y^2$

6.6   ALGEBRAIC SOLUTION: ADDITION METHOD

6.6.1   Eliminating One of the Variables

If one equation of a quadratic system is linear, then the solution is easily obtained by the substitution method. However, when both equations are quadratic, the substitution method generally yields a complicated equation involving radicals, and upon simplification, a fourth-degree equation, which is not always easy to solve.

For certain special cases it is possible to eliminate one of the variables by the *addition method*, which is similar to that used for linear systems. Each equation of the system is multiplied by an appropriate constant, and the resulting equations are added. The procedure is justified by reasoning similar to that used for the linear systems.

If (p, q) is a solution of both $A_1x^2 + B_1xy + C_1y^2 + D_1x + E_1y + F_1 = 0$ and $A_2x^2 + B_2xy + C_2y^2 + D_2x + E_2y + F_2 = 0$, then it is also a solution of any linear combination of the two quadratic equations—that is, (p, q) is a solution of

$$a(A_1x^2 + B_1xy + C_1y^2 + D_1x + E_1y + F_1)$$
$$+ b(A_2x^2 + B_2xy + C_2y^2 + D_2x + E_2y + F_2) = 0$$

since

$$a(A_1p^2 + B_1pq + C_1q^2 + D_1p + E_1q + F_1)$$
$$+ b(A_2p^2 + B_2pq + C_2q^2 + D_2p + E_2q + F_2) = 0$$

or

$$a(0) + b(0) = 0$$

EXAMPLE 6.6.1   Using the addition method, solve the system
$$3x^2 + 2y^2 = 23, \quad x^2 - 3y^2 = -7$$

Solution

1. Multiply first equation by 3: $9x^2 + 6y^2 = 69.$
   Multiply second equation by 2: $2x^2 - 6y^2 = -14.$
2. Add the new equations: $11x^2 \qquad = 55.$
3. Solve for x: $x^2 = 5$. Thus $x = \pm\sqrt{5}$.
4. Multiply first equation by 1: $3x^2 + 2y^2 = 23.$
   Multiply second equation by -3: $-3x^2 + 9y^2 = 21.$
5. Add the new equations: $11y^2 = 44.$
6. Solve for y: $y^2 = 4$. Thus $y = \pm 2$.
7. List the solution set: $\{(\sqrt{5}, 2), (\sqrt{5}, -2), (-\sqrt{5}, 2), (-\sqrt{5}, -2)\}$.
8. Check each solution in both equations of the original system. (The check is left for the student.)

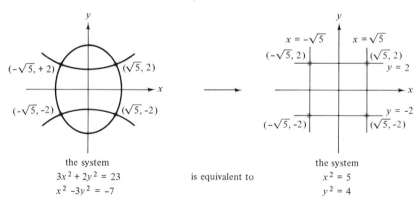

the system   the system
$3x^2 + 2y^2 = 23$ is equivalent to $x^2 = 5$
$x^2 - 3y^2 = -7$ $y^2 = 4$

FIGURE 6.6.1   Graphic illustration of the solution.

6.6.2   Eliminating the Second-Degree Terms

When it is not possible to eliminate one of the variables by adding multiples of the two equations, it may be possible to eliminate the second-degree terms and obtain a linear equation. Since each solution of the quadratic system must also be a solution of the linear equation, the linear equation may be solved simultaneously with either equation of the quadratic system.

EXAMPLE 6.6.2   Solve $x^2 + y^2 - 2x = 9$, $x^2 + y^2 - 4y = 1$.

Solution

1. Eliminate the second-degree terms:

$$(x^2 + y^2 - 2x = 9)(-1) \rightarrow -x^2 - y^2 + 2x = -9$$
$$(x^2 + y^2 - 4y = 1)(1) \quad \rightarrow \quad x^2 + y^2 - 4y = 1$$

Adding,
$$2x - 4y = -8$$
$$x - 2y = -4$$

2. Now solve the equivalent system:

$$x^2 + y^2 - 4y = 1$$
$$x - 2y = -4$$

Using the substitution method,

$$x = 2y - 4$$
$$(2y - 4)^2 + y^2 - 4y = 1$$
$$5y^2 - 20y + 15 = 0$$
$$y^2 - 4y + 3 = 0$$
$$(y - 3)(y - 1) = 0$$

Thus $y = 3$ or $y = 1$.

If $y = 3$, then $x = 2y - 4 = 2(3) - 4 = 2$.

If $y = 1$, then $x = 2y - 4 = 2(1) - 4 = -2$.

Thus the solution set is $\{(2, 3), (-2, 1)\}$.

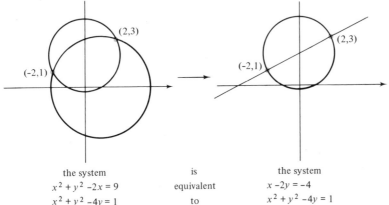

the system   is   the system
$x^2 + y^2 - 2x = 9$ equivalent $x - 2y = -4$
$x^2 + y^2 - 4y = 1$ to $x^2 + y^2 - 4y = 1$

FIGURE 6.6.2 Graphic illustration of the solution.

6.6.3   Eliminating the Constants

Sometimes the solution set of a quadratic system may be easily obtained by using the addition method to eliminate the constants.

EXAMPLE 6.6.3   Solve $x^2 + xy = 4$, $y^2 - xy = 6$.

Solution   Eliminating the constants,

$$(x^2 + xy = 4)(3) \rightarrow \quad 3x^2 + 3xy = \quad 12$$
$$(y^2 - xy = 6)(-2) \rightarrow -2y^2 + 2xy = -12$$

Adding, $3x^2 + 5xy - 2y^2 = 0$

Factoring, $(3x - y)(x + 2y) = 0$

$$3x - y = 0 \quad \text{or} \quad x + 2y = 0$$
$$y = 3x \quad \text{or} \quad x = -2y$$

Now either equation of the original system can be replaced by this set of two linear equations. Thus the original system is equivalent to the *union* of the two systems

$$\begin{array}{ccc} x^2 + xy = 4 & & x^2 + xy = 4 \\ & \text{and} & \\ y = 3x & & x = -2y \end{array}$$

Each of these systems can now be solved by the substitution method:

$$\begin{array}{cc} x^2 + x(3x) = 4 & (-2y)^2 + (-2y)y = 4 \\ 4x^2 = 4 & 2y^2 = 4 \\ x^2 = 1 & y^2 = 2 \\ x = \pm 1 & y = \pm\sqrt{2} \end{array}$$

If $x = 1$, then $y = 3x = 3$.   If $y = \sqrt{2}$, then $x = -2y = -2\sqrt{2}$.

If $x = -1$, then $y = -3$.   If $y = -\sqrt{2}$, then $x = 2\sqrt{2}$.

Thus the solution set is $\{(1, 3), (-1, -3), (-2\sqrt{2}, \sqrt{2}), (2\sqrt{2}, -\sqrt{2})\}$. Each of these solutions should be checked in both equations of the original system.

EXERCISES 6.6 A

Using the addition method, solve Exercises 1–4 by eliminating one of the variables. Check each solution.

1. $x^2 + y^2 = 20$
   $y^2 - x^2 = 12$

2. $x^2 + 3y^2 = 31$
   $x^2 = 8y^2 + 9$

3. $4x^2 + y^2 = 24$
   $2x^2 + y = 12$

4. $6x^2 - 5y^2 + 21 = 0$
   $7x^2 - 4y^2 + 8 = 0$

Solve Exercises 5–10 by eliminating either the second-degree terms or the constants. Check each solution.

5. $3xy + y^2 = 28$
$\quad 4x^2 + xy = 8$

6. $xy - 4x - 4y = 0$
$\quad xy = -8$

7. $x^2 + y^2 - 6x - 2y = 6$
$\quad x^2 + y^2 - 2x - 2y = 10$

8. $x^2 + 5xy + 2x = 8$
$\quad 2x^2 + 3xy + 4x = 9$

9. $xy = 48$
$\quad xy - 10x + 5y = 98$

10. $x^2 + xy = 24$
$\quad xy - y^2 = 4$

EXERCISES 6.6 B

Using the addition method, solve Exercises 1–4 by eliminating one of the variables. Check each solution.

1. $2x^2 + y^2 = 13$
$\quad 3x^2 + y^2 = 17$

2. $x^2 - y^2 + 7 = 0$
$\quad 3x^2 + 2y^2 = 24$

3. $x^2 - 5y^2 + 3 = 0$
$\quad 2x^2 + y^2 - 5 = 0$

4. $3x^2 - 2y^2 - 6 = 0$
$\quad 5x^2 - 3y^2 - 7 = 0$

Solve Exercises 5–10 by eliminating either the second-degree terms or the constants. Check each solution.

5. $y^2 + xy = 12$
$\quad 2x^2 - xy = 24$

6. $3xy + y^2 = 10$
$\quad xy = 3$

7. $x^2 - y^2 + 2x - y = -9$
$\quad x^2 - y^2 + 4x - 2y = -9$

8. $4x^2 - xy = 4$
$\quad 3y^2 + xy = 3$

9. $xy = 120$
$\quad xy + x - 4y = 124$

10. $xy - x = 3$
$\quad xy + y = 8$

6.7 ALGEBRAIC METHOD FOR SYMMETRIC EQUATIONS (Optional)

An equation in x and y is *symmetric in x and y* if the same equation is obtained when x and y are interchanged.

For example, $3x^2 - 2xy + 3y^2 + 5x + 5y = 9$ is symmetric in x and y.

If a quadratic system consists of two equations that are symmetric in x and y, then the substitutions $x = u + v$ and $y = u - v$ will yield a system which may be solvable by one of the previous methods.

EXAMPLE 6.7.1   Solve $x^2 + y^2 = 28$, $xy = x + y - 2$.

Solution

1. Let $x = u + v$ and let $y = u - v$.

$$x^2 + y^2 = 28 \qquad\qquad xy = x + y - 2$$

$$(u + v)^2 + (u - v)^2 = 28 \qquad (u + v)(u - v) = (u + v) + (u - v) - 2$$

$$u^2 + v^2 = 14 \qquad\qquad u^2 - v^2 = 2u - 2$$

2. Now solve the system $u^2 + v^2 = 14$, $u^2 - v^2 = 2u - 2$. Using the addition method,

$$2u^2 = 2u + 12$$

$$u^2 - u - 6 = 0$$

$$(u - 3)(u + 2) = 0$$

Thus $u = 3$ or $u = -2$.

If $u = 3$, then $v^2 = 14 - u^2 = 14 - 9 = 5$ and $v = \pm\sqrt{5}$.

If $u = -2$, then $v^2 = 14 - 4 = 10$ and $v = \pm\sqrt{10}$.

3. Now find the corresponding values for x and y.

u	v	$x = u + v$	$y = u - v$
3	$\sqrt{5}$	$3 + \sqrt{5}$	$3 - \sqrt{5}$
3	$-\sqrt{5}$	$3 - \sqrt{5}$	$3 + \sqrt{5}$
-2	$\sqrt{10}$	$-2 + \sqrt{10}$	$-2 - \sqrt{10}$
-2	$-\sqrt{10}$	$-2 - \sqrt{10}$	$-2 + \sqrt{10}$

Solution set: $\{(3 + \sqrt{5}, 3 - \sqrt{5}), (3 - \sqrt{5}, 3 + \sqrt{5}),$
$$(-2 + \sqrt{10}, -2 - \sqrt{10}), (-2 - \sqrt{10}, -2 + \sqrt{10})\}.$$

Each solution should be checked in each equation of the original system. The check for $(3 + \sqrt{5}, 3 - \sqrt{5})$ is illustrated below:

$$x^2 + y^2 = (3 + \sqrt{5})^2 + (3 - \sqrt{5})^2$$

$$= 9 + 6\sqrt{5} + 5 + 9 - 6\sqrt{5} + 5 = 28$$

$$xy - x - y = (3 + \sqrt{5})(3 - \sqrt{5}) - (3 + \sqrt{5}) - (3 - \sqrt{5})$$

$$= 9 - 5 - 3 - \sqrt{5} - 3 + \sqrt{5} = 4 - 6 = -2$$

EXERCISES 6.7 A

Solve and check each of the systems in Exercises 1–5.

1. $2xy + 13x + 13y + 56 = 0$
    $3x^2 + 2xy + 3y^2 - 8x - 8y = 88$

2. $x^2 + y^2 = 4$
    $xy = 10 - x - y$

3. $x^2 + y^2 = x + y$
    $xy = 3$

4. $3x^2 - 4xy + 3y^2 = 22$
    $x^2 + y^2 - 3xy + 3x + 3y = 3$

5. $(x + y)^2 + 9(x + y) = 10$
    $(x - y)^2 = 20$

EXERCISES 6.7 B

Solve and check each of the systems in Exercises 1–5.

1. $x^2 + y^2 = x + y + 2$
    $xy + 3x + 3y = 2$

2. $x^2 + y^2 - x - y = 2$
    $xy + x + y = 2$

3. $x^2 + y^2 + 3x + 3y + 2 = 0$
    $2xy + 5x + 5y + 10 = 0$

4. $x^2 + y^2 = 18$
    $xy = 4x + 4y + 1$

5. $x^2 - 2xy + y^2 = 36$
    $3(x + y)^2 - 22(x + y) = 16$

6.8 APPLICATIONS

There are many practical applications of quadratic relations and quadratic systems. Arches and dams are often parabolic, elliptical, or hyperbolic. Orbits of satellites and planets are elliptical, while the orbits of comets and meteors may be parabolic, elliptical, or hyperbolic. A focus of a conic may be used as an optical focus or as a sonar focus to concentrate light waves or sound waves.

A parabolic surface is obtained by rotating a parabola about its axis. Radar screens, searchlights, and headlights have parabolic reflectors. Other types of parabolic reflectors are used to concentrate rays of heat or light from a distant source; for example, rays of heat from the sun have been focused in this manner and used to drive machinery.

An elliptical surface is obtained by rotating an ellipse about its major axis. If a lamp is placed at one focus, rays of light striking the surface will be reflected toward the other focus. "Whispering galleries" are rooms with domed ceilings in the form of a semiellipse. A person at one focus of the ellipse can be heard distinctly by a person at the other focus, although he cannot be heard by people in between.

A hyperbolic surface is obtained by rotating one branch of a hyperbola about its transverse axis. If rays of light are directed toward one focus, the rays will be reflected toward the other focus. A Cassegrainian telescope is made by using both a hyperbolic mirror and a parabolic mirror.

Hyperbolas are used to determine positions of airplanes and to determine enemy positions by using properties of sound and the property that a hyperbola is the set of points such that the difference of the distances from any point on the hyperbola to the foci is a constant. The system of air navigation using this property is called *loran*.

The equation of the path of a projectile shot from the origin with initial velocity v feet per second directed along the line $y = mx$ is the parabola (Figure 6.8.1)

$$y = \frac{-16(m^2 + 1)}{v^2} x^2 + mx$$

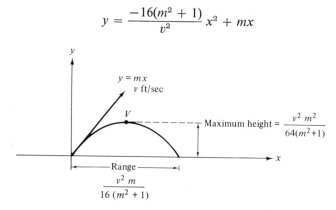

FIGURE 6.8.1

EXERCISES 6.8 A

1. A roadway 400 feet long is supported by a parabolic cable. The cable is 100 feet above the roadway at the ends and 4 feet above at the center. Find the lengths of the vertical supporting cables at 50-foot intervals along the roadway (Figure 6.8.2). (*Hint:* Let $y = kx^2 + 4$. Find k and then find y for $x = 50, 100, 150$.)

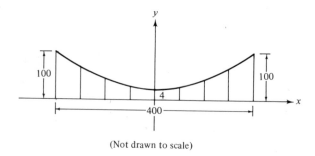

(Not drawn to scale)

FIGURE 6.8.2

2. The equation of the orbit of an artificial satellite of the earth is given by the ellipse

$$\frac{x^2}{(56.5)^2} + \frac{y^2}{(56)^2} = 1$$

where x and y are measured in hundreds of miles. The center of the earth is at $(7.5, 0)$.

a. Sketch the elliptical orbit, including a circular cross section of the earth, assuming that the radius of the earth is 40 hundred miles (4,000 miles).

b. From the graph, determine the minimum altitude and the maximum altitude of the satellite (the least distance and the greatest distance of the satellite from the surface of the earth).

3. Positions of airplanes and enemy positions can be determined by using the properties of sound and the mathematical curve, the hyperbola. An explosion is heard at A, with coordinates $(0, 0)$, $1\frac{1}{2}$ seconds after it is heard at B, with coordinates $(10, 0)$, and $2\frac{1}{2}$ seconds after it is heard at C, with coordinates $(0, 26)$. It can be shown that the place where the explosion occurred is (approximately) a point of intersection of two of the four hyperbolic asymptotes whose equations are

$$3y = 2(x - 5)$$
$$3y = -2(x - 5)$$
$$5x = 12(y - 13)$$
$$5x = -12(y - 13)$$

a. Graph these four equations on the same set of axes.

b. Find where the explosion occurred.

4. A rectangular piece of metal has an area of 200 square inches. A square whose side is 3 inches is cut from each of the four corners. The flaps are then turned up to form an open box whose volume is 168 cubic inches. Find the original dimensions of the sheet of metal.

5. A rectangular picture has a frame of uniform width and its area is equal to the area of the picture. If the width of the frame is 3 inches and the area of the picture is 240 square inches, find the dimensions of the picture.

6. A buyer bought some dresses for $480. If the price per dress had been $5 less, he could have bought 16 more dresses. How many dresses did he buy and what was the price per dress?

7. The cost of a club party was $72. If 20 more members had attended, the cost per person would have been 18 cents less. How many persons attended and what was the cost per person?

8. The equation of the path of a projectile shot from the origin with initial velocity 520 feet per second directed along the line $2y = 3x$ is

$$y = \frac{7800x - x^2}{5200}$$

a. Graph this relation.
b. Find the maximum height of the projectile.
c. Find the range of the projectile.
d. For what values of x is the height $y = 1625$ feet?

9. When two resistances x and y are connected in series, they have a joint resistance of 25 ohms—that is,

$$x + y = 25$$

When the same two resistances are connected in parallel, they have a joint resistance of 6 ohms—that is,

$$\frac{1}{x} + \frac{1}{y} = \frac{1}{6}$$

Find the number of ohms in each resistance.

10. A *demand law* is an equation relating the price y and the number of units x of a commodity demanded by consumers. A *supply law* is an equation relating the price y and the number of units x of the commodity that the manufacturer can supply. A point of intersection of a demand curve and a supply curve corresponds to "market equilibrium." The value of x and the value of y are called the equilibrium quantity and price, respectively. Find the equilibrium quantity and the equilibrium price for the following demand and supply laws:

> Demand law: $xy = 45$ $(x \geq 0, y \geq 0)$
> Supply law: $4x = y^2 + 3y - 4$ $(x \geq 0, y \geq 0)$

EXERCISES 6.8 B

1. An arch of a bridge across a river is in the shape of a semiellipse. The span of the arch at water level is 100 feet, and the greatest height of the arch above the water level is 30 feet. Find the equation of the arch.

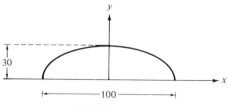

FIGURE 6.8.3

2. The equation of a comet moving in a parabolic orbit about the sun is

$$y^2 = 20x + 1000$$

where x and y are measured in millions of miles. The coordinates of the sun are $(0, 0)$.

a. Graph this equation.

b. Find how near the comet comes to the sun.

3. An apartment house contains 90 apartments. The owner finds that if he charges $80 per month rent, he can keep all of them rented. For every $1 he adds to the monthly rent, he has one vacancy.

a. Derive an equation relating the amount y he receives per month and the number of dollars x he charges over $80.

b. Graph the equation in (a) and find the value of x that yields the maximum amount y.

4. A mailman covers his 21-mile-long route by 11 a.m. each day. If his average rate of travel were half a mile faster each hour, he would cover the route by 10 a.m. What time does he start in the morning?

5. A boat takes 4 hours to go 18 miles downstream and then return to its starting point. The rate downstream is five times the rate of the current. Find the rate of the current and the speed of the boat in still water.

6. Determine the values of k so that the line $y = 2x + k$ will be tangent to the circle $x^2 + y^2 = 20$. (A tangent line intersects the circle in exactly one point.) (*Hint:* Use the substitution method to eliminate y. The roots of the resulting quadratic equation must be equal, and thus its discriminant must equal 0.)

7. A square park is surrounded by a sidewalk of uniform width. The sidewalk is surrounded by a paved road whose width is 8 times that of the sidewalk. If the area of the sidewalk is 4500 square feet and the area of the road is 43,200 square feet, find the length of a side of the park.

8. Letting P = power in watts of a transmission station

$\qquad E$ = output voltage in volts

$\qquad I$ = current in amperes

$\qquad V$ = input voltage in volts

$\qquad R$ = resistance of transmission wire in ohms

then $EI = P$ and $E = V - RI$.

Find E and I if $P = 75,000$ watts, $V = 650$ volts, and $R = 0.4$ ohm. (*Hint:* E must be near 650 volts for a practical solution to the problem.)

9. At 8 a.m. ship A was 125 miles due west of ship B. Ship A was sailing due north at 20 mph and ship B was sailing due west at 15 mph.

a. Derive an equation relating the distance y between them and the time t after 8 a.m.

b. Graph the equation in (a).

c. Determine the time of day when they will be nearest each other.

10. The total cost y in dollars of producing x units of a commodity is given by

$$y = x^2 + 2x$$

The total revenue in dollars received from the sale of x units of the commodity is given by

$$y = 20x - 2x^2$$

a. Graph each of the equations on the same set of axes. Note that $x \geq 0$ and $y \geq 0$.

b. Determine the values of x for which the revenue is greater than the cost.

c. The profit equation, where profit = revenue − cost, is given by

$$y = (20x - 2x^2) - (x^2 + 2x)$$

Graph this equation and determine the value of x for which the profit is a maximum.

SUMMARY

☐ The **solution set** of a quadratic equation in two variables x and y is $\{(x, y) \mid Ax^2 + Bxy + Cy^2 + Dx + Ey + F = 0$, where x and y are any complex numbers$\}$.

☐ The **equation of a circle** is $(x - a)^2 + (y - b)^2 = r^2$, where (a, b) is the *center* of the circle and r is its *radius*.

☐ The **equation of a parabola** having its axis (its line of symmetry) parallel to or on one of the coordinate axes is either $y = Ax^2 + Bx + C$ or $x = Ay^2 + By + C$.

☐ The **equation of an ellipse** having its center at the origin and its axes (its lines of symmetry) on the coordinate axes is

$$\frac{x^2}{a^2} + \frac{y^2}{b^2} = 1$$

☐ The **equation of a hyperbola** having its center at the origin and its axes (its lines of symmetry) on the coordinate axes is

$$\frac{x^2}{a^2} - \frac{y^2}{b^2} = 1 \quad \text{or} \quad \frac{y^2}{b^2} - \frac{x^2}{a^2} = 1$$

☐ The **equation of a hyperbola** having its center at the origin and its axes on the lines $y = \pm x$ is $xy = C$.

☐ The **graphic solution** of a quadratic system is obtained by graphing both equations on the same set of axes and then determining the points of intersection. This method yields estimates of the real solutions.

☐ The **algebraic solution by the substitution method** involves solving one of the equations for one of the variables and then replacing this variable by the expression obtained in the other equation. This method is most useful for linear-quadratic systems and for systems involving the equation of a parabola or a hyperbola, $xy = C$.

☐ The **algebraic solution** by the **addition method** involves the addition of constant multiples of the two equations in order to eliminate one of the variables, the second-degree terms, or the constant terms.

☐ **Symmetric equations** may often be replaced by simpler equations by the substitution $x = u + v$ and $y = u - v$.

REVIEW EXERCISES

Solve Exercises 1–6 algebraically and graphically.

1. $x^2 + y^2 = 25$ and $y^2 = x + 5$
2. $4y^2 - 4x^2 = 25$ and $x^2 + 4y = 2$
3. $x^2 + 4y^2 = 36$ and $x^2 - y^2 = 1$ **4.** $4x^2 + y^2 = 100$ and $xy = 24$
5. $x^2 - 4y^2 = 9$ and $2y^2 = 3 - x$ **6.** $4x^2 + y^2 = 25$ and $x^2 = y + 1$

Solve Exercises 7–12 algebraically.

7. $y^2 - x^2 = xy + 1$ and $x^2 - y^2 = xy - 5$
8. $x^2 - 3xy - y^2 = 9$ and $2x^2 - 4xy - 7y^2 = 9$
9. $12x^2 - y^2 = 12$ and $2y = x^2 + 8$
10. $x^2 + y^2 + 4y = 6$ and $x^2 + y^2 + 2x = 4$
11. $4x^2 - 3y^2 = 1$ and $2x^2 + y = 3$
12. $xy = 480$ and $xy + 16x - 5y = 560$

13. A dealer bought some television sets for $750. If they had each cost $25 less, he could have bought one more for the same amount. How many sets did he buy and what was the cost per set?

14. A train traveling at 8 mph less than its usual rate arrived at its destination 5 hours late. The destination was 800 miles from the starting point. What was the usual rate of the train? What was the usual time of the train?

15. Find the dimensions of a rectangle if its diagonal is 17 feet and its area is 120 square feet.

CHAPTER 7

SYSTEMS OF EQUATIONS, MATRICES, DETERMINANTS

7.1 LINEAR SYSTEMS

7.1.1 Two Variables

It has previously been stated that the solution set of a linear equation in two variables $ax + by = c$ consists of an infinite number of ordered pairs (x, y). The solution set of two linear equations in two variables is the intersection of the two solution sets of the equations—that is, the solution set of the system

$$a_1x + b_1y = c_1$$
$$a_2x + b_2y = c_2$$

is the set

$$\{(x, y) \mid a_1x + b_1y = c_1\} \cap \{(x, y) \mid a_2x + b_2y = c_2\}$$

One method for obtaining this solution set is based on the following theorem.

THEOREM

If (p, q) is a solution of both $a_1x + b_1y = c_1$ and $a_2x + b_2y = c_2$, then (p, q) is a solution of $A(a_1x + b_1y - c_1) + B(a_2x + b_2y - c_2) = 0$.

This theorem was shown to be valid because the replacement of x by p and y by q resulted in the statement $A(0) + B(0) = 0$, which is true for any real numbers for A and B.

EXAMPLE 7.1.1   Solve the system $3x - y = 9$, $x + 2y = 10$.

Solution

$$2(3x - y = 9) \qquad\qquad 6x - 2y = 18$$
$$1(x + 2y = 10) \qquad\qquad \underline{x + 2y = 10}$$
$$\qquad\qquad\qquad\qquad 7x = 28 \quad\text{and}\quad x = 4$$

$$1(3x - y = 9) \qquad\qquad 3x - y = 9$$
$$-3(x + 2y = 10) \qquad\qquad \underline{-3x - 6y = -30}$$
$$\qquad\qquad\qquad\qquad -7y = -21 \quad\text{and}\quad y = 3$$

Thus $\{(4, 3)\} = \{(x, y) \mid 3x - y = 9\} \cap \{(x, y) \mid x + 2y = 10\}$

It was also stated earlier that there are three possibilities for the solution set of a system of two linear equations in two variables:

1. There is exactly one solution.
2. The solution set is the empty set.
3. There are infinitely many solutions.

The geometric interpretations for these three possibilities are, respectively:

1. The lines intersect in exactly one point.
2. The lines are parallel.
3. The lines coincide.

7.1.2   Three Variables

DEFINITIONS

A **linear equation in three variables** x, y, and z is an equation of the form $ax + by + cz = d$, where a, b, c, and d are constants.

A **solution of an equation in three variables** is an ordered triple of numbers (x, y, z) such that the open equation becomes true when x is replaced by the first member of the ordered triple, y by the second member, and z by the third member.

EXAMPLE 7.1.2   Show that $(2, -1, 3)$ is a solution of $5x + 2y - z = 5$.

Solution   If $x = 2$, $y = -1$, and $z = 3$, then

$$5x + 2y - z = 5(2) + 2(-1) - 3 = 10 - 2 - 3 = 5$$

DEFINITION

The **solution set of an equation in three variables** is the set of all solutions of the equation.

As is the case with an equation in two variables, there are infinitely many solutions in the solution set.

DEFINITION

The **solution set of a system of three linear equations in three variables** is the intersection of the solution sets of each of the three equations in the system.

In other words, the solution set of the system

$$a_1x + b_1y + c_1z = d_1$$
$$a_2x + b_2y + c_2z = d_2$$
$$a_3x + b_3y + c_3z = d_3$$

is $\{(x, y, z) \mid a_1x + b_1y + c_1z = d_1\}$
$\cap\{(x, y, z) \mid a_2x + b_2y + c_2z = d_2\} \cap \{(x, y, z) \mid a_3x + b_3y + c_3z = d_3\}$

The solution set of a system of three linear equations in three variables can be obtained by a method similar to the one used for systems of two linear equations in two variables. By adding constant multiples of two equations of the system, one of the variables can be eliminated. Repeating this operation with a different pair of equations, the same variable can be eliminated again. There then results two linear equations in two variables which can be solved by the methods previously discussed. Again, there are three possibilities:

1. There is exactly one solution.
2. The solution set is the empty set.
3. There are infinitely many solutions.

Since there are three variables, the geometric interpretation is three-dimensional. It can be shown that a linear equation in three variables represents a plane. The solution set of a system of three equations in three variables is the geometrical intersection of the three planes. The possibilities are summarized below.

Algebraic Statement	*Geometric Statement*	*Illustration*
There is exactly one solution.	The three planes intersect in exactly one point P.	
The solution set is empty.	The three planes do not have a point in common.	
There are infinitely many solutions.	The three planes intersect on a line or the three planes are coincident.	

EXAMPLE 7.1.3   Solve the system
$$x + 2y - 3z = -11$$
$$x - y - z = 2$$
$$x + 3y + 2z = -4$$

Solution

1. Eliminate z by using the first and second equations:

$$-1(x + 2y - 3z = -11) \qquad -x - 2y + 3z = 11$$
$$3(x - y - z = 2) \qquad\quad \underline{3x - 3y - 3z = 6}$$
$$2x - 5y \qquad = 17$$

2. Eliminate z by using the second and third equations:

$$2(x - y - z = 2) \qquad\quad 2x - 2y - 2z = 4$$
$$1(x + 3y + 2z) = -4) \qquad \underline{x + 3y + 2z = -4}$$
$$3x + y \qquad\quad = 0$$

3. Now solve the system $2x - 5y = 17$, $3x + y = 0$:

$$1(2x - 5y = 17) \qquad\quad 2x - 5y = 17$$
$$5(3x + y = 0) \qquad\quad \underline{15x + 5y = 0}$$
$$17x \qquad\quad = 17 \quad \text{or} \quad x = 1$$

$$3(2x - 5y = 17) \qquad\quad 6x - 15y = 51$$
$$-2(3x + y = 0) \qquad \underline{-6x - 2y = 0}$$
$$-17y = 51 \quad \text{or} \quad y = -3$$

4. Use any one of the three original equations to find z. Using the third equation, $x + 3y + 2z = -4$, with $x = 1$ and $y = -3$,

$$1 + 3(-3) + 2z = -4$$
$$-8 + 2z = -4$$
$$2z = 4 \quad \text{or} \quad z = 2$$

5. State the solution set. The solution set is $\{(1, -3, 2)\}$.

6. Check the solution in each equation:

$$x + 2y - 3z = 1 + 2(-3) - 3(2)$$
$$= 1 - 6 - 6 = 1 - 12 = -11$$
$$x - y - z = 1 - (-3) - 2 = 1 + 3 - 2 = 4 - 2 = 2$$
$$x + 3y + 2z = 1 + 3(-3) + 2(2)$$
$$= 1 - 9 + 4 = -8 + 4 = -4$$

Thus if
$$A = \{(x, y, z) \mid x + 2y - 3z = -11\}$$
$$B = \{(x, y, z) \mid x - y - z = 2\}$$
$$C = \{(x, y, z) \mid x + 3y + 2z = -4\}$$

then the solution set of the system is $A \cap B \cap C = \{(1, -3, 2)\}$.

EXAMPLE 7.1.4   Solve the system
$$2x + 2y - 2z = 1$$
$$5x - 2y + z = 2$$
$$3x + 3y - 3z = 5$$

Solution

1. $1(2x + 2y - 2z = 1)$
    $2(5x - 2y + z = 2)$

$$2x + 2y - 2z = 1$$
$$10x - 4y + 2z = 4$$
$$\overline{}$$
$$12x - 2y = 5$$

2. $3(5x - 2y + z = 2)$
    $1(3x + 3y - 3z = 5)$

$$15x - 6y + 3z = 6$$
$$3x + 3y - 3z = 5$$
$$\overline{}$$
$$18x - 3y = 11$$

3. $3(12x - 2y = 5)$
    $-2(18x - 3y = 11)$

$$36x - 6y = 15$$
$$-36x + 6y = -22$$
$$\overline{}$$
$$0 = -7, \text{ a false statement}$$

4. The solution set is the empty set, $\varnothing$.

EXAMPLE 7.1.5   Solve the system
$$x - 2y + 3z = 4$$
$$2x + y - z = 1$$
$$3x - y + 2z = 5$$

Solution

1. $1(x - 2y + 3z = 4)$
    $3(2x + y - z = 1)$

$$x - 2y + 3z = 4$$
$$6x + 3y - 3z = 3$$
$$\overline{}$$
$$7x + y = 7$$

2. $2(2x + y - z = 1)$
    $1(3x - y + 2z = 5)$

$$4x + 2y - 2z = 2$$
$$3x - y + 2z = 5$$
$$\overline{}$$
$$7x + y = 7$$

Since the same equation was obtained in both cases, the solution set is infinite. Solving this equation for y, $y = 7 - 7x$. Now replacing y by $7 - 7x$ in the second equation,

$$2x + (7 - 7x) - z = 1$$
$$-5x - z = -6$$
$$z = 6 - 5x$$

The solution set is $\{(x, 7 - 7x, 6 - 5x) \mid x \text{ is any real number}\}$.

$$\text{Check:} \quad x - 2y + 3z = x - 2(7 - 7x) + 3(6 - 5x)$$

$$= x - 14 + 14x + 18 - 15x = 4$$

$$2x + y - z = 2x + (7 - 7x) - (6 - 5x)$$

$$= 2x + 7 - 7x - 6 + 5x = 1$$

$$3x - y + 2z = 3x - (7 - 7x) + 2(6 - 5x)$$

$$= 3x - 7 + 7x + 12 - 10x = 5$$

A system of four linear equations in four variables can be solved by eliminating the same variable using three different pairs of equations. Each of the four equations should be represented in at least one pair. A system of three linear equations in three variables is thus obtained, and this may be solved by the preceding technique.

EXERCISES 7.I A

Solve and check Exercises 1–12.

1. $x - 2y - 3z = 3$
$x + y - z = 2$
$2x - 3y - 5z = 5$

2. $x + y = 3$
$z - y = 3$
$2x + y = 8$

3. $x + y + 3z - 3 = 0$
$3x - y + 2z - 1 = 0$
$4x - 4y + z + 2 = 0$

4. $2r + s - t = 3$
$4r + 2s + 3t = 1$
$6r + 3s - 2t = 2$

5. $a + b + c = 5$
$2a - b - 2c = 3$
$3a - 3b - 5c = 1$

6. $x + y = z + 2$
$4x + 3z = y + 10$
$2x - 3y = 12 - z$

7. $x + 2y - 2z = 0$
$3x - 4y + 2z = 0$
$x + 12y - 10z = 0$

8. $x + 2y - z = 5$
$3x - 4y + 2z = 15$
$2x - y + z = 13$

9. $x + y + z = 9$
$x - y - z = 3$
$2x + 2y + 2z = 5$

10. $x + y + z = 9$
$x - y - z = 3$
$x + y - z = 5$

11. $x + y + z = 9$
$x - y - z = 3$
$x + 3y + 3z = 15$

12. $x + y - z = a$
$x - y + z = b$
$x - y - z = c$

13. Find the solution set of the system

$$x - 3y + z = 1$$
$$2x + y - 2z = -2$$

(*Hint:* Express y and z in terms of x.)

14. Find the solution set of the system

$$x + y + z = 4$$
$$2x - y + 3z = 14$$
$$3x - y - 2z = 1$$
$$6x - y + 2z = 19$$

(*Hint:* Find the solution set for the system consisting of the first three equations. Show that the solution also satisfies the fourth equation.)

15. Solve and check:

$$x + 2y - 3z + 2t = 6$$
$$2x + 3y + 4z + t = -5$$
$$3x - y + 2z - t = 0$$
$$2x + y + z - t = -4$$

16. Solve and check:

$$a + 2c = 11$$
$$a + 3b - 2d - 13 = 0$$
$$2a - 3c - d + 4 = 0$$
$$3a - 3b - 2c + 2d - 1 = 0$$

EXERCISES 7.1 B

Solve and check Exercises 1–12.

1. $x + y - z = 0$
$x - 2y + 3z = 0$
$2x - y - 2z = 0$

2. $x + 2y + z = 3$
$2x - y - 2z = 1$
$3x - 3y + 3z = 0$

3. $4a - b + 2c = 3$
$a + b - c = 4$
$a + 6b - 7c = 17$

4. $3x - y = z - 2$
$5x + 2y = 22 - z$
$2x + 4y - 5z = 0$

5. $x - y - z = 4$
$6x + y - 2z = 5$
$-3x + 3y + 3z = -7$

6. $x - y + 2z = 0$
$2x + y - z = 0$
$x + 5y - 8z = 0$

7. $2r + 2s - t = 28$
$2r - 2s - t = 8$
$2r - 2s + t = 20$

8. $2r + 2s - t = 28$
$2r - 2s - t = 8$
$2r + s - t = 20$

9. $2r + 2s - t = 28$
$2r - 2s - t = 8$
$6r - 2s - 3t = 44$

10. $2r + 2s - t = 4a$
$2r - 2s - t = 4b$
$2r - 2s + t = 4c$

11. $x - 2y - z = 0$
$2x - y + z = 0$
$3x + y - z = 20$

12. $x + y = 5$
$y + z = 10$
$x + z = 1$

13. Find the solution set of the system

$$x - 2y = 5$$
$$2x + 3y = 7$$
$$x + 5y = 2$$

14. Find the solution set of the system

$$x + y + z = 6$$
$$2x - y + z = 3$$
$$5x + 2y - 2z = 12$$
$$3x - y - z = 2$$

15. Solve and check:

$$x + y + z + t = 1$$
$$x + 2y \qquad - t = 10$$
$$2y - z + t = 1$$
$$x \qquad + 2z - 2t = 7$$

16. Solve and check:

$$a - b + c - d = 5$$
$$2a \qquad - 3c + d = 2$$
$$a \qquad - 3d = 6$$
$$2b - c + 2d = -3$$

7.2 MATRICES (Optional)

The *general solution* of a system of two linear equations in two variables is obtained as follows:

$$b_2(a_1x + b_1y = c_1)$$
$$-b_1(a_2x + b_2y = c_2)$$

$$a_1b_2x + b_1b_2y = b_2c_1$$
$$-a_2b_1x - b_1b_2y = -b_1c_2$$

$$\overline{(a_1b_2 - a_2b_1)x = b_2c_1 - b_1c_2}$$

$$-a_2(a_1x + b_1y = c_1)$$
$$a_1(a_2x + b_2y = c_2)$$

$$-a_1a_2x - a_2b_1y = -a_2c_1$$
$$a_1a_2x + a_1b_2y = a_1c_2$$

$$\overline{(a_1b_2 - a_2b_1)y = a_1c_2 - a_2c_1}$$

Now if $a_1b_2 - a_2b_1 \neq 0$, then

$$x = \frac{b_2c_1 - b_1c_2}{a_1b_2 - a_2b_1} \quad \text{and} \quad y = \frac{a_1c_2 - a_2c_1}{a_1b_2 - a_2b_1}$$

Although these equations provide a formula for the solution of the system, they are not easy to remember. Introducing the symbol $\begin{vmatrix} a & b \\ c & d \end{vmatrix}$ to designate the number $ad - bc$, the equations become

$$x = \frac{\begin{vmatrix} c_1 & b_1 \\ c_2 & b_2 \end{vmatrix}}{\begin{vmatrix} a_1 & b_1 \\ a_2 & b_2 \end{vmatrix}} \quad \text{and} \quad y = \frac{\begin{vmatrix} a_1 & c_1 \\ a_2 & c_2 \end{vmatrix}}{\begin{vmatrix} a_1 & b_1 \\ a_2 & b_2 \end{vmatrix}}$$

These arrays are more easily memorized, and moreover, they can be generalized to indicate the solution of a consistent system of three linear equations in three variables, or of four linear equations in four variables, or of n linear equations in n variables. The next sections are devoted to arrays of numbers, their properties, and their application to the solution of systems of equations.

DEFINITION

A matrix is a rectangular (or square) array of numbers (or elements) arranged in rows and columns enclosed by brackets, double vertical lines, or by large parentheses.

Some examples of matrices are:

$$\begin{bmatrix} 1 & 0 & 2 \\ 3 & -1 & 5 \\ 4 & 2 & 6 \end{bmatrix}, \quad \begin{Vmatrix} 5 & 1 & 2 \\ 3 & 2 & -7 \end{Vmatrix}, \quad \begin{pmatrix} 2 & 0 \\ 1 & 3 \end{pmatrix}, \quad \begin{bmatrix} 2 \\ 5 \end{bmatrix}, \quad [3 \quad 1]$$

DEFINITION

The **dimensions** of a matrix are $n \times k$ (read "n by k") if the matrix is an array of n rows and k columns.

If the number of rows is equal to the number of columns, then the matrix is called a **square matrix.**

If a matrix has only one row or one column, then it is called a **row matrix** or a **column matrix** (also called a **row vector** or a **column vector.**)

In the examples above, the first matrix is a 3×3 square matrix, the second is a 2×3 matrix, the third is a 2×2 square matrix, the fourth is a column matrix or column vector, and the fifth is a row matrix or a row vector.

HISTORICAL NOTE

Matrix algebra was originated by the English mathematician Arthur Cayley (1821–1895) in his work *A Memoir on the Theory of Matrices*, published in 1858. The term *matrix* was first used in 1850 by the English mathematician James Joseph Sylvester (1814–1897), a lifelong friend and collaborator of Cayley.

The nineteenth century witnessed the development of logical foundations of elementary algebra and the creation of modern abstract algebra. The discovery of noncommutative algebras such as Cayley's matrix algebra and Hamilton's quaternion algebra were events of major significance.

DEFINITION

Two **matrices are equal** if and only if they have the same dimensions and their corresponding elements are equal.

For example,

$$\begin{bmatrix} 2 & 1 \\ 3 & 0 \end{bmatrix} = \begin{bmatrix} \frac{6}{3} & \frac{3}{3} \\ \frac{9}{3} & 0 \end{bmatrix} \quad \text{and} \quad \begin{bmatrix} 4 & -1.5 & 2.5 \\ 1.5 & 3 & 5 \end{bmatrix} = \begin{bmatrix} 4 & -\frac{3}{2} & \frac{5}{2} \\ \frac{3}{2} & 3 & 5 \end{bmatrix}$$

$$\begin{bmatrix} 1 & 4 \\ 3 & 2 \end{bmatrix} \neq \begin{bmatrix} 1 & 3 \\ 4 & 2 \end{bmatrix} \quad \text{and} \quad \begin{bmatrix} 0 & 0 & 0 \\ 0 & 0 & 0 \end{bmatrix} \neq \begin{bmatrix} 0 & 0 \\ 0 & 0 \end{bmatrix}$$

DEFINITION

The **sum of two matrices** having the same dimensions is the matrix whose elements are the sums of the corresponding elements of the matrices being added.

For example,

$$\begin{bmatrix} a & b \\ c & d \end{bmatrix} + \begin{bmatrix} 1 & 2 \\ 3 & 4 \end{bmatrix} = \begin{bmatrix} a + 1 & b + 2 \\ c + 3 & d + 4 \end{bmatrix}$$

$$\begin{bmatrix} 1 & 0 & 2 \\ 2 & 1 & -3 \end{bmatrix} + \begin{bmatrix} 2 & 4 & -1 \\ 0 & 2 & 1 \end{bmatrix} = \begin{bmatrix} (1 + 2) & (0 + 4) & (2 - 1) \\ (2 + 0) & (1 + 2) & (-3 + 1) \end{bmatrix}$$

$$= \begin{bmatrix} 3 & 4 & 1 \\ 2 & 3 & -2 \end{bmatrix}$$

It should be noted that addition is defined only for matrices having the same dimensions. It is left for the student to verify that matrix addition is commutative and associative.

An element of a matrix is also called a **scalar**.

If the elements of a matrix are members of the set of real numbers, then a *scalar* is any real number.

DEFINITION

The **product of a scalar and a matrix** is the matrix each of whose elements is the product of the scalar and the corresponding element of the matrix being multiplied.

For example,

$$k\begin{bmatrix} 2 & 1 \\ 3 & 0 \end{bmatrix} = \begin{bmatrix} 2k & k \\ 3k & 0 \end{bmatrix} \quad \text{and} \quad 2\begin{bmatrix} 1 & -1 & 2 \\ 3 & 0 & 4 \end{bmatrix} = \begin{bmatrix} 2 & -2 & 4 \\ 6 & 0 & 8 \end{bmatrix}$$

Scalar multiplication may be shown to be commutative and associative.

DEFINITION

The **product of an** $n \times k$ **matrix** A **and a** $k \times m$ **matrix** B is an $n \times m$ matrix AB in which the element in the ath row and bth column is the sum of the products of the elements of the ath row of A multiplied by the corresponding elements of the bth row of B.

EXAMPLE 7.2.1 If $A = \begin{bmatrix} 2 & 3 \\ 4 & 5 \end{bmatrix}$ and $B = \begin{bmatrix} a & b & c \\ x & y & z \end{bmatrix}$, find the matrix product AB.

Solution $AB = \begin{bmatrix} 2a + 3x & 2b + 3y & 2c + 3z \\ 4a + 5x & 4b + 5y & 4c + 5z \end{bmatrix}$

EXAMPLE 7.2.2 Multiply $\begin{bmatrix} 2 & -1 & 3 \\ 3 & 4 & 1 \end{bmatrix}$ by $\begin{bmatrix} x \\ y \\ z \end{bmatrix}$.

Solution $\begin{bmatrix} 2 & -1 & 3 \\ 3 & 4 & 1 \end{bmatrix} \begin{bmatrix} x \\ y \\ z \end{bmatrix} = \begin{bmatrix} 2x - y + 3z \\ 3x + 4y + z \end{bmatrix}$

EXAMPLE 7.2.3 If $A = \begin{bmatrix} 2 & 0 \\ 1 & 3 \end{bmatrix}$ and $B = \begin{bmatrix} 1 & 5 \\ 3 & 2 \end{bmatrix}$, find AB and BA.

Solution

$AB = \begin{bmatrix} 2 & 0 \\ 1 & 3 \end{bmatrix} \begin{bmatrix} 1 & 5 \\ 3 & 2 \end{bmatrix} = \begin{bmatrix} 2 \cdot 1 + 0 \cdot 3 & 2 \cdot 5 + 0 \cdot 2 \\ 1 \cdot 1 + 3 \cdot 3 & 1 \cdot 5 + 3 \cdot 2 \end{bmatrix} = \begin{bmatrix} 2 & 10 \\ 10 & 11 \end{bmatrix}$

$BA = \begin{bmatrix} 1 & 5 \\ 3 & 2 \end{bmatrix} \begin{bmatrix} 2 & 0 \\ 1 & 3 \end{bmatrix} = \begin{bmatrix} 2 + 5 & 0 + 15 \\ 6 + 2 & 0 + 6 \end{bmatrix} = \begin{bmatrix} 7 & 15 \\ 8 & 6 \end{bmatrix}$

Example 7.2.3 illustrates that matrix multiplication is *not commutative*. However, the associative law is valid, and the right and left distributive laws are valid—that is,

$$(AB)C = A(BC)$$

$$A(B + C) = AB + AC$$

$$(B + C)A = BA + CA$$

It should be noted that the product of two matrices is defined *only* for the case where the number of columns of the left matrix is equal to the number of rows of the right matrix.

It may seem strange that matrix multiplication is defined as it is and not some other way; for example, as the matrix whose elements are the products of the corresponding elements. There are very few applications of such a definition, whereas there is a wide range of applications for the definition as stated, particularly in the solution of systems of equations. This will be shown in a later section.

DEFINITION

The **main diagonal** (or **principal diagonal**) of a square matrix is the diagonal from the upper-left element to the lower-right element.

For example, the main diagonal of the matrix $\begin{bmatrix} 1 & 2 & 3 \\ 4 & 5 & 6 \\ 7 & 8 & 9 \end{bmatrix}$ contains the

elements 1, 5, and 9.

DEFINITION

An **identity matrix** I is a square matrix whose elements along the main diagonal are ones and the rest of whose elements are zeros.

For example,

$$\begin{bmatrix} 1 & 0 \\ 0 & 1 \end{bmatrix}, \begin{bmatrix} 1 & 0 & 0 \\ 0 & 1 & 0 \\ 0 & 0 & 1 \end{bmatrix}, \quad \text{and} \quad \begin{bmatrix} 1 & 0 & 0 & 0 \\ 0 & 1 & 0 & 0 \\ 0 & 0 & 1 & 0 \\ 0 & 0 & 0 & 1 \end{bmatrix}$$

are identity matrices.

THEOREM

If I is an identity matrix and if A is any matrix having the same dimensions as I, then $IA = A = AI$.

For example,

$$\begin{bmatrix} 1 & 0 \\ 0 & 1 \end{bmatrix}\begin{bmatrix} a & b \\ c & d \end{bmatrix} = \begin{bmatrix} a & b \\ c & d \end{bmatrix} = \begin{bmatrix} a & b \\ c & d \end{bmatrix}\begin{bmatrix} 1 & 0 \\ 0 & 1 \end{bmatrix}$$

DEFINITION

If a matrix A^{-1} exists so that $AA^{-1} = I$, where I is an identity matrix and A is a matrix having the same dimensions as I, then A^{-1} is called the **inverse matrix** of A.

EXAMPLE 7.2.4   Find A^{-1} for $A = \begin{bmatrix} 2 & 1 \\ 4 & 5 \end{bmatrix}$.

Solution   Assume A^{-1} exists and let $A^{-1} = \begin{bmatrix} a & b \\ c & d \end{bmatrix}$.

Then
$$\begin{bmatrix} 2 & 1 \\ 4 & 5 \end{bmatrix}\begin{bmatrix} a & b \\ c & d \end{bmatrix} = \begin{bmatrix} 1 & 0 \\ 0 & 1 \end{bmatrix}$$

$$2a + c = 1 \qquad 2b + d = 0$$
$$4a + 5c = 0 \qquad 4b + 5d = 1$$

Solving for a, b, c, and d,
$$a = \tfrac{5}{6}, b = -\tfrac{1}{6}, c = -\tfrac{2}{3}, d = \tfrac{1}{3}$$

Therefore, $A^{-1} = \begin{bmatrix} \frac{5}{6} & -\frac{1}{6} \\ -\frac{2}{3} & \frac{1}{3} \end{bmatrix}$.

EXERCISES 7.2 A

Perform the indicated operations in Exercises 1–15, if possible. If not possible, state why.

1. $\begin{bmatrix} 3 & 5 \\ -2 & 4 \end{bmatrix} + \begin{bmatrix} 1 & 0 \\ -5 & -5 \end{bmatrix}$

2. $\begin{bmatrix} 2 & 4 & -3 \\ -1 & 0 & 1 \end{bmatrix} + \begin{bmatrix} 3 & 4 & 5 \\ 6 & 7 & 8 \end{bmatrix}$

3. $\begin{bmatrix} 4 & 2 & 9 \\ 2 & -1 & 0 \end{bmatrix} + \begin{bmatrix} 3 & 5 \\ 1 & -7 \end{bmatrix}$

4. $[5 \quad -2 \quad 7] + \begin{bmatrix} 3 \\ 8 \\ 2 \end{bmatrix}$

5. $\begin{bmatrix} 3 & -2 & 4 \\ 1 & 0 & 2 \\ 0 & 1 & 1 \end{bmatrix} + \begin{bmatrix} 2 & 5 & -8 \\ 1 & -3 & -2 \\ 3 & -2 & -2 \end{bmatrix}$

6. $\begin{bmatrix} 4 & 6 & -3 & 5 \\ 1 & -4 & 2 & 1 \\ 3 & 0 & 0 & 1 \end{bmatrix} + \begin{bmatrix} 7 & 0 & -5 \\ 2 & -1 & 0 \end{bmatrix}$

7. $[4 \quad 7 \quad 9] + [2 \quad 8 \quad 5]$

8. $-2\begin{bmatrix} \frac{1}{2} & 3 & \frac{5}{2} \\ 1 & 0 & -1 \\ 3 & -2 & 1 \end{bmatrix}$

9. $3[2 \quad -3 \quad 4]$

10. $5\begin{bmatrix} 2 & -3 & 5 \\ 1 & 4 & -2 \end{bmatrix}$

11. $\begin{bmatrix} 3 & 1 \\ 2 & 5 \end{bmatrix}\begin{bmatrix} -2 & 3 \\ 4 & 1 \end{bmatrix}$

12. $\begin{bmatrix} 2 & 3 \\ 1 & -2 \\ 4 & 3 \end{bmatrix}\begin{bmatrix} a & b & c \\ 4 & 2 & 3 \end{bmatrix}$

13. $\begin{bmatrix} 2 & 1 & 4 \\ 3 & -2 & 3 \end{bmatrix} \begin{bmatrix} a & b & c \\ 4 & 2 & 3 \end{bmatrix}$

14. $\begin{bmatrix} 1 & -1 & 1 \\ 2 & 1 & -3 \\ 3 & 2 & -1 \end{bmatrix} \begin{bmatrix} x \\ y \\ z \end{bmatrix}$

15. $\begin{bmatrix} 1 & -2 & 1 & 3 \\ 2 & 0 & 5 & 4 \end{bmatrix} \begin{bmatrix} a & 2 \\ b & -3 \\ c & -1 \\ d & 0 \end{bmatrix}$

Solve for the variables in Exercises 16–20.

16. $\begin{bmatrix} 1 & x & 3 \\ y & 2 & z \end{bmatrix} = \begin{bmatrix} 1 & 0 & 3 \\ 6 & 2 & 9 \end{bmatrix}$

17. $5 \begin{bmatrix} x & y \\ 3 & -2 \end{bmatrix} + \begin{bmatrix} 3 & -2 \\ -1 & 4 \end{bmatrix} = 2 \begin{bmatrix} -6 & 5 \\ 7 & -3 \end{bmatrix}$

18. $\begin{bmatrix} 1 & 0 & 0 \\ 0 & -1 & 0 \\ 0 & 0 & 2 \end{bmatrix} \begin{bmatrix} x \\ y \\ z \end{bmatrix} = \begin{bmatrix} 2 \\ 3 \\ 8 \end{bmatrix}$

19. $\begin{bmatrix} 2 & 0 & 0 \\ 0 & 3 & 0 \\ 0 & 0 & -4 \end{bmatrix} \begin{bmatrix} x & -1 \\ y & 2 \\ z & -3 \end{bmatrix} = 3 \begin{bmatrix} 3 & -\frac{2}{3} \\ -2 & 2 \\ 5 & 4 \end{bmatrix}$

20. $\begin{bmatrix} 1 & 2 & -1 \\ 0 & 1 & 1 \\ 0 & 0 & 1 \end{bmatrix} \begin{bmatrix} x \\ y \\ z \end{bmatrix} = \begin{bmatrix} -2 \\ 5 \\ 3 \end{bmatrix}$

In Exercises 21–30, let $A = \begin{bmatrix} a & b \\ c & d \end{bmatrix}$, $B = \begin{bmatrix} 1 & 1 \\ -1 & 1 \end{bmatrix}$, $C = \begin{bmatrix} 1 & 1 \\ 1 & -1 \end{bmatrix}$, *and let*

$S =$ *the set of* 2×2 *square matrices whose elements are real numbers.*

21. Show that $A + B \in S$ and $AB \in S$ (closure, addition and multiplication).

22. Show that $A + B = B + A$ (commutativity, addition).

23. Show that $BC \neq CB$ (multiplication is not commutative).

24. Show that $(A + B) + C = A + (B + C)$ and $(AB)C = A(BC)$ (associativity).

25. Find 0 so that $0 + A = A = A + 0$ (addition identity).

26. Find I so that $IA = A = AI$ (multiplication identity).

27. Find $-A$ so that $A + (-A) = 0$ (addition inverse).

28. Assuming $ad - bc \neq 0$, find A^{-1} so that $AA^{-1} = I$ (multiplication inverse). Does $A^{-1}A = AA^{-1}$? Why?

29. Find X so that $B + X = A$. Define subtraction.

30. Find Y so that $AY = B$. Define division.

EXERCISES 7.2 B

Perform the indicated operations in Exercises 1–10, if possible. If not possible, state why.

1. $3\begin{bmatrix} 5 & -6 & 7 \\ -2 & 8 & 4 \end{bmatrix} - 2\begin{bmatrix} -4 & 3 & -2 \\ 1 & -7 & -4 \end{bmatrix}$

2. $\begin{bmatrix} 2 & 4 & -3 \\ 9 & -5 & 6 \end{bmatrix} [x \ \ y \ \ z]$

3. $\begin{bmatrix} 1 & 3 & -4 \\ 5 & -1 & 2 \end{bmatrix} \begin{bmatrix} x \\ y \\ z \end{bmatrix}$

4. $\begin{bmatrix} 1 & 0 & 0 \\ 0 & 1 & 0 \\ 0 & 0 & 1 \end{bmatrix} \begin{bmatrix} a & b & c \\ d & e & f \\ g & h & i \end{bmatrix}$

5. $\begin{bmatrix} 1 & 2 & 1 & 3 \\ 0 & 1 & -2 & 1 \end{bmatrix} \begin{bmatrix} a & 0 \\ b & 1 \\ c & -1 \\ d & 0 \end{bmatrix}$

6. $\begin{bmatrix} 1 & 2 & 1 & 3 \\ 0 & 1 & -2 & 1 \end{bmatrix} + \begin{bmatrix} a & 0 \\ b & 1 \\ c & -1 \\ d & 0 \end{bmatrix}$

7. $\begin{bmatrix} 4 & 2 & 4 & -1 \\ 3 & -1 & 0 & 2 \end{bmatrix} \begin{bmatrix} 2 & x & 0 & 1 \\ y & 1 & -1 & 0 \end{bmatrix}$

8. $\begin{bmatrix} 4 & 2 & 4 & -1 \\ 3 & -1 & 0 & 2 \end{bmatrix} + \begin{bmatrix} 2 & x & 0 & 1 \\ y & 1 & -1 & 0 \end{bmatrix}$

9. $\begin{bmatrix} 1 & 2 \\ 2 & 4 \end{bmatrix} \begin{bmatrix} 6 & -2 \\ -3 & 1 \end{bmatrix}$

10. $\begin{bmatrix} 1 & 0 & 0 & 0 \\ 0 & 1 & 0 & 0 \\ 0 & 0 & 1 & 0 \\ 0 & 0 & 0 & 1 \end{bmatrix} \begin{bmatrix} w \\ x \\ y \\ z \end{bmatrix} = \begin{bmatrix} -2 \\ 3 \\ 1 \\ 5 \end{bmatrix}$

In Exercises 11–20, let $A = \begin{bmatrix} 1 & 2 \\ -1 & 1 \end{bmatrix}$ *and* $B = \begin{bmatrix} 1 & -1 \\ 2 & 1 \end{bmatrix}$.

11. Find $-B$ and $A - B$.   **12.** Find $A(-B)$.

13. Find $-BA$.   **14.** Find AB.

15. Find BA.   **16.** Find A^{-1}.

17. Find B^{-1}.   **18.** Find $(AB)^{-1}$.

19. Find $B^{-1}A^{-1}$.

20. Determine the relationship between (a) AB and BA, (b) $A(-B)$ and $(-B)A$, and (c) $(AB)^{-1}$ and $B^{-1}A^{-1}$.

21. Let $A = \begin{bmatrix} x & y \\ 2x & 2y \end{bmatrix}$ and $B = \begin{bmatrix} y & 3y \\ -x & -3x \end{bmatrix}$. Find AB. If $AB = 0$, does it follow that $A = 0$ or $B = 0$?

7.3 SOLUTION OF LINEAR SYSTEMS USING MATRICES (Optional)

7.3.1 Transformation to a Matrix Equation

The solution of linear systems by using matrices may be considered as an abbreviated form of the solution by the addition method previously discussed.

By applying the definition of equal matrices and the definition of matrix multiplication, it follows that the linear system

$$a_1x + b_1y = c_1$$
$$a_2x + b_2y = c_2$$

is equivalent to

$$\begin{bmatrix} a_1 & b_1 \\ a_2 & b_2 \end{bmatrix} \begin{bmatrix} x \\ y \end{bmatrix} = \begin{bmatrix} c_1 \\ c_2 \end{bmatrix}$$

and the system

$$a_1x + b_1y + c_1z = d_1$$
$$a_2x + b_2y + c_2z = d_2$$
$$a_3x + b_3y + c_3z = d_3$$

is equivalent to

$$\begin{bmatrix} a_1 & b_1 & c_1 \\ a_2 & b_2 & c_2 \\ a_3 & b_3 & c_3 \end{bmatrix} \begin{bmatrix} x \\ y \\ z \end{bmatrix} = \begin{bmatrix} d_1 \\ d_2 \\ d_3 \end{bmatrix}$$

Similar statements are valid for the cases involving four or more variables.

Each of the matrices, $\begin{bmatrix} a_1 & b_1 \\ a_2 & b_2 \end{bmatrix}$ and $\begin{bmatrix} a_1 & b_1 & c_1 \\ a_2 & b_2 & c_2 \\ a_3 & b_3 & c_3 \end{bmatrix}$, is called the *matrix of the coefficients.*

In solving a linear system by using matrices, it is convenient to work with the following matrices, which are called the **augmented matrices** of the linear system:

$$\begin{bmatrix} a_1 & b_1 & c_1 \\ a_2 & b_2 & c_2 \end{bmatrix}$$ for two variables and $$\begin{bmatrix} a_1 & b_1 & c_1 & d_1 \\ a_2 & b_2 & c_2 & d_2 \\ a_3 & b_3 & c_3 & d_3 \end{bmatrix}$$ for three variables

7.3.2 Equivalent Transformations of Matrices

Since the linear system

$$x = p$$
$$y = q$$
$$z = r$$

is equivalent to the open matrix equation

$$\begin{bmatrix} 1 & 0 & 0 \\ 0 & 1 & 0 \\ 0 & 0 & 1 \end{bmatrix} \begin{bmatrix} x \\ y \\ z \end{bmatrix} = \begin{bmatrix} p \\ q \\ r \end{bmatrix}$$

the solution set can be recognized immediately in either of these forms as $\{(p, q, r)\}$. Thus the objective is to transform the augmented matrix into one having zeros below and above the main diagonal of the matrix of the coefficients, if this is possible. Also, it is necessary that the linear system of the transformed matrix have the same solution set as that of the original matrix.

DEFINITION

Two **augmented matrices of linear systems are equivalent,** $A \approx B$, if and only if the associated linear systems of the two matrices are equivalent (that is, have the same solution set).

There are three elementary transformations which produce equivalent augmented matrices—interchanging rows, multiplying a row by a constant, and adding a constant multiple of a row to another row.

THEOREM I

Two augmented matrices are equivalent if one is obtained from the other by interchanging any two rows.

EXAMPLE 7.3.1　Show that $\begin{bmatrix} a_1 & b_1 & c_1 \\ a_2 & b_2 & c_2 \end{bmatrix} \approx \begin{bmatrix} a_2 & b_2 & c_2 \\ a_1 & b_1 & c_1 \end{bmatrix}$.

Solution　The associated linear systems are

$$\begin{array}{ll} a_1x + b_1y = c_1 & a_2x + b_2y = c_2 \\ \text{and} & \\ a_2x + b_2y = c_2 & a_1x + b_1y = c_1 \end{array}$$

Since the order in which the equations are written does not affect the solution set of the system, the two systems have the same solution set. Therefore, the two augmented matrices are equivalent.

EXAMPLE 7.3.2　Show that $\begin{bmatrix} a_1 & b_1 & c_1 & d_1 \\ a_2 & b_2 & c_2 & d_2 \\ a_3 & b_3 & c_3 & d_3 \end{bmatrix} \approx \begin{bmatrix} a_3 & b_3 & c_3 & d_3 \\ a_2 & b_2 & c_2 & d_2 \\ a_1 & b_1 & c_1 & d_1 \end{bmatrix}$.

Solution　The associated linear systems are

$$\begin{array}{lll} a_1x + b_1y + c_1z = d_1 & & a_3x + b_3y + c_3z = d_3 \\ a_2x + b_2y + c_2z = d_2 & \text{and} & a_2x + b_2y + c_2z = d_2 \\ a_3x + b_3y + c_3z = d_3 & & a_1x + b_1y + c_1z = d_1 \end{array}$$

are equivalent. Since the equations of a linear system may be written in any order without changing the solution set of the system, the augmented matrices are equivalent. A similar argument holds for the other possibilities.

THEOREM 2

Two augmented matrices are equivalent if one is obtained from the other by multiplying each element of a row by a nonzero constant.

Theorem 2 implies that an equivalent system is obtained if any equation of the system is multiplied by a constant.

EXAMPLE 7.3.3   Show that $\begin{bmatrix} a_1 & b_1 & c_1 \\ a_2 & b_2 & c_2 \end{bmatrix} \approx \begin{bmatrix} a_1 & b_1 & c_1 \\ ka_2 & kb_2 & kc_2 \end{bmatrix}$.

Solution   The associated linear systems are

$$\begin{array}{ll} a_1x + b_1y = c_1 \\ a_2x + b_2y = c_2 \end{array} \quad \text{and} \quad \begin{array}{ll} a_1x \ + b_1y = \ c_1 \\ ka_2x + kb_2y = kc_2 \end{array}$$

For $k \neq 0$, $ka_2x + kb_2y = kc_2$ is equivalent to $a_2x + b_2y = c_2$. Thus the two systems have the same solution set, and the augmented matrices are equivalent.

For the case of three variables,

$$\begin{array}{lll} a_1x + b_1y + c_1z = d_1 & & ka_1x + kb_1y + kc_1z = kd_1 \\ a_2x + b_2y + c_2z = d_2 & \text{and} & la_2x + lb_2y + lc_2z = ld_2 \\ a_3x + b_3y + c_3z = d_3 & & ma_3x + mb_3y + mc_3z = md_3 \end{array}$$

are equivalent for $klm \neq 0$, as was established previously. Thus the associated augmented matrices are equivalent—that is,

$$\begin{bmatrix} a_1 & b_1 & c_1 & d_1 \\ a_2 & b_2 & c_2 & d_2 \\ a_3 & b_3 & c_3 & d_3 \end{bmatrix} \approx \begin{bmatrix} ka_1 & kb_1 & kc_1 & kd_1 \\ la_2 & lb_2 & lc_2 & ld_2 \\ ma_3 & mb_3 & mc_3 & md_3 \end{bmatrix}$$

THEOREM 3

Two augmented matrices are equivalent if one is obtained from the other by adding a nonzero constant multiple of the elements of one row to the corresponding elements of another row.

Theorem 3 implies that any equation of a system may be replaced by a linear combination of two equations of the system. Since this equivalence has been established, it follows that the associated augmented matrices are equivalent.

EXAMPLE 7.3.4 Show that

$$\begin{bmatrix} 1 & 3 & 11 \\ 2 & -1 & 8 \end{bmatrix} \approx \begin{bmatrix} 1 & 3 & 11 \\ -2(1) + 2 & -2(3) - 1 & -2(11) + 8 \end{bmatrix}$$

Solution The associated linear systems are

$$\begin{aligned} x + 3y &= 11 \\ 2x - y &= 8 \end{aligned} \quad \text{and} \quad \begin{aligned} x + 3y &= 11 \\ (-2 + 2)x + (-6 - 1)y &= -2(11) + 8 \\ \text{or} \quad -7y &= -14 \end{aligned}$$

The solution set of $(x + 3y = 11$ and $-7y = -14)$ is $(5, 2)$.
The solution set of $(x + 3y = 11$ and $2x - y = 8)$ is $(5, 2)$.
Therefore, the two systems are equivalent, and their associated augmented matrices are equivalent.
As examples,

$$\begin{bmatrix} a_1 & b_1 & c_1 \\ a_2 & b_2 & c_2 \end{bmatrix} \approx \begin{bmatrix} a_1 & b_1 & c_1 \\ ka_1 + a_2 & kb_1 + b_2 & kc_1 + c_2 \end{bmatrix}$$

and

$$\begin{bmatrix} a_1 & b_1 & c_1 & d_1 \\ a_2 & b_2 & c_2 & d_2 \\ a_3 & b_3 & c_3 & d_3 \end{bmatrix} \approx \begin{bmatrix} a_1 & b_1 & c_1 & d_1 \\ ka_1 + a_2 & kb_1 + b_2 & kc_1 + c_2 & kd_1 + d_2 \\ a_3 & b_3 & c_3 & d_3 \end{bmatrix}$$

7.3.3 Gauss's Reduction Method

The method of solution illustrated in this section is known as Gauss's reduction method in honor of the German mathematician Karl Friedrich Gauss (1777–1855), who is generally acknowledged as the greatest mathematician of all time. Near the end of the eighteenth century, there were many persons who felt that all the important subject matter of mathematics had already been discovered. Gauss showed how wrong this attitude was as he opened the doors to the vast new area of modern mathematics.

The objective of this method is to obtain zeros below the main diagonal of the matrix of the coefficients.

EXAMPLE 7.3.5 By using matrices, solve the system

$$3x + 2y - z = 5$$
$$x - 3y + z = 2$$
$$2x - y - 2z = 1$$

Solution

1. Form the augmented matrix:
$$\begin{bmatrix} 3 & 2 & -1 & 5 \\ 1 & -3 & 1 & 2 \\ 2 & -1 & -2 & 1 \end{bmatrix}$$

2. Interchange rows to have a 1 in row 1, column 1:
$$\begin{bmatrix} 1 & -3 & 1 & 2 \\ 2 & -1 & -2 & 1 \\ 3 & 2 & -1 & 5 \end{bmatrix}$$

3. Multiply row 1 by -2, add to row 2:
   Multiply row 1 by -3, add to row 3:
$$\begin{bmatrix} 1 & -3 & 1 & 2 \\ 0 & 5 & -4 & -3 \\ 0 & 11 & -4 & -1 \end{bmatrix}$$

4. Multiply row 2 by $\frac{1}{5}$:
$$\begin{bmatrix} 1 & -3 & 1 & 2 \\ 0 & 1 & -\frac{4}{5} & -\frac{3}{5} \\ 0 & 11 & -4 & -1 \end{bmatrix}$$

5. Multiply row 2 by -11, add to row 3:
$$\begin{bmatrix} 1 & -3 & 1 & 2 \\ 0 & 1 & -\frac{4}{5} & -\frac{3}{5} \\ 0 & 0 & \frac{24}{5} & \frac{28}{5} \end{bmatrix}$$

6. Multiply row 3 by $\frac{5}{24}$:
$$\begin{bmatrix} 1 & -3 & 1 & 2 \\ 0 & 1 & -\frac{4}{5} & -\frac{3}{5} \\ 0 & 0 & 1 & \frac{7}{6} \end{bmatrix}$$

7. Write the associated linear system:
$$x - 3y + z = 2$$
$$y - \tfrac{4}{5}z = -\tfrac{3}{5}$$
$$z = \tfrac{7}{6}$$

8. Solve the system:
$$y = \tfrac{4}{5}(\tfrac{7}{6}) - \tfrac{3}{5} = \tfrac{14}{15} - \tfrac{9}{15} = \tfrac{1}{3}$$
$$x = 3(\tfrac{1}{3}) - \tfrac{7}{6} + 2 = \tfrac{18}{6} - \tfrac{7}{6} = \tfrac{11}{6}$$

9. State the solution set:
$$(\tfrac{11}{6}, \tfrac{1}{3}, \tfrac{7}{6})$$

10. Check the solution set: (Check left for student.)

7.3.4 Extension of Gauss's Method

It is sometimes possible to introduce zeros above the main diagonal of the matrix of the coefficients, as the following example illustrates.

EXAMPLE 7.3.6 By using matrices, solve the system

$$x - 2y + 3z = 1$$
$$2x + y - 2z = 13$$
$$x + 3y - z = 4$$

Solution

1. Multiply row 1 by
   -2, add to row 2:
   Multiply row 1 by
   -1, add to row 3:

$$\begin{bmatrix} 1 & -2 & 3 & 1 \\ 2 & 1 & -2 & 13 \\ 1 & 3 & -1 & 4 \end{bmatrix} \approx \begin{bmatrix} 1 & -2 & 3 & 1 \\ 0 & 5 & -8 & 11 \\ 0 & 5 & -4 & 3 \end{bmatrix}$$

2. Multiply row 2 by
   -1, add to row 3:

$$\approx \begin{bmatrix} 1 & -2 & 3 & 1 \\ 0 & 5 & -8 & 11 \\ 0 & 0 & 4 & -8 \end{bmatrix}$$

3. Multiply row 3 by
   2, add to row 2:
   Multiply row 3
   by $\frac{1}{4}$:

$$\approx \begin{bmatrix} 1 & -2 & 3 & 1 \\ 0 & 5 & 0 & -5 \\ 0 & 0 & 1 & -2 \end{bmatrix}$$

4. Multiply row 2
   by $\frac{1}{5}$:

$$\approx \begin{bmatrix} 1 & -2 & 3 & 1 \\ 0 & 1 & 0 & -1 \\ 0 & 0 & 1 & -2 \end{bmatrix}$$

5. Multiply row 2 by
   2, add to row 1:

$$\approx \begin{bmatrix} 1 & 0 & 3 & -1 \\ 0 & 1 & 0 & -1 \\ 0 & 0 & 1 & -2 \end{bmatrix}$$

6. Multiply row 3 by
   -3, add to row 1:

$$\approx \begin{bmatrix} 1 & 0 & 0 & 5 \\ 0 & 1 & 0 & -1 \\ 0 & 0 & 1 & -2 \end{bmatrix}$$

The associated linear system is $x = 5$, $y = -1$, $z = -2$. Thus the solution set is $\{(5, -1, -2)\}$. The solution should be checked in each of the three equations of the original system.

This technique, using matrices, is readily adaptable to the electronic computer. Moreover, it becomes increasingly effective as the number of equations and the number of variables increase.

EXERCISES 7.3 A

Solve Exercises 1–10 by using matrices.

1. $x - 2y = 7$
$3x + 2y = 5$

2. $5x + 6y = 8$
$3x - 4y = -18$

3. $2x - y = 0$
$2y - z = 0$
$x + 2y - z = 3$

4. $4x + 5y - 6z = 15$
$3x - 7y - 4z = -19$
$6x + 2y + z = 46$

5. $x - 5y + 2z - 9 = 0$
$4x + y + z - 1 = 0$
$2x - 7y + 3z - 13 = 0$

6. $2a - b + 2c - 3 = 0$
$-a - 3b + 4c - 5 = 0$
$3a - 5b + 8c - 11 = 0$

7. $x + 5y - 2z = 8$
$2x - 4y + 2z = 3$
$3x - 6y + 3z = 4$

8. $2a + b + c = 2$
$3a - b + c = 2$
$7a - 5b - 3c = 0$

9. $x - 2y + z = 0.5$
$-2x + 4y - 2z = 1$
$3x - 6y + 3z = 1.5$

10. $4r - 3s - t = 2$
$-r + 2s + 3t = 1$
$5r + s - 2t = 3$

EXERCISES 7.3 B

Solve Exercises 1–10 by using matrices.

1. $2x - 3y = 16$
$5x - 2y = -4$

2. $5x + 4y = 12$
$3x - 2y = 16$

3. $y - x = 1$
$z - x = 2$
$74x - 16y = 25z$

4. $x + y - z = 0$
$4x + 4y + 2z = 3$
$2x + 5y - z = 1$

5. $x + y + 2z - 7 = 0$
$4x - 2y + 3z - 1 = 0$
$9x - 3y + 8z - 4 = 0$

6. $2x + y - 3z - 2 = 0$
$3x - y - 2z - 3 = 0$
$x - z - 1 = 0$

7. $x + y + z = 0$
$x + y - 4z = 2$
$x - 2y + z = 1$
$2x - y - 3z = 3$

8. $x - y + z = 6$
$2x - y - 2z = 2$
$3x - 2y - z = 8$

9. $x + y + z + t = 0$
   $2x - y + t = 7$
   $x - 2z - t = -1$
   $2y - z - 2t = -3$

10. $x - y + 2z - t = 0$
    $2y - 3z + 2t = 5$
    $x - 2z - 3t = 2$
    $2x + y - 3z - 2t = 7$

7.4 DETERMINANTS (Optional)

DEFINITION

A **determinant of the second order**, denoted by the symbol $\begin{vmatrix} a_1 & b_1 \\ a_2 & b_2 \end{vmatrix}$, where a_1, b_1, a_2, and b_2 are real numbers, is the real number, $a_1b_2 - a_2b_1$—that is,

$$\begin{vmatrix} a_1 & b_1 \\ a_2 & b_2 \end{vmatrix} = a_1b_2 - a_2b_1$$

EXAMPLE 7.4.1 Evaluate the determinant $\begin{vmatrix} 3 & -1 \\ 2 & 5 \end{vmatrix}$.

Solution By the definition,

$$\begin{vmatrix} 3 & -1 \\ 2 & 5 \end{vmatrix} = 3(5) - (2)(-1) = 15 + 2 = 17$$

DEFINITION

A **determinant of the third order**, denoted by the symbol $\begin{vmatrix} a_1 & b_1 & c_1 \\ a_2 & b_2 & c_2 \\ a_3 & b_3 & c_3 \end{vmatrix}$, where the elements are real numbers, is the real number

$a_1 \begin{vmatrix} b_2 & c_2 \\ b_3 & c_3 \end{vmatrix} - a_2 \begin{vmatrix} b_1 & c_1 \\ b_3 & c_3 \end{vmatrix} + a_3 \begin{vmatrix} b_1 & c_1 \\ b_2 & c_2 \end{vmatrix}$ —that is,

$$\begin{vmatrix} a_1 & b_1 & c_1 \\ a_2 & b_2 & c_2 \\ a_3 & b_3 & c_3 \end{vmatrix} = a_1 \begin{vmatrix} b_2 & c_2 \\ b_3 & c_3 \end{vmatrix} - a_2 \begin{vmatrix} b_1 & c_1 \\ b_3 & c_3 \end{vmatrix} + a_3 \begin{vmatrix} b_1 & c_1 \\ b_2 & c_2 \end{vmatrix}$$

EXAMPLE 7.4.2   Evaluate the determinant $\begin{vmatrix} 1 & 2 & 3 \\ 3 & -1 & 2 \\ 4 & -3 & 1 \end{vmatrix}$

Solution   By the definition,

$$\begin{vmatrix} 1 & 2 & 3 \\ 3 & -1 & 2 \\ 4 & -3 & 1 \end{vmatrix} = 1\begin{vmatrix} -1 & 2 \\ -3 & 1 \end{vmatrix} - 3\begin{vmatrix} 2 & 3 \\ -3 & 1 \end{vmatrix} + 4\begin{vmatrix} 2 & 3 \\ -1 & 2 \end{vmatrix}$$

$$= 1[(-1)(1) - (-3)(2)] - 3[(2)(1) - (-3)(3)]$$
$$+ 4[(2)(2) - (-1)(3)]$$
$$= 1(5) - 3(11) + 4(7)$$
$$= 5 - 33 + 28$$
$$= 0$$

The **order of a determinant** is the number of rows (or columns) in the square array associated with the determinant.

The **expansion of a determinant** is the indicated calculation in the definition of the determinant.

The **minor** of an element of a determinant is the determinant obtained by deleting the row and column in which the element lies.

For example, for a determinant of order three,

$A_1 = \begin{vmatrix} b_2 & c_2 \\ b_3 & c_3 \end{vmatrix}$ is the minor of a_1.

$A_2 = \begin{vmatrix} b_1 & c_1 \\ b_3 & c_3 \end{vmatrix}$ is the minor of a_2.

$A_3 = \begin{vmatrix} b_1 & c_1 \\ b_2 & c_2 \end{vmatrix}$ is the minor of a_3.

$B_2 = \begin{vmatrix} a_1 & c_1 \\ a_3 & c_3 \end{vmatrix}$ is the minor of b_2, and so on.

Thus $D = \begin{vmatrix} a_1 & b_1 & c_1 \\ a_2 & b_2 & c_2 \\ a_3 & b_3 & c_3 \end{vmatrix} = a_1 A_1 - a_2 A_2 + a_3 A_3$

is called an expansion of D by the minors of the elements of the first column.

EXAMPLE 7.4.3   Evaluate the determinant $\begin{vmatrix} 1 & 2 & 3 \\ 3 & -1 & 2 \\ 4 & -3 & 1 \end{vmatrix}$ by expansion of the minors of the first column.

Solution　$A_1 = \begin{vmatrix} -1 & 2 \\ -3 & 1 \end{vmatrix} = (-1)(1) - (-3)(2) = 5$

$$A_2 = \begin{vmatrix} 2 & 3 \\ -3 & 1 \end{vmatrix} = (2)(1) - (-3)(3) = 11$$

$$A_3 = \begin{vmatrix} 2 & 3 \\ -1 & 2 \end{vmatrix} = (2)(2) - (-1)(3) = 7$$

$$D = a_1A_1 - a_2A_2 + a_3A_3$$
$$= 1(5) - 3(11) + 4(7)$$
$$= 5 - 33 + 28 = 0$$

(Compare this solution with Example 7.4.2.)

THEOREM I

Any row or column may be used to expand a determinant of the third order—that is,

$$D = a_1A_1 - a_2A_2 + a_3A_3$$
$$= -b_1B_1 + b_2B_2 - b_3B_3$$
$$= c_1C_1 - c_2C_2 + c_3C_3$$
$$= a_1A_1 - b_1B_1 + c_1C_1$$
$$= -a_2A_2 + b_2B_2 - c_2C_2$$
$$= a_3A_3 - b_3B_3 + c_3C_3$$

Whether a product in the expansion is to be multiplied by $+1$ or -1 is determined by the following array, called the checkerboard of signs:

$$\begin{vmatrix} + & - & + \\ - & + & - \\ + & - & + \end{vmatrix}$$

The proof given below shows that $a_1A_1 - a_2A_2 + a_3A_3 = -b_1B_1 + b_2B_2 - b_3B_3$. The other cases are proved similarly and are left for the student to verify.

$$a_1A_1 - a_2A_2 + a_3A_3 = a_1\begin{vmatrix} b_2 & c_2 \\ b_3 & c_3 \end{vmatrix} - a_2\begin{vmatrix} b_1 & c_1 \\ b_3 & c_3 \end{vmatrix} + a_3\begin{vmatrix} b_1 & c_1 \\ b_2 & c_2 \end{vmatrix}$$
$$= a_1(b_2c_3 - b_3c_2) - a_2(b_1c_3 - b_3c_1) + a_3(b_1c_2 - b_2c_1)$$
$$= a_1b_2c_3 + a_2b_3c_1 + a_3b_1c_2 - a_1b_3c_2 - a_2b_1c_3 - a_3b_2c_1$$
$$= -b_1(a_2c_3 - a_3c_2) + b_2(a_1c_3 - a_3c_1)$$
$$\quad - b_3(a_1c_2 - a_2c_1)$$
$$= -b_1\begin{vmatrix} a_2 & a_3 \\ c_2 & c_3 \end{vmatrix} + b_2\begin{vmatrix} a_1 & a_3 \\ c_1 & c_3 \end{vmatrix} - b_3\begin{vmatrix} a_1 & a_2 \\ c_1 & c_2 \end{vmatrix}$$
$$= -b_1B_1 + b_2B_2 - b_3B_3$$

EXAMPLE 7.4.4 Expand $\begin{vmatrix} 2 & 3 & -1 \\ 1 & -2 & 2 \\ -4 & -1 & 5 \end{vmatrix}$ by row 1.

Solution $D = 2 \begin{vmatrix} -2 & 2 \\ -1 & 5 \end{vmatrix} - 3 \begin{vmatrix} 1 & 2 \\ -4 & 5 \end{vmatrix} + (-1) \begin{vmatrix} 1 & -2 \\ -4 & -1 \end{vmatrix}$

$= 2(-10 - (-2)) - 3(5 - (-8)) - (-1 - 8)$

$= 2(-8) - 3(13) - (-9)$

$= -16 - 39 + 9 = -46$

EXAMPLE 7.4.5 Expand $\begin{vmatrix} 3 & 2 & -5 \\ 2 & 0 & 1 \\ 1 & 0 & 4 \end{vmatrix}$ by column 2.

Solution $D = -2 \begin{vmatrix} 2 & 1 \\ 1 & 4 \end{vmatrix} + 0 \begin{vmatrix} 3 & -5 \\ 1 & 4 \end{vmatrix} - 0 \begin{vmatrix} 3 & -5 \\ 2 & 1 \end{vmatrix}$

$= -2(8 - 1) + 0 - 0$

$= -14$

The two preceding examples illustrate that the determinant is much easier to evaluate when there are zeros in a row or column. Thus it is desirable to establish some theorems indicating the transformations that can be performed on determinants in order to obtain an equal determinant with zeros as some of its elements.

THEOREM 2

Two determinants of the same order are equal if one is obtained from the other by **interchanging the rows and columns.**

In symbols,

$$\begin{vmatrix} a_1 & b_1 \\ a_2 & b_2 \end{vmatrix} = \begin{vmatrix} a_1 & a_2 \\ b_1 & b_2 \end{vmatrix} \quad \text{and} \quad \begin{vmatrix} a_1 & b_1 & c_1 \\ a_2 & b_2 & c_2 \\ a_3 & b_3 & c_3 \end{vmatrix} = \begin{vmatrix} a_1 & a_2 & a_3 \\ b_1 & b_2 & b_3 \\ c_1 & c_2 & c_3 \end{vmatrix}$$

This is proved by applying the definition, regrouping the terms, and applying the definition again.

THEOREM 3

If two rows (or two columns) of a determinant are interchanged, then the resulting determinant is the negative of the original one.

For example, let $D = \begin{vmatrix} a_1 & b_1 & c_1 \\ a_2 & b_2 & c_2 \\ a_3 & b_3 & c_3 \end{vmatrix}$ and let $E = \begin{vmatrix} c_1 & b_1 & a_1 \\ c_2 & b_2 & a_2 \\ c_3 & b_3 & a_3 \end{vmatrix}$.

Then $E = -D$.

This is proved by expanding E, regrouping the terms, and applying the definition.

THEOREM 4

If each element of a row (or column) is multiplied by a constant, then the determinant is multiplied by a constant.

For example,

$$\begin{vmatrix} ka_1 & kb_1 & kc_1 \\ a_2 & b_2 & c_2 \\ a_3 & b_3 & c_3 \end{vmatrix} = k \begin{vmatrix} a_1 & b_1 & c_1 \\ a_2 & b_2 & c_2 \\ a_3 & b_3 & c_3 \end{vmatrix}$$

and

$$\begin{vmatrix} 6 & 9 & -12 \\ 2 & 1 & 5 \\ 3 & -1 & 2 \end{vmatrix} = 3 \begin{vmatrix} 2 & 3 & -4 \\ 2 & 1 & 5 \\ 3 & -1 & 2 \end{vmatrix}$$

The proof is left for the student.

THEOREM 5

If the corresponding elements of two rows (or columns) are equal, then the value of the determinant is 0.

For example,

$$\begin{vmatrix} 1 & 4 & 1 \\ 3 & 1 & 3 \\ -2 & 5 & -2 \end{vmatrix} = 0 \quad \text{and} \quad \begin{vmatrix} 2 & 1 & 5 \\ 2 & 1 & 5 \\ 3 & 7 & 9 \end{vmatrix} = 0$$

The proof is left for the student.

THEOREM 6

If each element of a row (or column) is multiplied by a constant and then added to the corresponding element of another row (or column), then the resulting determinant is equal to the original determinant.

For example,

$$\begin{vmatrix} a_1 + ka_3 & b_1 + kb_3 & c_1 + kc_3 \\ a_2 & b_2 & c_2 \\ a_3 & b_3 & c_3 \end{vmatrix} = \begin{vmatrix} a_1 & b_1 & c_1 \\ a_2 & b_2 & c_2 \\ a_3 & b_3 & c_3 \end{vmatrix}$$

and

$$\begin{vmatrix} a_1 & b_1 + kc_1 & c_1 \\ a_2 & b_2 + kc_2 & c_2 \\ a_3 & b_3 + kc_3 & c_3 \end{vmatrix} = \begin{vmatrix} a_1 & b_1 & c_1 \\ a_2 & b_2 & c_2 \\ a_3 & b_3 & c_3 \end{vmatrix}$$

This theorem may be proved by expanding the determinant by the altered row (or column) and expressing the expansion as the sum of two determinants, one of which is zero.

For example,

$$\begin{vmatrix} a_1 + ka_3 & b_1 + kb_3 & c_1 + kc_3 \\ a_2 & b_2 & c_2 \\ a_3 & b_3 & c_3 \end{vmatrix}$$

$$= (a_1 + ka_3)A_1 - (b_1 + kb_3)B_1 + (c_1 + kc_3)C_1$$

$$= a_1A_1 - b_1B_1 + c_1C_1 + k(a_3A_1 - b_3B_1 + c_3C_1)$$

$$= \begin{vmatrix} a_1 & b_1 & c_1 \\ a_2 & b_2 & c_2 \\ a_3 & b_3 & c_3 \end{vmatrix} + k\begin{vmatrix} a_3 & b_3 & c_3 \\ a_2 & b_2 & c_2 \\ a_3 & b_3 & c_3 \end{vmatrix}$$

$$= \begin{vmatrix} a_1 & b_1 & c_1 \\ a_2 & b_2 & c_2 \\ a_3 & b_3 & c_3 \end{vmatrix} + k(0) \text{ (By theorem 5)}$$

$$= \begin{vmatrix} a_1 & b_1 & c_1 \\ a_2 & b_2 & c_2 \\ a_3 & b_3 & c_3 \end{vmatrix}$$

EXAMPLE 7.4.6   Find a determinant equal to $\begin{vmatrix} -2 & 3 & 1 \\ 1 & 2 & 3 \\ 2 & 3 & 3 \end{vmatrix}$ having zeros everywhere in column 3 except the first row. Expand the resulting determinant.

Solution

$$
\begin{vmatrix} -2 & 3 & 1 \\ 1 & 2 & 3 \\ 2 & 3 & 3 \end{vmatrix} = \begin{vmatrix} -2 & 3 & 1 \\ 7 & -7 & 0 \\ 2 & 3 & 3 \end{vmatrix}
$$
(Multiply row 1 by -3 and add to row 2)

$$
= \begin{vmatrix} -2 & 3 & 1 \\ 7 & -7 & 0 \\ 8 & -6 & 0 \end{vmatrix}
$$
(Multiply row 1 by -3 and add to row 3)

$$
= 1 \begin{vmatrix} 7 & -7 \\ 8 & -6 \end{vmatrix} = 7(-6) - 8(-7) = -42 + 56 = 14
$$

EXAMPLE 7.4.7   Expand the determinant having zeros everywhere in row 2 except column 2 and equal to $\begin{vmatrix} 2 & -1 & 4 \\ 5 & 1 & -2 \\ 3 & -3 & -4 \end{vmatrix}$.

Solution

$$
\begin{vmatrix} 2 & -1 & 4 \\ 5 & 1 & -2 \\ 3 & -3 & -4 \end{vmatrix} = \begin{vmatrix} 7 & -1 & 4 \\ 0 & 1 & -2 \\ 18 & -3 & -4 \end{vmatrix}
$$
(Multiply column 2 by -5 and add to column 1)

$$
= \begin{vmatrix} 7 & -1 & 2 \\ 0 & 1 & 0 \\ 18 & -3 & -10 \end{vmatrix}
$$
(Multiply column 2 by 2 and add to column 3)

$$
= 1 \begin{vmatrix} 7 & 2 \\ 18 & -10 \end{vmatrix}
$$

$$
= 7(-10) - 18(2) = -70 - 36 = -106
$$

EXERCISES 7.4 A

Expand each given determinant in Exercises 1–6.

1. $\begin{vmatrix} 3 & 5 \\ 4 & 9 \end{vmatrix}$

2. $\begin{vmatrix} 2 & 3 \\ 4 & -5 \end{vmatrix}$

3. $\begin{vmatrix} 1 & 0 \\ 3 & -2 \end{vmatrix}$

4. $\begin{vmatrix} 2 & -5 \\ 5 & 2 \end{vmatrix}$

5. $\begin{vmatrix} -3 & -1 \\ -4 & -7 \end{vmatrix}$

6. $\begin{vmatrix} 0 & 2 \\ -9 & 10 \end{vmatrix}$

In Exercises 7–8, expand the given determinant about the indicated row or column.

7. $\begin{vmatrix} 2 & 3 & 5 \\ 1 & 4 & 2 \\ 3 & 1 & 1 \end{vmatrix}$

    a. Column 1

    b. Row 1

8. $\begin{vmatrix} 7 & -3 & 5 \\ 1 & 2 & -3 \\ 2 & -1 & 0 \end{vmatrix}$

    a. Column 1

    b. Column 3

In Exercises 9–12, find a determinant equal to the given one and satisfying the stated conditions. Expand the determinant.

9. $\begin{vmatrix} -2 & 3 & 1 \\ 1 & 2 & 3 \\ 2 & 3 & 3 \end{vmatrix}$ Zeros everywhere in row 2 except column 1

10. $\begin{vmatrix} 1 & 1 & 1 \\ 1 & 4 & 9 \\ 1 & 8 & 27 \end{vmatrix}$ Zeros everywhere in column 1 except row 1

11. $\begin{vmatrix} -1 & 4 & -4 \\ 2 & 3 & 2 \\ 1 & -1 & 2 \end{vmatrix}$ Zeros everywhere in column 3 except row 3

12. $\begin{vmatrix} 5 & -2 & 3 \\ -12 & 3 & 9 \\ 4 & 1 & -2 \end{vmatrix}$ Zeros everywhere in row 2 except column 2

In Exercises 13–14, find an equal determinant having zeros everywhere below the main diagonal.

13. $\begin{vmatrix} 1 & 2 & 3 \\ 1 & 4 & 9 \\ 1 & 8 & 27 \end{vmatrix}$

14. $\begin{vmatrix} 2 & 3 & 4 \\ 1 & -5 & 6 \\ 4 & -7 & -1 \end{vmatrix}$

For Exercises 15–16, the definition and theorems for third-order determinants can be generalized for fourth-order or higher-order determinants. Thus

$$\begin{vmatrix} a_1 & b_1 & c_1 & d_1 \\ a_2 & b_2 & c_2 & d_2 \\ a_3 & b_3 & c_3 & d_3 \\ a_4 & b_4 & c_4 & d_4 \end{vmatrix} = a_1 A_1 - a_2 A_2 + a_3 A_3 - a_4 A_4$$

15. Find an equal determinant having zeros everywhere in column 1 except row 1. Then expand the determinant.

$$\begin{vmatrix} 1 & 3 & 2 & -2 \\ 2 & 3 & -1 & 1 \\ -3 & 2 & 1 & 6 \\ -1 & 4 & 5 & 2 \end{vmatrix}$$

16. Find an equal determinant having zeros everywhere below the main diagonal. Then expand the determinant.

$$\begin{vmatrix} 1 & 1 & 1 & 1 \\ 1 & 2 & 3 & 4 \\ 1 & 4 & 9 & 16 \\ 1 & 8 & 27 & 64 \end{vmatrix}$$

EXERCISES 7.4 B

Expand each given determinant in Exercises 1–6.

1. $\begin{vmatrix} 2 & 5 \\ -1 & 3 \end{vmatrix}$ **2.** $\begin{vmatrix} 4 & 1 \\ 0 & -7 \end{vmatrix}$

3. $\begin{vmatrix} 1 & -1 \\ 5 & 4 \end{vmatrix}$ **4.** $\begin{vmatrix} 3 & 4 \\ -4 & 3 \end{vmatrix}$

5. $\begin{vmatrix} 11 & 6 \\ 2 & -5 \end{vmatrix}$ **6.** $\begin{vmatrix} 4 & -3 \\ -7 & 15 \end{vmatrix}$

In Exercises 7–14, find an equal determinant having zeros everywhere as indicated. Expand the resulting determinant.

7. $\begin{vmatrix} 10 & 25 \\ 8 & 12 \end{vmatrix}$ Row 2, except column 2

8. $\begin{vmatrix} -5 & -4 \\ 3 & -2 \end{vmatrix}$ Row 1, except column 2

9. $\begin{vmatrix} 3 & 1 & -1 \\ 2 & -1 & 2 \\ 8 & -2 & -1 \end{vmatrix}$ Row 1, except column 3

10. $\begin{vmatrix} 4 & 2 & -6 \\ 1 & -2 & 3 \\ 5 & -3 & -1 \end{vmatrix}$ Column 2, except row 1

11. $\begin{vmatrix} 2 & -4 & -3 \\ 1 & 5 & 2 \\ -5 & 2 & 1 \end{vmatrix}$ Row 3, except column 3

12. $\begin{vmatrix} 4 & 3 & -10 \\ 1 & -2 & 6 \\ 2 & -1 & 15 \end{vmatrix}$ Column 3, except row 2

13. $\begin{vmatrix} 1 & 2 & -2 & 3 \\ 2 & 1 & 1 & -4 \\ 4 & 3 & -5 & 1 \\ 3 & -3 & 2 & -2 \end{vmatrix}$ Row 1, except column 1

14. $\begin{vmatrix} 2 & -4 & 7 & 1 \\ 1 & 3 & 1 & 2 \\ 5 & -1 & 2 & -1 \\ -3 & 2 & 6 & -6 \end{vmatrix}$ Column 4, except row 1

15. Without expanding, show that

$$\begin{vmatrix} 1 & 1 & 2x - 2y \\ 0 & x & x^2 - xy \\ x & y & x^2 - y^2 \end{vmatrix} = 0$$

State the reason for each step.

16. Without expanding, show that

$$\begin{vmatrix} 1 & 1 & 1 & 1 \\ 1 & a & a^2 & a^3 \\ 1 & b & b^2 & b^3 \\ 1 & c & c^2 & c^3 \end{vmatrix} = (a - 1)(b - 1)(c - 1)(b - a)(c - a)(c - b)$$

7.5 SOLUTION OF LINEAR SYSTEMS USING DETERMINANTS (Optional)

Case 1: Two equations, two variables. In solving the system

$$a_1 x + b_1 y = k_1$$
$$a_2 x + b_2 y = k_2$$

by the addition method, the following equations were obtained:

$$(a_1 b_2 - a_2 b_1)x = b_2 k_1 - b_1 k_2$$
$$(a_1 b_2 - a_2 b_1)y = a_1 k_2 - a_2 k_1$$

Now letting

$$D = \begin{vmatrix} a_1 & b_1 \\ a_2 & b_2 \end{vmatrix}, \quad X = \begin{vmatrix} k_1 & b_1 \\ k_2 & b_2 \end{vmatrix}, \quad \text{and} \quad Y = \begin{vmatrix} a_1 & k_1 \\ a_2 & k_2 \end{vmatrix}$$

These equations become $Dx = X$, $Dy = Y$.

The determinant D is called the *determinant of the coefficients*.

If $D \neq 0$, then there is a unique solution, $(X/D, Y/D)$ (the case of two intersecting lines).

If $D = 0$ and $X = Y = 0$, then the solution set is infinite (the case of coincident lines).

If $D = 0$ and either $X \neq 0$ or $Y \neq 0$, then the solution set is empty (the case of parallel lines).

EXAMPLE 7.5.1   Use determinants to solve the system

$$3x + 2y = 5$$
$$x - 3y = 2$$

Solution $D = \begin{vmatrix} 3 & 2 \\ 1 & -3 \end{vmatrix} = (3)(-3) - (1)(2) = -9 - 2 = -11$

$$X = \begin{vmatrix} 5 & 2 \\ 2 & -3 \end{vmatrix} = (5)(-3) - (2)(2) = -15 - 4 = -19$$

$$Y = \begin{vmatrix} 3 & 5 \\ 1 & 2 \end{vmatrix} = (3)(2) - (1)(5) = 6 - 5 = 1$$

$$\frac{X}{D} = \frac{-19}{-11} = \frac{19}{11}, \frac{Y}{D} = \frac{1}{-11} = -\frac{1}{11}$$

Thus $(\frac{19}{11}, -\frac{1}{11})$ is the unique solution.

Check: $3(\frac{19}{11}) + 2(-\frac{1}{11}) = \frac{57}{11} - \frac{2}{11} = \frac{55}{11} = 5$

                   $\frac{19}{11} - 3(-\frac{1}{11}) = \frac{19}{11} + \frac{3}{11} = \frac{22}{11} = 2$

Case 2: Three equations, three variables. If in the system

$$a_1 x + b_1 y + c_1 z = k_1$$
$$a_2 x + b_2 y + c_2 z = k_2$$
$$a_3 x + b_3 y + c_3 z = k_3$$

the first equation is multiplied by A_1, the minor of a_1 for the determinant of the coefficients, and the second equation is multiplied by $-A_2$, and the third equation is multiplied by A_3, and the resulting three equations are added, then

$$(a_1 A_1 - a_2 A_2 + a_3 A_3)x + (b_1 A_1 - b_2 A_2 + b_3 A_3)y$$
$$+ (c_1 A_1 - c_2 A_2 + c_3 A_3)z = k_1 A_1 - k_2 A_2 + k_3 A_3$$

The coefficients of the variables and the constant term can be recognized as the expansions of determinants as follows:

$$\begin{vmatrix} a_1 & b_1 & c_1 \\ a_2 & b_2 & c_2 \\ a_3 & b_3 & c_3 \end{vmatrix} x + \begin{vmatrix} b_1 & b_1 & c_1 \\ b_2 & b_2 & c_2 \\ b_3 & b_3 & c_3 \end{vmatrix} y + \begin{vmatrix} c_1 & b_1 & c_1 \\ c_2 & b_2 & c_2 \\ c_3 & b_3 & c_3 \end{vmatrix} z = \begin{vmatrix} k_1 & b_1 & c_1 \\ k_2 & b_2 & c_2 \\ k_3 & b_3 & c_3 \end{vmatrix}$$

Thus
$$\begin{vmatrix} a_1 & b_1 & c_1 \\ a_2 & b_2 & c_2 \\ a_3 & b_3 & c_3 \end{vmatrix} x + 0 \cdot y + 0 \cdot z = \begin{vmatrix} k_1 & b_1 & c_1 \\ k_2 & b_2 & c_2 \\ k_3 & b_3 & c_3 \end{vmatrix}$$

since the determinants which are the coefficients of y and z each contain two identical columns and thus equal 0.

Similar equations may be obtained where x and y or x and z are eliminated. Finally,

$$\begin{vmatrix} a_1 & b_1 & c_1 \\ a_2 & b_2 & c_2 \\ a_3 & b_3 & c_3 \end{vmatrix} x = \begin{vmatrix} k_1 & b_1 & c_1 \\ k_2 & b_2 & c_2 \\ k_3 & b_3 & c_3 \end{vmatrix}$$

$$\begin{vmatrix} a_1 & b_1 & c_1 \\ a_2 & b_2 & c_2 \\ a_3 & b_3 & c_3 \end{vmatrix} y = \begin{vmatrix} a_1 & k_1 & c_1 \\ a_2 & k_2 & c_2 \\ a_3 & k_3 & c_3 \end{vmatrix}$$

$$\begin{vmatrix} a_1 & b_1 & c_1 \\ a_2 & b_2 & c_2 \\ a_3 & b_3 & c_3 \end{vmatrix} z = \begin{vmatrix} a_1 & b_1 & k_1 \\ a_2 & b_2 & k_2 \\ a_3 & b_3 & k_3 \end{vmatrix}$$

The determinant of the coefficients D may be readily obtained from the equations when they are expressed in the form stated at the beginning of this section. The determinants on the right of each equation above can be obtained from the determinant of the coefficients by replacing the column containing the coefficients of the variable in the equation by the constant terms of the system.

Now designating the determinants on the right of each equation by X, Y, and Z, respectively, these equations can be expressed as follows:

$$Dx = X$$
$$Dy = Y$$
$$Dz = Z$$

If $D \neq 0$, then there is exactly one solution: $(X/D, Y/D, Z/D)$.

If $D = 0$ and $X = Y = Z = 0$, then the solution set is infinite.

If $D = 0$ and either $X \neq 0$ or $Y \neq 0$ or $Z \neq 0$, then the solution set is empty.

This solution of linear systems using determinants is known as *Cramer's rule* in honor of the Swiss mathematician Gabriel Cramer (1704–1752).

EXAMPLE 7.5.2 Solve by using determinants, and check:

$$
\begin{aligned}
x + y + z &= 4 \\
2x - y - 2z &= -1 \\
x - 2y - z &= 1
\end{aligned}
$$

Solution

$$
D = \begin{vmatrix} 1 & 1 & 1 \\ 2 & -1 & -2 \\ 1 & -2 & -1 \end{vmatrix} = \begin{vmatrix} 1 & 1 & 2 \\ 2 & -1 & 0 \\ 1 & -2 & 0 \end{vmatrix} = 2 \begin{vmatrix} 2 & -1 \\ 1 & -2 \end{vmatrix} = 2(-4 + 1) = -6
$$

$$
X = \begin{vmatrix} 4 & 1 & 1 \\ -1 & -1 & -2 \\ 1 & -2 & -1 \end{vmatrix} = \begin{vmatrix} 4 & -3 & -7 \\ -1 & 0 & 0 \\ 1 & -3 & -3 \end{vmatrix} = \begin{vmatrix} -3 & -7 \\ -3 & -3 \end{vmatrix} = \begin{vmatrix} 3 & 7 \\ 3 & 3 \end{vmatrix}
$$

$$
= 9 - 21 = -12
$$

$$
Y = \begin{vmatrix} 1 & 4 & 1 \\ 2 & -1 & -2 \\ 1 & 1 & -1 \end{vmatrix} = \begin{vmatrix} 1 & 4 & 1 \\ 0 & -3 & 0 \\ 1 & 1 & -1 \end{vmatrix} = -3 \begin{vmatrix} 1 & 1 \\ 1 & -1 \end{vmatrix} = -3(-1 - 1)
$$

$$
= 6
$$

$$
Z = \begin{vmatrix} 1 & 1 & 4 \\ 2 & -1 & -1 \\ 1 & -2 & 1 \end{vmatrix} = \begin{vmatrix} 1 & 1 & 4 \\ 2 & -1 & -1 \\ 3 & -3 & 0 \end{vmatrix} = \begin{vmatrix} 1 & 2 & 4 \\ 2 & 1 & -1 \\ 3 & 0 & 0 \end{vmatrix} = 3 \begin{vmatrix} 2 & 4 \\ 1 & -1 \end{vmatrix} = -18
$$

Thus
$$
x = \frac{X}{D} = \frac{-12}{-6} = 2
$$

$$
y = \frac{Y}{D} = \frac{6}{-6} = -1
$$

$$
z = \frac{Z}{D} = \frac{-18}{-6} = 3
$$

The solution set is $(2, -1, 3)$.

Check: $x + y + z = 2 - 1 + 3 = 4$

$2x - y - 2z = 2(2) - (-1) - 2(3) = 4 + 1 - 6 = -1$

$x - 2y - z = 2 - 2(-1) - 3 = 2 + 2 - 3 = 1$

EXAMPLE 7.5.3 Solve by using determinants:

$$2x - 4y + 2z = 3$$
$$x + y - z = 2$$
$$3x - 6y + 3z = 2$$

Solution

$$D = \begin{vmatrix} 2 & -4 & 2 \\ 1 & 1 & -1 \\ 3 & -6 & 3 \end{vmatrix} = \begin{vmatrix} 4 & -2 & 2 \\ 0 & 0 & -1 \\ 6 & -3 & 3 \end{vmatrix} = -(-1) \begin{vmatrix} 4 & -2 \\ 6 & -3 \end{vmatrix} = -12 + 12 = 0$$

$$X = \begin{vmatrix} 3 & -4 & 2 \\ 2 & 1 & -1 \\ 2 & -6 & 3 \end{vmatrix} = \begin{vmatrix} 7 & -2 & 2 \\ 0 & 0 & -1 \\ 8 & -3 & 3 \end{vmatrix} = -(-1) \begin{vmatrix} 7 & -2 \\ 8 & -3 \end{vmatrix} = -21 + 16$$

$$= -5$$

Since $D = 0$ and $X \neq 0$, the solution set is the empty set, $\emptyset$.

EXAMPLE 7.5.4 Solve by using determinants, and check:

$$x + 2y - 3z = 4$$
$$2x - y + z = 1$$
$$3x + y - 2z = 5$$

Solution

$$D = \begin{vmatrix} 1 & 2 & -3 \\ 2 & -1 & 1 \\ 3 & 1 & -2 \end{vmatrix} = \begin{vmatrix} 1 & 0 & 0 \\ 2 & -5 & 7 \\ 3 & -5 & 7 \end{vmatrix} = 0$$

$$X = \begin{vmatrix} 4 & 2 & -3 \\ 1 & -1 & 1 \\ 5 & 1 & -2 \end{vmatrix} = \begin{vmatrix} 6 & 2 & -1 \\ 0 & -1 & 0 \\ 6 & 1 & -1 \end{vmatrix} = -1 \begin{vmatrix} 6 & -1 \\ 6 & -1 \end{vmatrix} = 0$$

$$Y = \begin{vmatrix} 1 & 4 & -3 \\ 2 & 1 & 1 \\ 3 & 5 & -2 \end{vmatrix} = \begin{vmatrix} 1 & 1 & -3 \\ 2 & 2 & 1 \\ 3 & 3 & -2 \end{vmatrix} = 0$$

$$Z = \begin{vmatrix} 1 & 2 & 4 \\ 2 & -1 & 1 \\ 3 & 1 & 5 \end{vmatrix} = \begin{vmatrix} 1 & 2 & 4 \\ 3 & 1 & 5 \\ 3 & 1 & 5 \end{vmatrix} = 0$$

Thus the solution set is infinite. Now try to solve two equations for which the determinant of the coefficients of two of the variables, say x and y, is *not* zero. Using the first and second equations,

$$x + 2y = 4 + 3z$$
$$2x - y = 1 - z$$

$$D = \begin{vmatrix} 1 & 2 \\ 2 & -1 \end{vmatrix} = -1 - 4 = -5$$

$$X = \begin{vmatrix} 4 + 3z & 2 \\ 1 - z & -1 \end{vmatrix} = -6 - z$$

$$Y = \begin{vmatrix} 1 & 4 + 3z \\ 2 & 1 - z \end{vmatrix} = -7 - 7z$$

Thus $x = \dfrac{X}{D} = \dfrac{-6 - z}{-5} = \dfrac{6 + z}{5}$ and $y = \dfrac{Y}{D} = \dfrac{-7 - 7z}{-5} = \dfrac{7 + 7z}{5}$

The solution set is $\left\{ \left(\dfrac{6 + z}{5}, \dfrac{7 + 7z}{5}, z \right) \middle| z \text{ is any real number} \right\}$

Check: $x + 2y - 3z = \dfrac{6 + z}{5} + 2\left(\dfrac{7 + 7z}{5} \right) - 3z$

$$= \dfrac{6 + 14 + z + 14z - 15z}{5} = 4$$

$$2x - y + z = 2\left(\dfrac{6 + z}{5} \right) - \dfrac{7 + 7z}{5} + z$$

$$= \dfrac{12 - 7 + 2z - 7z + 5z}{5} = 1$$

$$3x + y - 2z = 3\left(\dfrac{6 + z}{5} \right) + \dfrac{7 + 7z}{5} - 2z$$

$$= \dfrac{18 + 7 + 3z + 7z - 10z}{5} = 5$$

EXERCISES 7.5 A

Solve each of the systems in Exercises 1–10 by using determinants, and check.

1. $3x + 5y = 2$
   $x + 2y = 0$

2. $2x - 3y = 7$
   $-5x + 4y = -14$

3. $x - 4y + 6 = 0$
   $-3x - 7y + 1 = 0$

4. $10x + 3y - 4 = 0$
   $7x + 2y - 1 = 0$

5. $x - 2y - 3z = -20$
$2x + 4y - 5z = 11$
$3x + 7y - 4z = 33$

6. $4x + y - 3z = 0$
$x - y + z = -7$
$3x + 2y - z = -5$

7. $8x + 2y = 5$
$4y - 3z = 0$
$4x + 6z = -1$

8. $x + y = z$
$2x + 2z = 3 - 2y$
$5x + 2y = z + 2$

9. $x + 2y - 4 = 0$
$4y - z + 2 = 0$
$2x + z = 10$

10. $6x + 3y - 3z = 2$
$2x + 2y - 2z = 5$
$3x - 3y + 3z = 7$

In Exercises 11–12,

If $D = \begin{vmatrix} a_1 & b_1 & c_1 & d_1 \\ a_2 & b_2 & c_2 & d_2 \\ a_3 & b_3 & c_3 & d_3 \\ a_4 & b_4 & c_4 & d_4 \end{vmatrix}$, $X = \begin{vmatrix} k_1 & b_1 & c_1 & d_1 \\ k_2 & b_2 & c_2 & d_2 \\ k_3 & b_3 & c_3 & d_3 \\ k_4 & b_4 & c_4 & d_4 \end{vmatrix}$, $Y = \begin{vmatrix} a_1 & k_1 & c_1 & d_1 \\ a_2 & k_2 & c_2 & d_2 \\ a_3 & k_3 & c_3 & d_3 \\ a_4 & k_4 & c_4 & d_4 \end{vmatrix}$,

$Z = \begin{vmatrix} a_1 & b_1 & k_1 & d_1 \\ a_2 & b_2 & k_2 & d_2 \\ a_3 & b_3 & k_3 & d_3 \\ a_4 & b_4 & k_4 & d_4 \end{vmatrix}$, *and* $T = \begin{vmatrix} a_1 & b_1 & c_1 & k_1 \\ a_2 & b_2 & c_2 & k_2 \\ a_3 & b_3 & c_3 & k_3 \\ a_4 & b_4 & c_4 & k_4 \end{vmatrix}$, *then the system*

$$a_1 x + b_1 y + c_1 z + d_1 t = k_1$$
$$a_2 x + b_2 y + c_2 z + d_2 t = k_2$$
$$a_3 x + b_3 y + c_3 z + d_3 t = k_3$$
$$a_4 x + b_4 y + c_4 z + d_4 t = k_4$$

is equivalent to $Dx = X$, $Dy = Y$, $Dz = Z$, $Dt = T$.

11. $2x + y - z + t = 1$
$x - y + 2z - t = 2$
$-x - y + z - t = -1$
$3x + y - z + 2t = 0$

12. $x - 2z = 4$
$y + z + t = 6$
$x + 3t = 6$
$x + y + z + t = 0$

EXERCISES 7.5 B

Solve each of the systems in Exercises 1–12 by using determinants, and check.

1. $4x - 3y = 0$
$2x - 3y = 6$

2. $3x + 8y = 14$
$x + 7y = 22$

3. $2x + 3y - 5 = 0$
$x - 2y + 8 = 0$

4. $6x - 2y = 9$
$9x - 8y = 1$

5. $2x + y + 3z = 15$
$2x + 7z = 25$
$3x + 2y + 6z = 35$

6. $x + 2y = 2 + z$
$x - 4y = 5z - 7$
$x + 3y + 4z = 5$

7. $x + 2y = z + 7$
$4x + 3y = 1 - 2z$
$9x + 8y = 4 - 3z$

8. $5x + 2y - z + 1 = 0$
$2x - y + z = 0$
$3x + z - 1 = 0$

9. $2x - 4y + 7z = 5$
$3x + 2y - z = 2$
$x - 10y + 15z = 8$

10. $x - 2y + 3 = 0$
$4x + z - 2 = 0$
$3x + 2y - 2z = 5$

11. $x + y + z + t = 2$
$2x + 2y - z = -3$
$y - 2z - 2t = 1$
$x + y - t = 5$

12. $x + y + z = 14$
$x + z + t = 13$
$y + z + t = 10$
$x - y + z - t = 9$

7.6 VERBAL PROBLEMS

There are many modern applications of linear systems involving three or more variables. Some problems require the solution of 100 linear equations in 100 variables, with the electronic computer programmed to perform the arithmetical calculations.

Matrices, determinants, and linear systems are used to solve problems in modern physics and chemistry related to quantum mechanics and atomic structure, in biology related to genetics and heredity, in sociology and psychology requiring data analysis, in economics and industry pertaining to cost analysis, research, and management, and in military tactics.

EXERCISES 7.6 A

1. There are 47 coins in a collection of nickels, dimes, and quarters. The total value of the collection is $5. The total number of nickels and quarters is 3 less than the number of dimes. How many coins of each kind are there?

2. Three machines, working together, require 20 minutes to complete a certain job. If the first two machines require 30 minutes to complete the job when working together, and if the first and the third require 36 minutes, find the time it would take for each machine alone.

3. Applying Kirchhoff's laws to the electrical circuit at the right, the following equations are obtained:

$$
\begin{aligned}
I_1 - I_2 - I_3 &= 0 \\
5I_1 \qquad + 20I_3 &= 100 \\
15I_2 - 20I_3 &= 15 \\
5I_1 + 15I_2 \qquad &= 115
\end{aligned}
$$

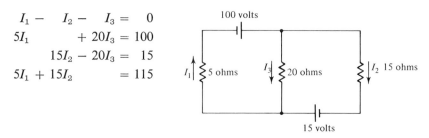

FIGURE 7.6.1

Solve this system by showing that the common solution of the first three equations is also a solution of the fourth equation.

4. A man invested $40,000 in three different investments. The first yielded 6 percent, the second 5 percent, and the third 10 percent. The total income was $2660. If the last two amounts had been interchanged, his income would have been $2860. Find the amount of money invested at each rate.

5. A factory has 2400 pounds of A, 310 pounds of B, and 28 pounds of C. Product P requires 25 pounds of A and 5 pounds of B. Product Q requires 20 pounds of A, 2 pounds of B, and 1 pound of C. Product R requires 150 pounds of A, 10 pounds of B, and $\frac{1}{2}$ pound of C. How many items of each product should be produced in order to use all of the raw material?

6. Find a, b, and c for $y = ax^2 + bx + c$ if the points $P:(2, 6)$, $Q:(-3, 21)$, and $R:(1, 5)$ are on the parabola.

7. Find the amounts of the following three foods that will provide precisely the daily minimum vitamin and mineral requirements indicated in the last column of the array. The content of each food is given in milligrams per ounce.

Vitamin	Food			Daily Minimum Requirement
	I	I	III	
Thiamine	0.2	0.2	0.5	2.00
Niacin	90	30	30	450.00
Iron	1.5	2	1.5	11.0

8. A problem in genetics is to determine the probable genetic structure of a future generation if the offspring are repeatedly bred with hybrids. The solution is obtained by determining a unique fixed-point probability vector—that is, by solving the matrix equation

$$\begin{bmatrix} \frac{1}{2} & \frac{1}{4} & 0 \\ \frac{1}{2} & \frac{1}{2} & \frac{1}{2} \\ 0 & \frac{1}{4} & \frac{1}{2} \end{bmatrix} \begin{bmatrix} d \\ h \\ r \end{bmatrix} = \begin{bmatrix} d \\ h \\ r \end{bmatrix}$$

Solve this equation for (d, h, r) if $d + h + r = 1$.

EXERCISES 7.6 B

1. Tickets to a certain campus show were 50 cents for students, 75 cents for faculty, and $1 for the general public. The total sales from 550 people amounted to $337.50. If there were 4 times as many student tickets sold as general public tickets, how many tickets of each type were sold?

2. Two men A and B when working together can do a certain job in 4 days. If B and a young boy C work together 6 days, then B can complete the job alone in 2 more days. If A and C work together, they can complete the job in 6 days.

   a. How long would it take for each person working alone to complete the job?

   b. If A can be hired for $4.80 per hour, what should be the equivalent rate for each of the other two workers?

3. Three machines A, B, and C each are able to produce three different products P, Q, and R. The following table indicates the number of hours required by each machine to make each product. If machine A

Machine	Product		
	P	Q	R
A	2	1	3
B	4	2	2
C	1	3	2

can operate 14 hours a day; B, 16 hours, and C, 11 hours, find the number of items of each product the factory is able to produce when all three machines work to full capacity.

4. Applying Kirchhoff's laws to the electrical circuit shown, the following equations are obtained. Solve for I_1, I_2, I_3.

$$I_1 - I_2 - I_3 = 0$$
$$2I_1 + 6I_2 = 22$$
$$6I_2 - 4I_3 = 0$$
$$2I_1 + 4I_3 = 22$$

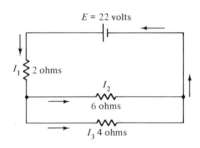

FIGURE 7.6.2

5. The copper content of U.S. coins is 95 percent for the cent, 75 percent for the nickel, and 10 percent for the dime. For 82 pounds of coins, all of the nickels and one third of the dimes were made from 50 pounds of ore containing 50 percent copper and the rest of the dimes and all of the cents were made from 50 pounds of ore containing 42 percent copper. How many pounds of each type of coin were there?

6. A man went to a bus station in a cab averaging 25 mph and took a bus averaging 40 mph to an airport. At the airport he boarded a plane, which flew him to his destination. The plane averaged 600 mph. The entire trip of 1469 miles required 3 hours and 12 minutes. The plane trip took 3 times as long as the other two trips combined. How much time was spent in each type of travel?

7. Find the equation of the plane $Ax + By + Cz = D$ determined by the three points $P:(2, -1, 0)$, $Q:(0, 3, 4)$, $R:(-3, 0, 5)$.

8. Find all possible amounts of the following four foods that will provide precisely the amounts of nutrients indicated in the last column of the array if each contains the amounts per unit as indicated.

Nutrient	Food				
	I	II	III	IV	Total
A	1	3	3	0	12
B	2	2	0	3	26
C	3	1	5	4	44
D	3	9	1	2	32

If food *I* costs 40 cents per unit, food *II* 40 cents per unit, food *III* 10 cents per unit, and food *IV* 20 cents per unit, is there a solution costing exactly $1? Exactly $3?

SUMMARY

☐ The **solution set** of a **linear equation in three variables** x, y, and z is $\{(x, y, z) \mid ax + by + cz = d$, where x, y, and z are real numbers$\}$.

☐ The **solution set of a system of three linear equations in three variables** is the intersection of the solution sets of each of the equations in the system.

☐ **Methods for solving a system** of linear equations in three variables are:

    1. Elimination of a variable by adding linear combinations of pairs of equations

    2. Transformation to a matrix equation

    3. Application of determinants

☐ **Matrix transformations** that produce equivalent associated linear systems are:

    1. Interchanging rows

    2. Multiplying each element of a row by a constant

    3. Adding a constant multiple of a row to another row

☐ The **expansion of a determinant** can be accomplished by using any row or any column.

☐ If a **constant multiple of any row (or column) is added to another** row (or column), then the value of the determinant is unchanged.

☐ **Cramer's rule**

$$\text{If } D = \begin{vmatrix} a_1 & b_1 & c_1 \\ a_2 & b_2 & c_2 \\ a_3 & b_3 & c_3 \end{vmatrix}, \ X = \begin{vmatrix} k_1 & b_1 & c_1 \\ k_2 & b_2 & c_2 \\ k_3 & b_3 & c_3 \end{vmatrix}, \ Y = \begin{vmatrix} a_1 & k_1 & c_1 \\ a_2 & k_2 & c_2 \\ a_3 & k_3 & c_3 \end{vmatrix},$$

$$Z = \begin{vmatrix} a_1 & b_1 & k_1 \\ a_2 & b_2 & k_2 \\ a_3 & b_3 & k_3 \end{vmatrix}$$

$$\text{then } \begin{cases} a_1x + b_1y + c_1z = k_1 \\ a_2x + b_2y + c_2z = k_2 \\ a_3x + b_3y + c_3z = k_3 \end{cases} \text{ is equivalent to } \begin{cases} Dx = X \\ Dy = Y \\ Dz = Z \end{cases}$$

REVIEW EXERCISES

1. Add:

$$\begin{bmatrix} 2 & -4 & 6 \\ 4 & 5 & -4 \\ -1 & 2 & 7 \end{bmatrix} + \begin{bmatrix} 4 & 2 & -3 \\ 6 & -8 & -3 \\ 1 & 6 & 3 \end{bmatrix}$$

2. Multiply:

$$\begin{bmatrix} 2 & 5 & 3 & 0 \\ 3 & 1 & 2 & -1 \\ 1 & 7 & 0 & -2 \end{bmatrix} \begin{bmatrix} 2 & x \\ 3 & y \\ -1 & z \\ 1 & t \end{bmatrix}$$

3. By using elementary equivalence transformations, transform the matrix so that the entries below the main diagonal are zeros.

$$\begin{bmatrix} 2 & -2 & 4 & -2 \\ 1 & -3 & 2 & -5 \\ -3 & 6 & 1 & 2 \end{bmatrix}$$

4. Find an equal determinant having zeros below the main diagonal.

$$\begin{vmatrix} 2 & -4 & 6 \\ -3 & 1 & 1 \\ 5 & 0 & -3 \end{vmatrix}$$

5. Solve, by using matrices:

$$\begin{aligned} x + y + z &= 1 \\ -x - 2y + 3z &= 9 \\ 3x + 4y - 2z &= -10 \end{aligned}$$

6. Solve, using Cramer's Rule:

$$\begin{aligned} 2x - 4y - 2z &= 5 \\ 3x + 2y - z &= 3 \\ 4x - 3y - 2z &= 1 \end{aligned}$$

7. Solve:

$$\begin{bmatrix} 5 & 1 & 2 \\ 3 & -2 & 6 \\ 2 & 3 & -4 \end{bmatrix} \begin{bmatrix} x \\ y \\ z \end{bmatrix} = \begin{bmatrix} 0 \\ 0 \\ 0 \end{bmatrix}$$

8. The amounts per unit and the total amounts on hand of three different chemicals needed to manufacture three different products are given in the following table. How many units of each product should be made if all of the chemicals are used?

Chemical	Product			
	P	Q	R	Total
A	3	4	1	330
B	5	0	2	330
C	1	2	3	240

CHAPTER **8**

SEQUENCES AND SERIES

8.1 BASIC CONCEPTS

8.1.1 Sequences

There are many applications of sets of numbers that are arranged in order. One familiar example is the ordered set of natural numbers, 1, 2, 3, 4, 5, ..., used for counting. An ordered set of numbers is called a *sequence*.

DEFINITION

A **finite sequence of n terms** is a function c whose domain is the ordered set of natural numbers 1, 2, 3, ..., n and whose range is the ordered set $c(1)$, $c(2)$, $c(3)$, ..., $c(n)$, also written as $c_1, c_2, c_3, ..., c_n$.

An **infinite sequence** is a function c whose domain is the ordered set of all natural numbers, 1, 2, 3, ..., and whose range is $c(1)$, $c(2)$, $c(3)$, ..., $c(n)$, ..., also written as $c_1, c_2, c_3, ..., c_n, ...$.

The *terms* of the sequence are the elements of the range—that is, the values of $c(n)$. The first term is c_1, the second term is c_2, the third term is c_3, and the nth term or general term is c_n.

Examples of Sequences

a. 3, 7, 11, 15, ... $c(n) = 4n - 1$ Domain $= \{1, 2, 3, ...\}$
b. 1, 3, 9, 27, 81, 243. $c(n) = 3^{n-1}$ Domain $= \{1, 2, 3, 4, 5, 6\}$
c. 12, 9, 6, 3, 0. $c(n) = 15 - 3n$ Domain $= \{1, 2, 3, 4, 5\}$
d. 5, -1, $\frac{1}{5}$, $-\frac{1}{25}$, ... $c(n) = 5(-\frac{1}{5})^{n-1}$ Domain $= \{1, 2, 3, ...\}$

e. 1, $\frac{1}{2}$, $\frac{1}{3}$, $\frac{1}{4}$, $\frac{1}{5}$, ... $c(n) = \dfrac{1}{n}$ Domain $= \{1, 2, 3, ...\}$

Examples (b) and (c) above are finite sequences and examples (a), (d), and (e) are infinite.

When the rule is stated, then other terms may be found by using the rule. In example (a) above, where $c(n) = 4n - 1$, the 8th term is $c(8) = 4(8) - 1 = 32 - 1 = 31$.

In applications, the rule of a sequence function may be stated specifically or it may have to be inferred from given data. In general, if the rule is guessed from the first terms of a sequence, there is no guarantee that the rule guessed is the correct one. For example, let 1, 2, 4, ... be the first three terms of a sequence. Suppose one guesses that the rule is $c(n) = 2^{n-1}$. Then the fourth term would be $c(4) = 2^{4-1} = 2^3 = 8$. However, it is possible that the rule might be $c(n) = (n^2 - n + 2)/2$. Then $c(4) = (16 - 4 + 2)/2 = 7$. Thus there is no unique (exactly one) solution to the problem of finding the rule from the first few terms of a sequence.

8.1.2   Series

For each sequence, there is an associated *series* which is obtained by replacing the commas in the ordered set of terms by plus symbols.

DEFINITION

A **series** is the indicated sum of the terms of a sequence.
The **finite series** $c_1 + c_2 + \cdots + c_n$ is obtained from the finite sequence $c_1, c_2, ..., c_n$.
The **infinite series** $c_1 + c_2 + \cdots + c_n + \cdots$ is obtained from the infinite sequence $c_1, c_2, ..., c_n, ...$.

The nth term of a series is the nth term of its corresponding sequence.

For a finite series, there is always a finite number which is the sum obtained by adding the terms of the series. Thus the sum of the series $1 + 3 + 9 + 27 + 81 + 243$ is 364. On the other hand, it is important that this sum, a number, is not confused with the series which indicates the addition. This is especially important for infinite series. For example, there is no finite number that can be designated as the sum of the infinite series $3 + 7 + 11 + 15 + 19 + \cdots$. However, there are some infinite series which can be assigned a number, called the sum. This shall be seen later.

EXAMPLE 8.1.1   Write the first five terms of the sequence where the nth term is given by $c(n) = 2^{n-1} + 2n$.

Solution $c(1) = 2^0 + 2(1) = 3$

$c(2) = 2^1 + 2(2) = 6$

$c(3) = 2^2 + 2(3) = 10$

$c(4) = 2^3 + 2(4) = 16$

$c(5) = 2^4 + 2(5) = 26$

EXAMPLE 8.1.2   By using a trial and error process, (a) find an expression for c_n, the general term, of the series $5 - \frac{5}{2} + \frac{5}{4} - \frac{5}{8} + \frac{5}{16} - \cdots$ and (b) add two more terms to the series.

Solution $5 - \frac{5}{2} + \frac{5}{4} - \frac{5}{8} + \frac{5}{16} = 5 - \frac{5}{2} + \frac{5}{2^2} - \frac{5}{2^3} + \frac{5}{2^4}$

Comparing with values for n,

$$\text{for } n = 1, c_1 = \frac{5}{2^0}$$

$$\text{for } n = 2, c_2 = \frac{-5}{2^1}$$

$$\text{for } n = 3, c_3 = \frac{5}{2^2}$$

$$\text{for } n = 4, c_4 = \frac{-5}{2^3}$$

$$\text{for } n = 5, c_5 = \frac{5}{2^4}$$

A possibility for c_n is:

$$c_n = \frac{5(-1)^{n+1}}{2^{n-1}}$$

Then $c_6 = \frac{5(-1)^7}{2^5} = \frac{-5}{32}$ and $c_7 = \frac{5(-1)^8}{2^6} = \frac{5}{64}$

EXERCISES 8.1 A

In Exercises 1–10, write the first five terms of the sequence whose nth term is given.

1. $c_n = 2n - 1$ **2.** $c_n = n^2$

3. $c_n = 5(10)^{-n}$ **4.** $c(n) = n$

5. $c(n) = \dfrac{(-1)^n}{n^2}$ **6.** $c(n) = 3(-\tfrac{1}{2})^n$

7. $c(n) = \dfrac{1 + (-1)^n}{n}$ **8.** $c(n) = \dfrac{1 - n}{1 + n}$

9. $c_n = \dfrac{1}{2^n}$ **10.** $c_n = \dfrac{1}{n^2 + 1}$

By using a trial and error process, find an expression for the general term of each of the series in Exercises 11–20 and add two more terms to the series. (Remember that there is no unique solution.)

11. $\frac{1}{2} + \frac{2}{3} + \frac{3}{4} + \frac{4}{5} + \cdots$ **12.** $6 + 12 + 18 + 24 + \cdots$

13. $4 + 7 + 10 + 13 + \cdots$ **14.** $23 + 19 + 15 + 11 + \cdots$

15. $5 - \frac{5}{3} + \frac{5}{9} - \frac{5}{27} + \cdots$ **16.** $\frac{1}{2} - \frac{1}{4} + \frac{1}{8} - \frac{1}{16} + \cdots$

17. $4 + \frac{5}{4} + \frac{6}{9} + \frac{7}{16} + \cdots$ **18.** $\frac{1}{2} + \frac{1}{5} + \frac{1}{10} + \frac{1}{17} + \cdots$

19. $0 - \frac{1}{3} - \frac{2}{4} - \frac{3}{5} - \cdots$ **20.** $3 - 9 + 27 - 81 + \cdots$

EXERCISES 8.1 B

In Exercises 1–10, write the first five terms of the sequence whose nth term is given.

1. $c_n = 4n - 15$ **2.** $c_n = 5(2^n)$

3. $c_n = \dfrac{(-1)^n}{2n + 1}$ **4.** $c_n = \dfrac{n - 1}{n + 1}$

5. $c(n) = 43(0.01)^n$ **6.** $c(n) = 100(1.01)^n$

7. $c(n) = (-1)^n(2^n)$ **8.** $c(n) = \dfrac{n}{n + 2}$

9. $c_n = \dfrac{1}{2(-1)^n}$ **10.** $c_n = \dfrac{n^2}{2^n}$

By using a trial and error process, find an expression for the general term of each of the series in Exercises 11–20 and add two more terms to the series. (Remember that there is no unique solution.)

11. $5 + 10 + 15 + 20 + \cdots$ **12.** $10 + 9 + 8 + 7 + \cdots$

13. $1 - \frac{1}{4} + \frac{1}{9} - \frac{1}{16} + \cdots$

14. $(1 \times 2) + (2 \times 3) + (3 \times 4) + (4 \times 5) + \cdots$

15. $\sqrt[3]{1} - \sqrt[3]{4} + \sqrt[3]{9} - \sqrt[3]{16} + \cdots$ **16.** $32 - 16 + 8 - 4 + \cdots$

17. $\frac{1}{5} - \frac{1}{10} + \frac{1}{15} - \frac{1}{20} + \cdots$ **18.** $\frac{1}{2} + 1 + \frac{9}{8} + 1 + \frac{25}{32} + \cdots$

19. $0 + \frac{1}{3} + \frac{2}{4} + \frac{3}{5} + \cdots$ **20.** $\frac{1}{3} - \frac{1}{9} + \frac{1}{27} - \frac{1}{81} + \cdots$

8.2 SIGMA NOTATION

The Greek letter Σ (sigma) is used in many branches of mathematics in designating a series. This sigma notation greatly reduces the amount of writing necessary. For example, the finite series $3 + 7 + 11 + 15 + 19$ can be written as

$$\sum_{k=1}^{5} (4k - 1)$$

where it is understood that k is to be replaced by each of the numbers 1, 2, 3, 4, and 5 in the expression $4k - 1$, and then the sum of the resulting values is to be indicated.

If $k = 1$, then $4k - 1 = 3$

$\quad\quad k = 2 \quad\quad 4k - 1 = 7$

$\quad\quad k = 3 \quad\quad 4k - 1 = 11$

$\quad\quad k = 4 \quad\quad 4k - 1 = 15$

$\quad\quad k = 5 \quad\quad 4k - 1 = 19$

Thus $\displaystyle\sum_{k=1}^{5} (4k - 1) = 3 + 7 + 11 + 15 + 19.$

DEFINITION

$$\sum_{k=1}^{n} c(k) = c(1) + c(2) + c(3) + \cdots + c(n)$$

The symbol Σ is called the *summation symbol*, the letter k is called the *index of summation*, and the replacement set of integers for k is called the *range of summation.* The letter k is often referred to as a *dummy variable*, since any other letter could be used and still the same sum would be indicated. Thus

$$\sum_{k=1}^{5} (4k - 1) = \sum_{i=1}^{5} (4i - 1) = 3 + 7 + 11 + 15 + 19$$

EXAMPLE 8.2.1   Express in expanded notation $\displaystyle\sum_{k=1}^{4} \frac{k}{k + 1}$.

Solution

$$\sum_{k=1}^{4} \frac{k}{k + 1} = \sum_{k=1}^{4} c(k) = c(1) + c(2) + c(3) + c(4)$$

where
$$c(k) = \frac{k}{k + 1}$$

Thus $\displaystyle\sum_{k=1}^{4} c(k) = \frac{1}{2} + \frac{2}{3} + \frac{3}{4} + \frac{4}{5}.$

EXAMPLE 8.2.2   Express in expanded notation $\displaystyle\sum_{i=1}^{5} (-1)^i 2^i.$

Solution $c(i) = (-1)^i 2^i$

$$\sum_{i=1}^{5} (-1)^i 2^i = (-1)^1 2^1 + (-1)^2 2^2 + (-1)^3 2^3 + (-1)^4 2^4 + (-1)^5 2^5$$
$$= -2 + 4 - 8 + 16 - 32$$

EXAMPLE 8.2.3   Express in sigma notation

$$1 + 4 + 9 + 16 + 25 + 36 + 49$$

Solution   First find $c(k)$. A possibility is $c(k) = k^2$. To see that this is possible, check the terms:

$$c(1) = 1^2 = 1$$
$$c(2) = 2^2 = 4$$
$$c(3) = 3^2 = 9$$
$$c(4) = 4^2 = 16$$
$$c(5) = 5^2 = 25$$
$$c(6) = 6^2 = 36$$
$$c(7) = 7^2 = 49$$

Thus k is taken from 1 to 7, and in sigma notation

$$1 + 4 + 9 + 16 + 25 + 36 + 49 = \sum_{k=1}^{7} k^2$$

An infinite series is designated by the notation

$$\sum_{k=1}^{\infty} c(k)$$

For example, $\displaystyle\sum_{k=1}^{\infty} \frac{1}{k} = 1 + \frac{1}{2} + \frac{1}{3} + \cdots$

It is not necessary for the index to start with 1. For instance, $\displaystyle\sum_{k=3}^{5} c(k)$ merely indicates that the first term of the summation is $c(3)$. Thus $\displaystyle\sum_{k=3}^{5} c(k) = c(3) + c(4) + c(5)$.

EXAMPLE 8.2.4　Write in expanded notation $\displaystyle\sum_{k=2}^{4} (2k + 1)$.

Solution　$\displaystyle\sum_{k=2}^{4} c(k) = c(2) + c(3) + c(4)$

where $c(k) = 2k + 1$.

Thus $\displaystyle\sum_{k=2}^{4} (2k + 1) = 5 + 7 + 9$.

EXAMPLE 8.2.5　Write in expanded notation $\displaystyle\sum_{i=3}^{\infty} (i^2 - i)$.

Solution　$\displaystyle\sum_{i=3}^{\infty} c(i) = c(3) + c(4) + c(5) + \cdots$

where $c(i) = i^2 - i$.

Therefore, $\displaystyle\sum_{i=3}^{\infty} (i^2 - i) = 6 + 12 + 20 + \cdots$.

EXERCISES 8.2 A

Write Exercises 1–10 in expanded notation.

1. $\displaystyle\sum_{k=1}^{4} (2k - 1)$　　　　　　　2. $\displaystyle\sum_{i=1}^{5} i^3$

3. $\displaystyle\sum_{j=1}^{3} 3(10)^{-j}$　　　　　　　4. $\displaystyle\sum_{i=1}^{5} (-1)^i 2^i$

5. $\displaystyle\sum_{k=1}^{6} \frac{1}{k+1}$

6. $\displaystyle\sum_{k=1}^{5} (2 - 7k)$

7. $\displaystyle\sum_{i=3}^{7} \frac{i}{i+1}$

8. $\displaystyle\sum_{k=98}^{100} (-1)^k(5k + 7)$

9. $\displaystyle\sum_{k=1}^{\infty} \left(\frac{1}{2k-1}\right)$

10. $\displaystyle\sum_{k=1}^{\infty} 2400(10)^{-2k}$

Write each of the series in Exercises 11–20 in sigma notation.

11. $1 + 3 + 5 + 7 + \cdots + 17$

12. $1 + \frac{1}{5} + \frac{1}{25} + \frac{1}{125} + \frac{1}{625}$

13. $1 + 4 + 9 + 16 + 25 + \cdots$

14. $1 - \frac{1}{4} + \frac{1}{9} - \frac{1}{16} + \frac{1}{25}$

15. $7 - 7^2 + 7^3 - 7^4 + \cdots$

16. $2 + 5 + 10 + 17 + 26 + 37 + 50$

17. $20 + 16 + 12 + 8 + \cdots$

18. $\frac{1}{2} + \frac{1}{3} + \frac{1}{4} + \frac{1}{5} + \cdots + \frac{1}{18}$

19. $1 + (\frac{1}{2})^2 + (\frac{1}{3})^3 + (\frac{1}{4})^4 + \cdots$

20. $(1 \times 3) + (3 \times 5) + (5 \times 7)$

Evaluate Exercises 21–24 for the given information by expanding the expression first and then finding the sum.

21.

i	z_i
1	-2
2	0
3	2
4	4

$\displaystyle\sum_{i=1}^{4} z_i$

22.

i	z_i
1	-1
2	0
3	1
4	2
5	3

$\displaystyle\sum_{i=1}^{5} 3z_i$

23.

i	z_i
1	-1
2	1
3	-1
4	1

$\displaystyle\sum_{i=1}^{4} (z_i + 2)$

24.

i	f_i	z_i
1	1	2
2	2	4
3	3	8
4	4	16

$\displaystyle\sum_{i=1}^{4} f_i z_i$

EXERCISES 8.2 B

Write Exercises 1–10 in expanded notation.

1. $\displaystyle\sum_{k=1}^{5} (3k + 2)$

2. $\displaystyle\sum_{i=1}^{5} \frac{1}{i^3}$

3. $\displaystyle\sum_{j=1}^{4} 7(10)^{-j}$

4. $\displaystyle\sum_{i=1}^{3} \frac{1}{2(-1)^i}$

5. $\displaystyle\sum_{k=1}^{4} \frac{k-1}{k+1}$

6. $\displaystyle\sum_{k=1}^{5} (6-5k)$

7. $\displaystyle\sum_{i=5}^{10} \frac{i+1}{i}$

8. $\displaystyle\sum_{k=18}^{20} (-1)^k k$

9. $\displaystyle\sum_{k=1}^{\infty} (-1)^k k^{1/2}$

10. $\displaystyle\sum_{k=0}^{\infty} (-1)^k \frac{1}{(k+1)(k+2)}$

Write each of the series in Exercises 11–20 in sigma notation.

11. $1 + 2 + 3 + 4 + \cdots + 20$ **12.** $1 - \frac{1}{2} + \frac{3}{4} - \frac{7}{8} + \cdots$

13. $3 + 6 + 9 + 12 + 15 + 18$

14. $\frac{3}{2} + \frac{4}{3} + \frac{5}{4} + \frac{6}{5} + \frac{7}{6} + \cdots + \frac{21}{20}$

15. $-5 + 0.5 - 0.05 + 0.005 - 0.0005$

16. $3 - 3^2 + 3^3 - 3^4 + 3^5 - 3^6$ **17.** $10 + 9 + 8 + 7 + \cdots$

18. $\frac{1}{2} - \frac{1}{4} + \frac{1}{8} - \frac{1}{16} + \frac{1}{32} + \frac{1}{64} + \frac{1}{128}$

19. $\sqrt{1} + \sqrt{3} + \sqrt{5} + \sqrt{7} + \sqrt{9} + \cdots$

20. $(\frac{1}{3} + 3) + (\frac{1}{4} + 4) + (\frac{1}{5} + 5)$

Evaluate Exercises 21–24 for the given information by expanding the expression first and then finding the sum.

21.

i	z_i
1	2
2	4
3	6

$\displaystyle\sum_{i=1}^{3} z_i$

22.

i	z_i
1	3
2	5
3	7
4	9
5	12

$\displaystyle\sum_{i=1}^{5} 2z_i$

23.

i	z_i
1	$\frac{1}{2}$
2	1
3	2
4	4

$\displaystyle\sum_{i=1}^{4} (z_i + i)$

24.

i	f_i	z_i
1	0	4
2	2	3
3	4	2
4	6	1
5	8	0

$\displaystyle\sum_{i=1}^{5} f_i z_i$

8.3   ARITHMETIC PROGRESSIONS

DEFINITION

An **arithmetic progression** is a sequence in which the difference between any two consecutive terms is a constant d called the common difference.

If the first term of an arithmetic progression is designated by the letter a, then the definition states that the arithmetic progression can be written as follows:

$$a, a + d, a + 2d, a + 3d, a + 4d, \ldots$$

Examples of arithmetic progressions are:

a. $1, 2, 3, 4, 5, \ldots$, where $a = 1$ and $d = 1$
b. $2, 4, 6, 8, 10, 12, \ldots$, where $a = 2$ and $d = 2$
c. $20, 30, 40, \ldots, 500$, where $a = 20$ and $d = 10$
d. $20, 18, 16, 14, 12, 10, \ldots$, where $a = 20$ and $d = -2$

The nth or general term $c_n = a + (n - 1)d$.

EXAMPLE 8.3.1   Find the 35th term of the arithmetic progression $5, 9, 13, \ldots$.

Solution $a = 5, d = 9 - 5 = 4$, and $n = 35$

The 35th term $= c(35) = 5 + (35 - 1)4 = 5 + 34(4) = 141$

DEFINITION

The **arithmetic means** between two given numbers are the terms of an arithmetic progression with the two given numbers assigned as the first and the last of the sequence.

EXAMPLE 8.3.2   Insert three arithmetic means between 4 and 20.

Solution   The arithmetic progression is $4, c_2, c_3, c_4, 20$.

$$c_5 = 4 + (5 - 1)d = 20$$
$$4d = 16 \text{ and } d = 4$$

Thus $c_2 = 8, c_3 = 12$, and $c_4 = 16$. The three arithmetic means are $8, 12, 16$.

If one arithmetic mean is inserted between two numbers, then it is called the average or the arithmetic mean of the two numbers.

DEFINITION

The **sum S_n of an arithmetic progression** is the sum of the terms of the series associated with the arithmetic sequence. In symbols,

$$S_n = a + (a + d) + (a + 2d) + \cdots + (a + [n - 1]d)$$

where a is the first term of the progression and d is the common difference.

An arithmetic progression which has a finite number (n) of terms can also be written in reverse order, listing the last term first, then the next-to-last term, and so on, and listing the first term last.

If l designates the last term or nth term of the series, then S_n can also be expressed as

$$S_n = l + (l - d) + (l - 2d) + \cdots + (l - [n - 1]d)$$

By adding the corresponding terms of these two equations for S_n, one obtains

$$2S_n = (a + l) + (a + l) + (a + l) + \cdots + (a + l)$$

Noting that $a + l$ occurs n times,

$$2S_n = n(a + l)$$

$$S_n = \frac{n(a + l)}{2} = \frac{n}{2}(2a + [n - 1]d)$$

For example, consider the sum of the arithmetic progression

$$S_5 = 2 + 5 + 8 + 11 + 14$$

Writing this progression in reverse order yields

$$S_5 = 14 + 11 + 8 + 5 + 2$$

Adding,

$$2S_5 = (2 + 14) + (5 + 11) + (8 + 8) + (11 + 5) + (14 + 2)$$

$$2S_5 = 16 + 16 + 16 + 16 + 16$$

$$2S_5 = 5(16)$$

$$2S_5 = 80$$

$$S_5 = 40$$

Now $a = 2$ and $l = 14$, $n = 5$, $d = 3$.

Thus

$$S_n = \frac{n(a + l)}{2} = \frac{5(2 + 14)}{2} = 40$$

Also,

$$S_n = \frac{n}{2}(2a + [n - 1]d) = \frac{5}{2}(4 + 4(3))$$

$$= \tfrac{5}{2}(16)$$

$$= 40$$

The above illustration does not constitute a proof but is intended as an aid to understanding the summation formula for an arithmetic progression.

EXAMPLE 8.3.3   Find the sum of the first twelve terms of the arithmetic progression 1, 3, 5, 7, ...

Solution $a = 1, d = 2, n = 12$

$$l = a + (n - 1)d = 1 + (12 - 1)2 = 1 + (11)2 = 23$$

$$S_n = \frac{12(1 + 23)}{2} = 6(24) = 144$$

EXAMPLE 8.3.4   Find the sum of the first 100 positive integers.

Solution $1 + 2 + 3 + \cdots + 100$ is an arithmetic series, with $a = 1$, $d = 1, n = 100$, and $l = 100$.

$$S_n = \frac{n(a + l)}{2} = \frac{100(1 + 100)}{2} = 50(101) = 5050$$

EXERCISES 8.3 A

Find the indicated terms of each of the arithmetic progressions in Exercises 1–5.

1. 20th term of 3, 8, 13, . . .   **2.** 38th term of 2, 2.5, 3.0, . . .
3. 51st term of 87, 81, 75, . . .   **4.** 47th term of 8, $7\frac{1}{3}$, $6\frac{2}{3}$, . . .
5. 16th term of $2\sqrt{5}$, $4\sqrt{5}$, $6\sqrt{5}$, . . .

Find the sum of each of the arithmetic series in Exercises 6–15.

6. $3 + 5 + 7 + 9 + 11 + 13$ **7.** $\sum_{k=1}^{10} (2k + 1)$

8. $\sum_{i=1}^{8} 3i$ **9.** $\sum_{k=1}^{15} \frac{1}{2}(k + 2)$

10. $\sum_{j=1}^{10} (3j - 2)$ **11.** $1 + 2 + 3 + \cdots + 200$

12. $30 + 29.5 + 29 + \cdots + 5$ **13.** $1 + 3 + 5 + \cdots + (2n - 1)$

14. $\sum_{k=1}^{10} (5k - 3)$ **15.** $\sum_{i=0}^{n} (i + 1)$

Find the sum of the arithmetic progression for each of Exercises 16–20.

16. $a = 5, d = 2, n = 10$ **17.** $a = 0, d = 5, n = 20$

18. $a = 30, d = -2, n = 10$ **19.** $a = 4, d = \frac{1}{2}, n = 6$

20. $l = 34, d = 5, n = 14$

In Exercises 21–25, insert the indicated number of arithmetic means between the two given numbers.

21. Three between 80 and 90   **22.** One between 60 and 75

23. Two between 24 and -6 **24.** Four between 1 and 2

25. One between 3 and 6

26. How many positive integers between 5 and 100 are divisible by 7? (*Hint:* $a = 7, d = 7$; find l and n.)

27. A man is offered two jobs: one with a beginning salary of $6000 per year and a $1200 raise at the end of each year, the other with a starting salary of $3000 for the first six months and a raise of $500 for every six months thereafter. What is the salary for each job for the sixth year? Which job pays the most and why?

28. A debt of $2000 is paid by paying $100 at the end of each month plus $\frac{1}{2}$ percent interest on the amount unpaid at the end of that month. What is the total payment?

29. A certain charity raffles a car in the following way. Tickets are numbered from 1 to 200, each ticket is placed in a sealed envelope, and a ticket marked n is sold for n dollars. If all of the tickets are sold, how much money is received from the sale of the tickets?

30. A certain auditorium has 40 rows of seats. The first row has 16 seats, the second row 18 seats, with each row having 2 more seats than the row in front of it. How many seats are in the auditorium?

EXERCISES 8.3 B

Find the indicated term of each of the arithmetic progressions in Exercises 1–5.

1. 12th term of $9, 3, -3, \ldots$ **2.** 40th term of $1, 3, 5, \ldots$

3. 26th term of $0.5, 0.75, 1, \ldots$

4. 17th term of $10 + \sqrt{2}, 10 - \sqrt{2}, \ldots$

5. 100th term of $\frac{1}{6}, \frac{1}{2}, \frac{5}{6}, \ldots$

Find the sum of each of the arithmetic series in Exercises 6–15.

6. $10 + 13 + 16 + 19 + 21$

7. $\displaystyle\sum_{k=1}^{12} (1 - 2k)$

8. $\displaystyle\sum_{k=1}^{40} 2k$

9. $\displaystyle\sum_{k=1}^{12} \frac{1}{4}(k + 4)$

10. $\displaystyle\sum_{j=1}^{200} (2j + 3)$

11. $2 + 4 + 6 + \cdots + 100$

12. $12 + 8 + 4 + \cdots + (-20)$

13. $3 + 6 + 9 + \cdots + 3n$

14. $\displaystyle\sum_{k=1}^{50} \frac{k+1}{2}$

15. $\displaystyle\sum_{i=1}^{n} (2 - 3i)$

Find the sum of the arithmetic progression for each of Exercises 16–20.

16. $a = 12, d = 4, n = 16$

17. $a = -6, d = 5, n = 20$

18. $a = 0, d = -\frac{1}{2}, n = 100$

19. $l = 320, d = 4, n = 32$

20. $l = 2, d = -4, n = 1000$

In Exercises 21–25, insert the indicated number of arithmetic means between the two given numbers.

21. Two between 45 and 50

22. One between 78 and 65

23. One between 1 and 10

24. Three between 10 and -10

25. Two between 0 and 4

26. The fare charged by a certain taxicab company is 30 cents for the first $\frac{1}{2}$ mile and 15 cents for each $\frac{1}{4}$ mile thereafter. What is the fare from the center of a city to an airport 10 miles away?

27. If a body falls from rest in a vacuum, it falls 16 feet the first second and 32 feet more each second thereafter. How far does the body fall in 20 seconds? How far in k seconds?

28. A man contributes to his savings account by depositing $100 the first month and by increasing his deposit by $5 each month thereafter. How much has he saved at the end of 5 years?

29. There are 80 watermelons lying on a straight line, each watermelon 5 feet from the next one. A picker starts with a crate at the first watermelon, places it in the crate, then carries each of the other watermelons back to the crate, one at a time. How far does the picker travel?

30. A certain grocery display consists of cans arranged in the form of a pyramid, with 1 can on the top and 17 cans on the bottom row. If each row contains 2 more cans than the one above it, how many cans are in the display?

8.4 GEOMETRIC PROGRESSIONS

8.4.1 Finite Geometric Progressions

DEFINITION

A **geometric progression** is a sequence in which the ratio between any two consecutive terms is a constant r called the common ratio.

If the first term of a geometric progression is designated by the letter a, then the definition states that the geometric progression can be expressed as follows:

$$a, ar, ar^2, ar^3, ar^4, \ldots, ar^{n-1}$$

Examples of geometric progressions are:

a. $2, 4, 8, 16, \ldots, 128$, where $a = 2$ and $r = 2$
b. $5, -15, 45, -135$, where $a = 5$ and $r = -3$
c. $30, 10, 3\frac{1}{3}, 1\frac{1}{9}, \frac{10}{27}$, where $a = 30$ and $r = \frac{1}{3}$

Since r to the first power does not occur until the second term of the progression, the nth term is ar^{n-1}, or

$$c(n) = ar^{n-1}$$

EXAMPLE 8.4.1 Find the 6th term of the geometric progression $3, 15, 75, \ldots$.

Solution $c(n) = ar^{n-1}$, $a = 3$, $r = \frac{15}{3} = 5$, and $n = 6$
$$c(6) = 3(5)^{6-1} = 3(5)^5 = 3(3125) = 9375$$

DEFINITION

The **geometric means** between two given numbers are the terms of a geometric progression with the two given numbers assigned as the first and last of the sequence.

EXAMPLE 8.4.2 Insert two geometric means between 1 and 64.

Solution The geometric progression is $1, c_2, c_3, 64$.
$$a = 1, \quad c_4 = ar^{4-1} = 1(r)^3 = 64$$

Thus

$$r = 4 \qquad c_2 = 4 \qquad c_3 = 4 \cdot 4 = 16$$

The two geometric means are 4 and 16.

EXAMPLE 8.4.3   Insert three real geometric means between 3 and 48.

Solution   The geometric progression is 3, c_2, c_3, c_4, 48.

$$a = 3 \quad \text{and} \quad c_5 = 3r^4 = 48, \quad \text{or} \quad r^4 = 16$$

Thus, $r^2 = \pm 4$ and $r = \pm 2$ or $r = \pm 2i$. For the means to be real, $r = 2$ or $r = -2$. There are two possibilities: the real means are 6, 12, 24, or -6, 12, -24.

If one geometric mean is inserted between two numbers, then it is called the **mean proportional**, or the *geometric mean* of the two numbers.

DEFINITION

The **sum S_n of a geometric progression** is the sum of the terms of the series associated with the geometric sequence. In symbols,

$$S_n = a + ar + ar^2 + ar^3 + \cdots + ar^{n-1}$$

If each term of the indicated sum is multiplied by $-r$, then

$$-rS_n = -ar - ar^2 - ar^3 - \cdots - ar^{n-1} - ar^n$$

Adding these two equations,

$$S_n - rS_n = a - ar^n$$
$$S_n(1 - r) = a(1 - r^n)$$

Now if $r \neq 1$, then

$$S_n = \frac{a(1 - r^n)}{1 - r}$$

$$= \frac{a(r^n - 1)}{r - 1}$$

EXAMPLE 8.4.4   Find the sum of the geometric progression 5, 10, ..., 320.

Solution $a = 5$ and $r = \frac{10}{5} = 2$

$$c_n = ar^{n-1} = 5(2)^{n-1} = 320$$
$$2^{n-1} = 64 = 2^6$$
$$n - 1 = 6$$
$$n = 7$$
$$S_7 = \frac{a(r^7 - 1)}{r - 1} = \frac{5(2^7 - 1)}{2 - 1} = 5(128 - 1) = 5(127) = 635$$

8.4.2   Infinite Geometric Progressions

The sum of a finite geometric progression

$$S_n = \frac{a(1 - r^n)}{1 - r}$$

can also be written as

$$S_n = \frac{a}{1 - r}(1 - r^n)$$

If $|r| < 1$ (or $-1 < r < 1$), then r^n becomes smaller and smaller as n becomes larger and larger. For example, let $r = \frac{1}{10} = 0.1$. Then

$$r^2 = (0.1)^2 = 0.01$$

$$r^3 = (0.1)^3 = 0.001$$

$$r^4 = (0.1)^4 = 0.0001$$

$$\vdots$$

$$r^9 = (0.1)^9 = 0.000000001$$

By taking n large enough, r^n can be made as close to 0 as one wants. Then $1 - r^n$ will be as close to 1 as one wants and, as a result, S_n can be made as close to $a/(1 - r)$ as one wants. Therefore, S, the sum of an infinite geometric progression with $|r| < 1$, is defined as this value, $a/(1 - r)$.

DEFINITION

If $|r| < 1$ and $S = a + ar + ar^2 + \cdots$, then

$$S = \frac{a}{1 - r}$$

A more precise way of stating this result symbolically is as follows:

$$\lim_{n \to \infty} S_n = \frac{a}{1 - r}$$

which is read "the limit of S_n as n increases without bound is $a/(1 - r)$."

EXAMPLE 8.4.5   Find the sum of the infinite geometric series 12, 3, $\frac{3}{4}, \ldots$.

Solution $a = 12, r = \frac{1}{4}$

$$S = \frac{a}{1-r} = \frac{12}{1-(1/4)} = \frac{12(4)}{4-1} = \frac{48}{3} = 16$$

EXAMPLE 8.4.6   Find the sum of the infinite geometric series $1 - \frac{1}{2} + \frac{1}{4} - \frac{1}{8} + \cdots$.

Solution $a = 1, r = -\frac{1}{2}$

$$S = \frac{a}{1-r} = \frac{1}{1-(-1/2)} = \frac{1}{3/2} = \frac{2}{3}$$

EXERCISES 8.4 A

Find the indicated term of each of the geometric progressions in Exercises 1–4.

1. 8th term of 18, 12, 8, . . .   **2.** 5th term of 1, 7, 49, . . .

3. 9th term of -16, 20, -25, . . .   **4.** 15th term of 9, $\frac{9}{10}$, $\frac{9}{100}$, . . .

In Exercises 5–8, insert the indicated number of real geometric means between the two given numbers.

5. Two between 54 and 16   **6.** Three between 2 and 162

7. One between 4 and 9   **8.** Two between 5 and 10

Find the sum of each of the geometric series in Exercises 9–14.

9. $25 + 10 + \cdots + 0.256$ **10.** $1 + 7 + \cdots + 7^5$

11. $\displaystyle\sum_{i=1}^{6} 5\left(\frac{1}{3}\right)^i$ **12.** $\displaystyle\sum_{k=2}^{7} 3(-2)^k$

13. $1 + 2 + 2^2 + \cdots + 2^8$ **14.** $1 - 2 + 2^2 - 2^3 + \cdots + 2^8$

If the sum exists, find the sum of each of the geometric series in Exercises 15–18. If the sum does not exist, state why.

15. $10 + 2 + 0.4 + \cdots$ **16.** $\frac{1}{50} + \frac{1}{40} + \frac{1}{32} + \cdots$

17. $\frac{1}{75} - \frac{1}{90} + \frac{2}{108} - \cdots$ **18.** $\displaystyle\sum_{k=1}^{\infty} 100\left(\frac{3}{4}\right)^k$

19. Each arc length of the path of the bob of a swinging pendulum is 95 percent of the preceding arc length. If its initial length is 18 inches, find the total distance the bob travels before the pendulum comes to rest.

20. How many ancestors has an individual had in the 12 generations preceding him if it is assumed that there have been no intermarriages?

21. When the oil used in operating certain machinery is refined so it can be used again, 25 percent of the oil is lost each time it is refined. If there is originally 200 gallons of refined oil, which is refined each time it becomes dirty, find approximately the total amount of oil used in operating the machinery before all of the oil is lost.

22. A piece of equipment costing $12,500 depreciates by 20 percent of its value each year. At the end of each year, the amount the equipment has depreciated is placed in a fund. In how many years will the fund contain $8404?

EXERCISES 8.4 B

Find the indicated term of each of the geometric progressions in Exercises 1–4.

1. 6th term of $4, 2\sqrt{2}, 2, \ldots$ **2.** 7th term of $\sqrt{2}, -\sqrt{6}, 3\sqrt{2}, \ldots$

3. 10th term of $-4, 2, -1, \ldots$

4. 8th term of $2\sqrt[3]{2}, 4, 4\sqrt[3]{4}, \ldots$

In Exercises 5–8, insert the indicated number of real geometric means between the two given numbers.

5. One between 6 and 30   **6.** Two between 56 and 189

7. Three between 1 and 16   **8.** Two between 3 and 24

Find the sum of each of the geometric series in Exercises 9–12.

9. $2 - 12 + \cdots + 2592$ **10.** $10 + 5 + \cdots + 10(0.5)^6$

11. $\displaystyle\sum_{k=1}^{7} 4\left(-\frac{3}{2}\right)^k$ **12.** $\displaystyle\sum_{i=0}^{8} 2(3^i)$

Express each of Exercises 13–14 as a quotient.

13. a. $1 + x + x^2 + \cdots + x^{10}$
    b. $1 + x + x^2 + \cdots + x^n$

14. a. $1 - x + x^2 - x^3 + \cdots + x^{10}$
    b. $1 - x + x^2 - x^3 + \cdots - x^9$

If the sum exists, find the sum of each of the infinite geometric series in Exercises 15–18. If the sum does not exist, state why.

15. $12 - 2 + \cdots$

16. $(3)^{-1/2} + (6)^{-1/2} + \cdots$

17. $\frac{1}{54} - \frac{1}{36} + \frac{1}{24} - \cdots$

18. $\displaystyle\sum_{i=1}^{\infty} 32\left(-\frac{5}{8}\right)^{i}$

19. On each rebound, a certain ball bounces back to $\frac{5}{8}$ of its preceding height. If the ball is dropped from a height of 4 feet, find the total distance it travels before coming to rest.

20. If \$100 is deposited at the end of each month in a savings fund and if $\frac{1}{2}$ percent interest is paid on the money in the fund each month, how much money is in the fund at the end of one year?

21. If an air pump removes 60 percent of the air in a container with each stroke, find the percentage of air left in the container after 4 strokes. How many strokes would it require to reduce the air to less than $\frac{1}{2}$ percent?

22. In a certain culture, a bacterium divides into two bacteria every 30 minutes. If there were 50 bacteria in the culture at the beginning, how many bacteria are there at the end of 3 hours?

8.5 MATHEMATICAL INDUCTION (Optional)

An important property of the set of natural numbers, $N = \{1, 2, 3, 4, 5, \ldots\}$, is the *axiom of mathematical induction.*

THE AXIOM OF MATHEMATICAL INDUCTION

Let S be a set of natural numbers. If (1) $1 \in S$ and (2) $k + 1 \in S$ whenever $k \in S$, then $S = N$. (In other words, S is the set of *all* natural numbers.)

This axiom provides a very powerful method of proof called *proof by mathematical induction.* If it can be shown that a certain open sentence is true for $n = 1$ and also for $n = k + 1$ whenever it is true for $n = k$, then the axiom of mathematical induction states that the set of natural numbers for which the open sentence is true consists of *all* the natural numbers.

It is important to observe that a proof by mathematical induction consists of *two* parts: part (1), the proof for $n = 1$; and part (2), the proof for $n = k + 1$, assuming the statement is true for $n = k$.

The axiom of mathematical induction is often called the *domino property* because an analogy can be made with a row of dominos. If some dominos are stood on edge and spaced along a row so that felling one will cause the next one to fall, and if the first one is felled, then all of the dominos must fall.

EXAMPLE 8.5.1 Using mathematical induction, prove that the sum of the first n odd positive integers is n^2.

Solution Let S be the set of natural numbers for which

$$1 + 3 + 5 + \cdots + (2n - 1) = n^2$$

Part 1: If $n = 1$, then $1 = 1^2$. Thus $1 \in S$.

Part 2: Assume $k \in S$. Then $1 + 3 + 5 + \cdots + (2k - 1) = k^2$.

For $n = k + 1$, then $2n - 1 = 2(k + 1) - 1$.
Therefore, adding $2(k + 1) - 1$ to each side,

$$1 + 3 + 5 + \cdots + (2k - 1) + [2(k + 1) - 1] = k^2 + 2(k + 1) - 1$$

$$1 + 3 + \cdots + 2(k + 1) - 1 = k^2 + 2k + 1 = (k + 1)^2$$

Thus the open sentence is true for $n = k + 1$ whenever it is true for $n = k$. In other words, if $k \in S$, then $k + 1 \in S$.

Combining parts (1) and (2), $S = N$ or $1 + 3 + \cdots + (2n - 1) = n^2$ is true for *all* natural numbers n.

EXAMPLE 8.5.2 Prove by mathematical induction that

$$1 + 2 + 3 + \cdots + n < \frac{(n + 1)^2}{2}$$

for all natural numbers n.

Solution Let S be the set of natural numbers for which the statement is true.

Part 1: If $n = 1$, then $\dfrac{(n + 1)^2}{2} = \dfrac{(1 + 1)^2}{2} = 2$ and $1 < 2$. Thus $1 \in S$.

Part 2: Assume true for $n = k$. Then $1 + 2 + \cdots + k < \dfrac{(k + 1)^2}{2}$.

Adding $k + 1$ to both sides,

$$1 + 2 + \cdots + k + (k + 1) < \frac{(k + 1)^2}{2} + (k + 1)$$

Now $\dfrac{(k + 1)^2}{2} + k + 1 = \dfrac{k^2 + 4k + 3}{2} < \dfrac{k^2 + 4k + 4}{2} = \dfrac{(k + 2)^2}{2}$

Thus $1 + 2 + \cdots + (k + 1) < \dfrac{k^2 + 4k + 3}{2} < \dfrac{([k + 1] + 1)^2}{2}$

Therefore, if $k \in S$, then $k + 1 \in S$. Combining parts (1) and (2),

$$S = N$$

EXERCISES 8.5 A

Prove by mathematical induction that each of the statements in Exercises 1–6 is true for all natural numbers n.

1. $2 + 4 + 6 + \cdots + 2n = n(n + 1)$

2. $1^2 + 2^2 + 3^2 + \cdots + n^2 = \dfrac{n(n + 1)(2n + 1)}{6}$

3. $\dfrac{1}{1(2)} + \dfrac{1}{2(3)} + \dfrac{1}{3(4)} + \cdots + \dfrac{1}{n(n + 1)} = \dfrac{n}{n + 1}$

4. $n < 2^n$

5. $x - y$ is a factor of $x^n - y^n$
   [*Hint:* $x^{k+1} - y^{k+1} = x^k(x - y) + y(x^k - y^k)$.]

6. $n(n + 1)(n + 2)$ is exactly divisible by 3

EXERCISES 8.5 B

Prove by mathematical induction that each of the statements in Exercises 1–6 is true for all natural numbers n.

1. $5 + 10 + 15 + \cdots + 5n = \dfrac{5n(n + 1)}{2}$

2. $1^3 + 2^3 + 3^3 + \cdots + n^3 = \dfrac{n^2(n + 1)^2}{4}$

3. $1(2) + 2(3) + 3(4) + \cdots + n(n + 1) = \dfrac{n(n + 1)(n + 2)}{3}$

4. $\dfrac{1}{1(3)} + \dfrac{1}{3(5)} + \dfrac{1}{5(7)} + \cdots + \dfrac{1}{(2n - 1)(2n + 1)} = \dfrac{n}{2n + 1}$

5. If $0 < x < y$, then $\left(\dfrac{x}{y}\right)^{n+1} < \left(\dfrac{x}{y}\right)^{n}$

6. $x + y$ is a factor of $x^{2n+1} + y^{2n+1}$

8.6   THE BINOMIAL THEOREM (Optional)

The expansions of the following powers of the binomial $(a + b)^n$ may be obtained by performing the indicated multiplications:

$$(a + b)^1 = a + b$$
$$(a + b)^2 = a^2 + 2ab + b^2$$
$$(a + b)^3 = a^3 + 3a^2b + 3ab^2 + b^3$$
$$(a + b)^4 = a^4 + 4a^3b + 6a^2b^2 + 4ab^3 + b^4$$
$$(a + b)^5 = a^5 + 5a^4b + 10a^3b^2 + 10a^2b^3 + 5ab^4 + b^5$$

By examining these special cases, the following properties may be observed for $n = 1, 2, 3, 4$, or 5.

1. Each expansion has $(n + 1)$ terms.
2. The first term is a^n and the last term is b^n.
3. The coefficient of the second term and the next to the last term is n.
4. The exponent of a is one less than that of the preceding term.
5. The exponent of b is one more than that of the preceding term.
6. The sum of the exponents of a and b in each term is n.
7. The coefficient of any term after the first can be obtained by multiplying the coefficient of the preceding term by its exponent of a and dividing by the number of this preceding term.

Assuming that these properties are valid for any natural number n, an expansion can be written for $(a + b)^n$. This statement is called the *binomial theorem*.

THE BINOMIAL THEOREM

Let n be any natural number. Then

$$(a + b)^n = a^n + na^{n-1}b + \frac{n(n-1)}{1(2)} a^{n-2}b^2$$

$$+ \frac{n(n-1)(n-2)}{1(2)(3)} a^{n-3}b^3 + \cdots$$

$$+ \frac{n(n-1)(n-2)\cdots(n-r+1)}{1(2)(3)\cdots(r)} a^{n-r}b^r + \cdots + b^n$$

Proof: By mathematical induction on n. Let $S =$ the set of natural numbers for which the statement is true. We have already observed that $1, 2, 3, 4$, and 5 are in S.

Part 1: If $n = 1$, then $(a + b)^1 = a + b$. Thus $1 \in S$.

Part 2: Assume $k \in S$. Then

$$(a + b)^k = a^k + ka^{k-1}b + \cdots + \frac{k(k-1)\cdots(k-r+2)}{1(2)\cdots(r-1)} a^{k-r+1}b^{r-1}$$

$$+ \frac{k(k-1)\cdots(k-r+2)(k-r+1)}{1(2)\cdots(r-1)r} a^{k-r}b^r + \cdots + b^k$$

Multiplying both sides of this equation by $a + b$, and writing the terms of the product of $a(a + b)^k$ above the corresponding terms of the product of $b(a + b)^k$, then

$$(a + b)^{k+1} = a(a + b)^k + b(a + b)^k$$

$$= a^{k+1} + ka^k b + \cdots + \frac{k(k - 1) \cdots (k - r + 1)}{1(2) \cdots (r)} a^{k-r+1} b^r + \cdots$$

$$+ a^k b + \cdots + \frac{k(k - 1) \cdots (k - r + 2)}{1(2) \cdots (r - 1)} a^{k-r+1} b^r + \cdots + b^{k+1}$$

$$(a + b)^{k+1} = a^{k+1} + (k + 1)a^k b + \cdots$$

$$+ \frac{(k + 1)k(k - 1) \cdots (k + 1 - r + 1)}{1(2)(3) \cdots (r)} a^{k+1-r} b^r + \cdots + b^{k+1}$$

Since this expansion is the same as that stated in the theorem for $n = k + 1$, it follows that $k + 1 \in S$ whenever $k \in S$. Thus $S = N$, or the binomial theorem is valid for every natural number n.

The addition of the terms involving b^r in the proof of the binomial theorem is verified as follows:

$$\frac{k(k - 1) \cdots (k - r + 2)}{1(2) \cdots (r - 1)} = \frac{k(k - 1) \cdots (k - r + 2)r}{1(2) \cdots (r - 1)r}$$

and

$$\frac{k(k - 1) \cdots (k - r + 2)(k - r + 1) + k(k - 1) \cdots (k - r + 2)r}{1(2) \cdots (r)}$$

$$= \frac{k(k - 1) \cdots (k - r + 2)(k - r + 1 + r)}{1(2) \cdots (r)}$$

$$= \frac{(k + 1)k \cdots (k - r + 2)}{1(2) \cdots (r)}$$

$$= \frac{(k + 1)k \cdots (k + 1 - r + 1)}{1(2) \cdots (r)}$$

The term that involves b^r is the $(r + 1)$st term in the expansion of $(a + b)^n$.

The $(r + 1)$st term of $(a + b)^n = \dfrac{n(n - 1) \cdots (n - r + 1)}{1(2)(3) \cdots (r)} a^{n-r} b^r$.

EXAMPLE 8.6.1   Expand and simplify $(x - 2)^6$.

Solution $a = x$, $b = -2$, and $n = 6$. Substituting in $(a + b)^6$,

$$(x - 2)^6 = x^6 + 6x^5(-2) + \frac{6(5)}{2}x^4(-2)^2 + \frac{6(5)(4)}{1(2)(3)}x^3(-2)^3$$

$$+ \frac{6(5)(4)(3)}{1(2)(3)(4)}x^2(-2)^4 + \frac{6(5)(4)(3)(2)}{1(2)(3)(4)(5)}x(-2)^5 + (-2)^6$$

Simplifying,

$$(x - 2)^6 = x^6 - 12x^5 + 60x^4 - 160x^3 + 240x^2 - 192x + 64$$

EXAMPLE 8.6.2   Find the 9th term of $(5y + 0.2)^{15}$.

Solution $(r + 1)$st term $= \dfrac{n(n - 1) \cdots (n - r + 1)}{1(2)(3) \cdots (r)} a^{n-r}b^r$

Since $r + 1 = 9$, then $r = 8$. Also $a = 5y$, $b = 0.2$, $n = 15$, and $n - r + 1 = 15 - 8 + 1 = 8$.
Substituting,

$$9\text{th term} = \frac{15(14)(13)(12)(11)(10)(9)(8)}{1(2)(3)(4)(5)(6)(7)(8)}(5y)^{15-8}(0.2)^8$$

$$= 1287\, y^7$$

Pascal's Triangle

By including $(a + b)^0 = 1$, the coefficients of the terms of the binomial expansion form an interesting pattern, known as *Pascal's triangle:*

$(a + b)^0$	1
$(a + b)^1$	1 1
$(a + b)^2$	1 2 1
$(a + b)^3$	1 3 3 1
$(a + b)^4$	1 4 6 4 1
$(a + b)^5$	1 5 10 10 5 1
$(a + b)^6$	1 6 15 20 15 6 1

Each of the numbers different from 1 may be obtained by adding the two numbers to its left and right in the row immediately above. Thus $10 = 4 + 6$ and $15 = 5 + 10$ and so on.

HISTORICAL NOTE

The French mathematician Blaise Pascal (1623–1662) discovered many properties of this triangular array which appear in his *Traité de triangle arithmétique* written in 1653. In this work is to be found the binomial theorem for positive integral exponents and one of the first acceptable uses of the method of mathematical induction.

Although the "arithmetical triangle" is named after Pascal because of his work on it, Pascal did not discover this array of numbers. It appears in one of the works of the great Chinese algebraist Chu Shï-kié written around 1303. The array appeared in European publications more than 100 years before Pascal's time.

The great English mathematician Sir Isaac Newton (1642–1727) generalized the binomial theorem for rational values of the exponent. His work appears in letters he wrote in 1676.

EXERCISES 8.6 A

Expand Exercises 1–16 by using the binomial theorem and simplify.

1. $(x + 1)^3$

2. $(x + 1)^4$

3. $(x - 1)^3$

4. $(x - 1)^4$

5. $(x + 2)^5$

6. $(y - 2)^5$

7. $(x + 2)^7$

8. $(y - 1)^8$

9. $(b - 1)^6$

10. $(x + y)^5$

11. $(x^2 + 3)^3$

12. $(1 - y^2)^5$

13. $(x + \frac{1}{2})^4$

14. $(a^2 - b^2)^3$

15. $(2x + 5)^3$

16. $(3x - 2)^4$

Express in simplified form the specified term in the expansion of each of Exercises 17–20.

17. 5th term of $(x - 3)^{10}$

18. 4th term of $\left(\dfrac{x}{2} + 1\right)^{16}$

19. 6th term of $(1 - 0.02y)^8$

20. 7th term of $\left(x + \dfrac{1}{x}\right)^{12}$

21. Approximate $(1.02)^{12}$ by finding the sum of the first four terms of the expansion of $(1 + 0.02)^{12}$.

22. Approximate $(0.98)^8$ correct to the nearest hundredth. Use $(1 - 0.02)^8$.

23. Evaluate $(1 - i)^8$ where $i^2 = -1$.

EXERCISES 8.6 B

Expand Exercises 1–8 by using the binomial theorem and simplify.

1. $(t - 1)^9$

2. $\left(x + \dfrac{y}{2}\right)^5$

3. $(2t^2 - 3)^6$

4. $(2y + 1)^8$

5. $(x^2 - 2)^7$

6. $(r - 2)^5$

7. $(3t - 1)^3$

8. $\left(\dfrac{x}{2} + 1\right)^4$

Write the first four terms of the expansions of each of Exercises 9–12.

9. $(x - y)^{34}$

10. $(t + 1)^{100}$

11. $(y^2 - \sqrt{2})^{15}$

12. $\left(\dfrac{2x + 5}{10}\right)^5$

Express in simplified form the specified term in the expansion of each of Exercises 13–16.

13. 4th term of $(y + 4)^{12}$

14. 6th term of $(x^2 - \tfrac{1}{2})^{20}$

15. 4th term of $\left(r^3 + \dfrac{1}{s^3}\right)^9$

16. 8th term of $\left(\dfrac{2}{x} - \dfrac{y}{2}\right)^{15}$

17. Approximate $(0.97)^{10}$ by using the first four terms of $(1 - 0.03)^{10}$.

18. Approximate $(1.01)^{10}$ correct to the nearest thousandth.

19. Evaluate

$$\left(\frac{1}{2} + \frac{\sqrt{3}}{2}i\right)^9$$

where $i^2 = -1$.

SUMMARY

☐ A **finite sequence of n terms** is a function c whose domain is the ordered set of natural numbers, $1, 2, 3, \ldots, n$, and whose range is the ordered set, $c(1), c(2), \ldots, c(n)$ also written as $c_1, c_2, \ldots, c_n$.

☐ An **infinite sequence** is a function whose domain is the ordered set of all natural numbers and whose range is $c(1), c(2), \ldots, c(n), \ldots$, also written as $c_1, c_2, \ldots, c_n, \ldots$.

☐ A **series** is the indicated sum of the terms of a sequence:

$$\sum_{k=1}^{n} c(k) = c(1) + c(2) + \cdots + c(n)$$

☐ An **arithmetic progression** is a sequence in which the difference between any two consecutive terms is a constant d called the **common difference**.

☐ The **arithmetic means** between two given numbers are the terms of an arithmetic progression with the two given numbers assigned as the first and the last of the sequence.

☐ The **nth or last term** of an arithmetic progression: $l = a + (n - 1)d$.

☐ The **sum of an arithmetic progression:**

$$S = \frac{n(a + l)}{2} = \frac{n}{2}(2a + [n - 1]d)$$

☐ A **geometric progression** is a sequence in which the ratio between any two consecutive terms is a constant r called the *common ratio*.

☐ The **geometric means** between two given numbers are the terms of a geometric progression with the two given numbers assigned as the first and last of the sequence.

☐ The **nth or last term** of a geometric progression: $l = ar^{n-1}$.

☐ The **sum of a geometric progression:**

$$S_n = \frac{a(1 - r^n)}{1 - r}$$

☐ The **sum of an infinite geometric progression:**

$$\text{If } |r| < 1, \text{ then } S = \frac{a}{1 - r}$$

☐ The **axiom of mathematical induction.** Let S be a set of natural numbers. If $1 \in S$ and $k + 1 \in S$ whenever $k \in S$, then $S = N$.

☐ **The binomial theorem.** Let n be any natural number. Then

$$(a + b)^n = a^n + na^{n-1}b + \cdots + \frac{n(n - 1)\cdots(n - r + 1)}{1(2)\cdots(r)} a^{n-r}b^r + \cdots + b^n$$

REVIEW EXERCISES

Each of the series in Exercises 1–10 is either an arithmetic progression (A), *a geometric progression* (G), *or a binomial expansion* (B). *Determine the type of series and find the next two terms of each.*

1. $\frac{2}{5} + 1 + \frac{5}{2} + \cdots$

2. $\dfrac{x + 1}{x} + \dfrac{x + 2}{x} + \dfrac{x + 3}{x} + \cdots$ 3. $3 - 6 - 15 - \cdots$

4. $3 - 6 + 12 - \cdots$

5. $1 - \frac{7}{2} + \frac{21}{4} - \cdots$

6. $\dfrac{-1}{1 + \sqrt{2}} + 1 + \dfrac{1}{1 - \sqrt{2}} + \cdots$

7. $\log 2 + \log 8 + \log 32 + \cdots$

8. $\log 9 + \log 3 + \log \sqrt{3} + \cdots$

9. $(1 - i) + 2 + (2 + 2i) + \cdots$

10. $4\sqrt{2} + 60 + 180 \sqrt{2} + \cdots$

Expand Exercises 11–15.

11. $\displaystyle\sum_{i=1}^{6} (2i - 1)$

12. $\displaystyle\sum_{i=3}^{9} i^2$

13. $\displaystyle\sum_{k=0}^{4} 3(10)^{-k}$

14. $\displaystyle\sum_{k=1}^{\infty} \frac{k}{k + 3}$

15. $\displaystyle\sum_{k=8}^{14} \frac{k - 1}{k + 1}$

Find the sum of each of the arithmetic series in Exercises 16–20.

16. $2 + 4 + 6 + \cdots + 50$

17. $\displaystyle\sum_{i=1}^{10} (2 - 3i)$

18. $3 + 6 + 9 + \cdots + 3n$

19. $\displaystyle\sum_{k=1}^{25} (5k - 3)$

20. $\displaystyle\sum_{k=1}^{50} \frac{k + 1}{2}$

Find the sum of each of the geometric series in Exercises 21–25.

21. $24 + 12 + 6 + 3$

22. $\displaystyle\sum_{i=1}^{12} 3(2^i)$

23. $\displaystyle\sum_{k=1}^{7} \frac{1}{2^k}$

24. $\displaystyle\sum_{k=1}^{\infty} 16\left(\frac{1}{2}\right)^{k-1}$

25. $\displaystyle\sum_{k=1}^{10} 5(-2)^{k-1}$

Find the sum of an arithmetic series with the information given in Exercises 26–27.

26. $a = 14, d = 5, n = 12$

27. $l = 10, d = -\frac{1}{2}, n = 20$

Find the sum, if it exists, of each geometric series in Exercises 28–29.

28. $2 + 4 + 8 + \cdots$

29. $10 + 4 + \frac{8}{5} + \frac{16}{25} + \frac{32}{125} + \cdots$

30. Solve for x: $\frac{7}{8} = 1 + x + x^2 + x^3 + \cdots$

31. Solve for x: $x^2 - x^4 + x^6 - x^8 + \cdots = \dfrac{1 - x^2}{x^2}$.

32. Expand by using the binomial theorem:
 a. $(1 - x)^3$
 b. $(2 - t)^4$
 c. $(a + 3)^5$

33. Approximate $(0.95)^6$ correct to the nearest thousandth.

34. Find the 4th term of $\left(5\sqrt{2} - \dfrac{x}{5}\right)^9$

35. Express $5.6363\ldots$ as a simplified common fraction.

36. Approximate $(1.005)^{12}$ correct to four decimal places.

37. Prove by using mathematical induction:

$$\frac{1}{2(5)} + \frac{1}{5(8)} + \frac{1}{8(11)} + \cdots + \frac{1}{(3n - 1)(3n + 2)} = \frac{n}{6n + 4}$$

38. Prove that the geometric mean of two positive integers is never larger than their arithmetic mean.

39. Find the cost of digging a well 200 yards deep if the charges are $4 for the first yard, and each successive yard costs 50 cents more than the previous one.

40. An instructor receives a beginning salary of $6000 a year and a yearly 4 percent raise. What is his salary for his 10th year of teaching?

ANSWERS

Exercises 0.1, page 6

1. $5 + N$
2. $5N$
3. $N - 5$
4. N^2
5. $5/N$
6. N^3
7. $\sqrt{5}$
8. $\sqrt[3]{N}$
9. $5(x + 8)$
10. $6x - 4$
11. $(y/5)^2$
12. $\frac{1}{2}(x^2 - \sqrt{y})$
13. $n^3 + (n - 1)^2$
14. $2xy/(x + y)$
15. $3x - 10 = 7$
16. $(x + 4)^2 < x/4$
17. $6(9 - x) \geq 8$
18. $\sqrt[3]{x + y} \neq \sqrt[3]{x} + \sqrt[3]{y}$
19. $4(3x + 6) - 12x = 24$
20. $(4x/5) - 5 = 3(x + 2)$
21. 4
22. 14
23. 9
24. 9
25. 1
26. 12
27. 36
28. 1
29. 2
30. 5
31. 5
32. 4
33. 3
34. 252
35. 28
36. Undefined
37. 77
38. 10
39. 18
40. 30
41. 5
42. 15
43. 0
44. 2
45. 5
46. True
47. False
48. True
49. False
50. False
51. False
52. True
53. False
54. True
55. True

Exercises 0.2, page II

1. $\{1, 2, 3, 4, 5, 6, 7\}$
2. $\{1, 2, 3, 5, 6, 10, 15, 30\}$
3. $\{30, 60, 90, 120, 150, \ldots\}$
4. $\{16, 17, 18, 19, 20, \ldots\}$
5. $\{1, 4, 9, 16, 25, \ldots\}$
6. $\{1, 8, 27, 64, 125, \ldots\}$
7. $\{12, 20, 30\}$
8. $\{2, 3, 5\}$
9. $\{31, 37, 41, 43, 47\}$
10. $\{31, 33, 35, 37, 39\}$
11. $\{18, 20, 22, 24\}$
12. $\{20, 21, 22, 24, 25, 26\}$
13. $\{1, 2, 4, 7, 14, 28\}$
14. $\{13, 26, 39, 52, 65, \ldots\}$
15. $\{1, 2, 3, 6, 9, 18\}$
16. $\{25, 50, 75, 100, 125, \ldots\}$
17. a. $\{1, 5, 12, 15, 20, 25\}$ b. $\{5, 20\}$
18. a. $\{1, 2, 3, 4, 5, \ldots\} = N$ b. $\varnothing$
19. a. $\{2, 3, 4, 5, 6, 7, 8, 9\}$ b. $\varnothing$
20. a. $\{31, 33, 35, 36, 37, 39\}$ b. $\{33, 39\}$
21. a. $\{1, 2, 3, 5, 6, 7, 10, 11, 13, 15, 17, 19, 23, 29, 30\}$ b. $\{2, 3, 5\}$
22. $\{\ \}, \{2\}, \{3\}, \{5\}, \{2, 3\}, \{2, 5\}, \{3, 5\}, \{2, 3, 5\}$
23. $\{\ \}, \{1\}, \{3\}, \{5\}, \{7\}, \{1, 3\}, \{1, 5\}, \{1, 7\}, \{3, 5\}, \{3, 7\}, \{5, 7\}, \{1, 3, 5\}, \{1, 3, 7\},$
    $\{1, 5, 7\}, \{3, 5, 7\}, \{1, 3, 5, 7\}$
24. True
25. False
26. False
27. True
28. False
29. False, because 1 is not prime or composite
30. True
31. True
32. True
33. True

Exercises 0.3, page 14

1. -5, integer, rational, real
2. $\frac{1}{4}$, rational, real
3. 1, natural, integer, rational, real
4. 3, natural, integer, rational, real
5. $\sqrt{5}$, real
6. 2, natural, integer, rational, real
7. -1, integer, rational, real
8. $-\frac{2}{3}$, rational, real
9. -2, integer, rational, real
10. 0, integer, rational, real
11. Commutative, addition
12. Inverse, addition
13. Inverse, multiplication
14. Distributive
15. Associative, addition
16. Commutative, multiplication
17. Identity, multiplication
18. Associative, multiplication
19. Closure, addition, multiplication
20. Identity, addition
21. x
22. x
23. x
24. x
25. x
26. x
27. $x + 5$
28. $4x - 9$

29. -100

30. 1000

31. -30

32. -5

33. 10

34. -7

35. -3

36. 6

37. 0

38. 5

39. -5

40. -3

41. -6

42. 0

43. -6

44. -6

45. 0

46. 3

47. 0

48. -1

49. -100

50. 121

51. -2

52. 0

53. 1

54. 77

55. $-\frac{5}{16}$

56. -7000

57. $(-35)(-\frac{2}{3})(-\frac{3}{2}) = (-35)(1) = -35$

58. $(-57)(-\frac{5}{3} + \frac{2}{3}) = (-57)(-\frac{3}{3}) = 57$

59. $[\frac{3}{4} + (-\frac{3}{4})] + \frac{7}{12} = 0 + \frac{7}{12} = \frac{7}{12}$

60. $-12(\frac{2}{3}) + (-12)(-\frac{3}{4}) = -8 + 9 = 1$

61. $[(-9)(4)(-25)]/-25 = -36$ **62.** $79(-25)(-4) = 7900$

63. $(-8)(-\frac{7}{9} - \frac{2}{9}) = (-8)(-\frac{9}{9}) = (-8)(-1) = 8$

64. $(64 + 36) - (75 + 25) = 100 - 100 = 0$

Exercises 0.4, page 20

1. Commutative, addition

2. Inverse, addition

3. Inverse, multiplication

4. Distributive

5. Associative, addition

6. Commutative, multiplication

7. Identity, multiplication

8. Associative, multiplication

9. Closure, addition, multiplication

10. Identity, addition

11. Addition, division

12. Multiplication, subtraction

13. Subtraction, multiplication

14. Subtraction, division

15. Subtraction, division

16. Addition, subtraction, division

17. Addition (twice), division

18. Subtraction (twice)

19. Multiplication, subtraction, division

20. Addition, division

21. -12

22. 1

23. -5

24. $\frac{3}{2}$

25. 8

26. 6

27. -7

28. 4

29. 1

30. $\varnothing$

31. R

32. 4

33. $\frac{1}{2}$

34. -2 **35.** -5

36. 5 **37.** -3

38. 7 **39.** $\varnothing$

40. 3 **41.** 0

42. -9 **43.** 2

44. -3 **45.** R

46. -5 **47.** 3

48. -5 **49.** 2

50. 4 **51.** $6x - 4$

52. $x/2 + 9$ **53.** $10x + 5(2x) = 20x$

54. $5000 - x$ **55.** $5x + 5(3x) = 20x$

56. $24 - x$ **57.** $x + 40$

58. $x + \frac{1}{2}$ **59.** $0.05x + 0.065(2000)$

60. $x + (x + 1) + (x + 2) = 3x + 3$ **61.** $6 + 2x = 5(x - 3)$
                                                            7

62. $3(2x - 15) = 90 + 3x$ **63.** $x + 0.4(18) = 0.5(x + 18)$
    Bus, 45 mph; train, 75 mph   3.6 ounces

64. $2x + 2(2x - 2) = 50$ $[3(9)(16)]/400 = 1.08$ gallons
    $x = 9$

65. $29x + 25(2x) + 10(2x - 1) = 221$
    $2\frac{1}{3}$ pounds peaches, $4\frac{2}{3}$ pounds plums, $3\frac{2}{3}$ pounds bananas

66. $\frac{1}{2}x(x + 11) - \frac{1}{2}x(x - 10) = 2310$ **67.** $x + (x + 3) + (x/2) + 17 = 70$
    $x + 11 = 231$ feet   20 years

68. $3.2(2x + 5 - 88) = 67.2$
    52 pounds, 57 pounds

69. $65x + 50(x + 1) = 280$
    $x = 2$ hours; 2 hours after 11 a.m. = 1 p.m.

70. $2(x - 9) = x + 8$
    26

Exercises 0.5, page 27

1. $8x^2 - 4x - 4$ **2.** $5a$

3. $-6y - 8$ **4.** $-6a + 13$

5. $y^2 - y + 1$ **6.** $a - 2b - 2c$

7. $-2r - 2s + 2t$ **8.** -1

9. $-1 + x - x^4$ **10.** $14a - 7b$

11. $7x + 4y - 5z$ **12.** $2x^2$

13. $10x - 20$ **14.** $2y - 23$

15. $2a^2 - 2a^2b^2 + 4b^2$ **16.** $-2t^2 + 5t - 8$

17. $2y - 2$ **18.** $6x - 2$

19. 0 **20.** $5t - 23$

21. $2x^5 + x^4 - 2x^3$

22. $-3y^7 + 2y^5 + 4y^4$

23. $2x^3y^2 - 6x^2y^3 + 2xy^4$

24. $x^{n+2} + x^{n+1} + x^n$

25. $x^2 - 3x - 28$

26. $6x^2 - 7xy - 5y^2$

27. $2x^3 + 9x^2 - 18x$

28. $x^3 - x^2 - 14x + 8$

29. $18y^3 + 48y^2 + 32y$

30. $x^3 + 216y^3$

31. $49x^2 - 1$

32. $4t^3 - 20t^2 - 81t + 405$

33. $n^3 - 125$

34. $27r^3 + 54r^2s + 36rs^2 + 8s^3$

35. $6x^3 - x^2 + 8x + 3$

36. $x^4 + 2x^3 - 8x^2 - 13x + 6$

37. $x^4 - 16$

38. $x^4 - x^2y^2 + 2xy^3 - y^4$

39. $x^6 - 1$

40. $25x^4 - 60x^2 + 36$

41. $3x(x + 8y - 2)$

42. $(t + 3)(t + 8)$

43. $(x + 5)^2$

44. $(7y^2 - 1)(7y^2 + 1)$

45. $(x - 1)(x^2 + x + 1)$

46. $(a + 1)(a^2 - a + 1)$

47. $-3x(x + 3)(x - 3)$

48. $(3a + 1)(2a - 3)$

49. $-3x(x - 2)(x + 2)(x^2 + 3)$

50. $25(2p - q)(2p + q)$

51. $4y(4y - 3)(4y + 3)$

52. $(x - y)(x + y)(x^2 - xy + y^2)(x^2 + xy + y^2)$

53. $2y(3x + 2)(2x - 5)$

54. $(8a - 3b)(a + b)$

55. $2y(y - 3)(y + 3)(y^2 + 9)$

56. $3ax^4(x + 3)(5x - 1)$

57. $(5r + 1)^2$

58. $(6x - 1)^2$

59. $(2p - 5q)^2$

60. $(x - 5)(x + 5)(x^2 + 4)$

Exercises 0.6, page 34

1. $\frac{2}{5}$

2. $5x^2/9yz$

3. -1

4. $[9(x + 2y)]/(x - 2y)$

5. $(2x - 15)/(2x - 3)$

6. $[-4x^3(x - 1)]/[3(x + 1)]$

7. 1

8. $(x - 1)/3$

9. $-1/[3(x + 3)]$

10. $(xy + 3)/(xy - 3)$

11. $(5 - x)/(x + 3)$

12. $(6x^2 - 3x)/(x^2 - 9)$

13. $(4x - 15)/(x^2 - 6x + 5)$

14. $4/[x(x + 1)(x - 1)]$

15. $(x^2 + 14x + 4)/[2x(x + 2)]$

16. $1/[(x + 3)(x + 4)]$

17. $16/[(x - 2)^2(x + 2)^2]$

18. $[5(5x + 2)]/[12(3x + 1)]$

19. $(9x - 2)/3$

20. Undefined

21. $x^2 - x + [-1/(x - 3)]$

22. $x^2 - 2 + [4/(x^2 + 3)]$

23. $3x^2 - 2x + 4 - [1/(5x + 10)]$

24. $a^2 + a + 1$

25. $5x^2 - 2x - 3 + [-7/(2x + 1)]$

26. a. -3 b. 3 c. 2

27. $x \neq 6, x \neq -6; \{-\frac{1}{2}\}; -\frac{32}{143} = -\frac{32}{143}$

28. $x \neq -2; \{0\}; \frac{3}{2} = \frac{3}{2}$

29. $x \neq 3, x \neq -3; \{9\}; -\frac{1}{6} = -\frac{1}{6}$

30. $x \neq 1, x \neq -3; \{-\frac{1}{4}\}; -2 = -2$

31. $x \neq 0, x \neq -\frac{1}{3}; \{\frac{1}{7}\}; 0 = 0$

32. $x \neq 4, x \neq -4; \varnothing$

33. All real numbers except $x = 1$.

34. $x = (a - b)/2$; $x \neq a$, $x \neq -b$, $a \neq -b$

35. $x = (p - 1)/(c + 1)$; $c \neq -1$, $x \neq 0$

36. $2/\frac{7}{2} + 2/x = 1$; $4\frac{2}{3}$ hours

37. $860/(x - 35) = 1140/(x + 35)$; 250 mph

38. $(x/80) + (x/120) = 1$; 48 minutes   **39.** $(x - 3 - 2)/(x + 1) = \frac{1}{4}$; $\frac{4}{7}$

40. $120/(32 + x) + 2 = 120/(32 - x)$; 8 mph

41. $(x/10) + (x/15) = 1$; 6 hours

42. $500/(t + 5) = 400/t$; 20 hours, 25 hours

43. 8   **44.** $\frac{720}{16} = 450/x$; 10 gallons

Exercises 0.7, page 40

1. $(5, -2)$ **2.** $(2, 1)$

3. $(-3, -7)$ **4.** $(-6, 4)$

5. $(\frac{1}{2}, -3)$ **6.** $(7, -4)$

7. $(-1, 2)$ **8.** $(\frac{7}{3}, \frac{5}{2})$

9. $(0, -\frac{36}{5})$ **10.** $(-\frac{35}{2}, -\frac{75}{2})$

11. $(2, 4)$ **12.** $(3, 5)$

13. $(-3, -3)$ **14.** $(5, 3)$

15. $(-2, -11)$ **16.** $(4, -\frac{1}{2})$

17. $(7, -1)$ **18.** $(4, -\frac{1}{2})$

19. $\emptyset$ (no solution)

20. $\{(x, y) \mid 3y = x - 1\}$ (infinitely many solutions)

Chapter 0   Review Exercises, page 41

1. $5a/b$ **2.** $x^2 + 3x - 10$

3. $x^2 + 6x + 9$ **4.** $1 - x + y$

5. $4/(4 - x)$ **6.** $(x + 4)(x - 7)$

7. $5x - 4x = 60$ **8.** $9x^2 - 12x + 4$

9. 1   **10.** $(x + 3y)(x + 4y)$

11. $2n - 5n + 5 = 3$ **12.** $(x + 3)/(x + 2)$

13. $2(5^2) = 50$ **14.** $2(y + 5)(y - 5)$

15. $x = 0$ or $x = 5$ **16.** $25x^2$

17. $14x/5$ **18.** $x = \frac{7}{3}$

19. $x - 3x + 6$ **20.** $x - 3x + 6$

21. $x/(x + 3)$ **22.** $2x = 5$

23. $x = -\frac{5}{3}$ **24.** $x^2 + 25$

25. No solutions   **26.** -4

27. Origin   **28.** -3

29. $(1, 3)$ **30.** 6

31. $x^4 + 3x^2 + 10x - 21$ **32.** $-2x - 31$

33. $20x^4 - 45x^2$ **34.** $3x - 6$

35. x **36.** $20 - x$

37. $1/x + 1/(x + 1)$ **38.** $(x - 3)/(x + 3)$

39. $900/(200 - x)$ **40.** $(30 - 2x)/2$

41. $1/x + 1/\frac{5}{2}$ **42.** $0.80x + 1.20y$

43. $x + (x + 7) - 20$ **44.** $0.35x/3$

45. $V = \frac{1}{3}h(a^2 + ab + b^2)$

Multiple-Choice Examination, page 43

1. b   **2.** c

3. d   **4.** b

5. d   **6.** a

7. c   **8.** e

9. a   **10.** c

11. e   **12.** b

13. b   **14.** e

15. d   **16.** c

17. d   **18.** a

19. a   **20.** b

21. d   **22.** a

23. b   **24.** c

25. a   **26.** e

27. c   **28.** e

29. d   **30.** e

Exercises 1.1 A, page 53

1. x^8, theorem 2   **2.** x^7, theorem 1

3. x^5, theorem 3   **4.** $-8y^3$, theorem 4

5. $81/y^4$, theorem 5   **6.** $1/x^4$, theorem 3

7. $10^4 = 10,000$, theorem 4   **8.** $-5(2^6) = -320$, theorem 2

9. $2^6 = 64$, theorem 5   **10.** $(-a)^{10} = a^{10}$, theorem 1

11. $-125x^6$ **12.** $-25x^6$

13. a^3b^8 **14.** $27/x^3$

15. $3y^8$ **16.** 64

17. $-x^4$ **18.** 1

19. $125x^3y^6$ **20.** x^7y^6

21. $36a^2c^2/25b^4$ **22.** $-32a^{10}b^{25}$

23. -1 **24.** $100,000,000$

25. 1000 **26.** $\frac{8}{25}$

27. $40,000$ **28.** 1

29. 32

30. x^{n-1}

31. y^{4n}

32. $x^{2n}/5^{2n}$

33. 1

34. x^n

35. x^{2n+1}

Exercises 1.2 A, page 58

1. $\frac{1}{2}$

2. $\frac{1}{125}$

3. 16

4. 100,000

5. 9

6. 625

7. $\frac{1}{1000} = 0.001$

8. $\frac{5}{3}$

9. $\frac{81}{10,000} = 0.0081$

10. 27

11. $\frac{1}{16}$

12. $\frac{1}{100} = 0.01$

13. $\frac{1}{100,000} = 0.00001$

14. 1

15. 1,000,000

16. $\frac{1}{100,000} = 0.00001$

17. 1

18. $\frac{1}{1000} = 0.001$

19. $\frac{1}{64}$

20. $\frac{81}{25}$

21. -4

22. $\frac{25}{4}$

23. $24\frac{3}{2}$

24. 1600

25. 1

26. 6

27. -15

28. 64

29. 10

30. $\frac{10}{7}$

31. $\frac{1}{1000} = 0.001$

32. $\frac{7}{1000} = 0.007$

33. $1/8x^3$

34. $2/x^3$

35. $16/x^2$

36. $x^3/125$

37. $4x^2$

38. $4x^2/(x^2 + 4x + 4)$

39. y^2

40. $x^2/2$

41. 1

42. x^2/y

43. 1

44. $xy/(x + y)$

45. 2

46. $xy^9/4$

47. $-1/x$

48. a^5/b

49. $y^6/9$

50. $3/y^5$

51. x

52. $(x - 1)/x$

53. x^n/y^n

54. $1/x^{10n}$

55. y^{3n}/x^{3n}

Exercises 1.3 A, page 67

1. 4

2. -6

3. 3

4. $\frac{2}{5}$

5. 20

6. 3

7. 1

8. 13

9. 17

10. $\frac{5}{7}$

11. 6.71

12. 3.16

13. 80.62

14. 0.22

15. 4.47

16. 3.27

17. 4.06

18. 1.72

19. -0.26

20. 3.46

21. 13

22. 12

23. $\sqrt{2}$

24. 8

25. $\sqrt{8} \approx 2.83$

26. $\sqrt{2} \approx 1.41$

27. $\sqrt{32} \approx 5.66$

28. $t^2 - 1$

29. $4u^2 + v^2$

30. $x^2 + 3$

31. $\{5\}$

32. $\{5, -5\}$

33. $\{6, -6\}$

34. $\{-6\}$

35. $\varnothing$

36. $|x|$

37. $2|x|$

38. $|x|$

39. $3|x|$

40. $|x - y|$

Exercises 1.4 A, page 73

1. 4

2. -3

3. 2

4. -8

5. 7

6. -1

7. 10

8. 15

9. 5

10. $\frac{1}{2}$

11. $\frac{3}{5}$

12. 10

13. 100

14. 0.1

15. $25x^2$

16. 4.48

17. 3.14

18. 3.85

19. 2.29

20. 3.42

21. 0.79

22. 0.56

23. $\frac{3}{4}$

24. 3

25. 2.52

26. $\{4\}$

27. $\{-4\}$

28. $\{4\}$

29. $\{-4\}$

30. $\{2\}$

31. $\{2, -2\}$

32. $\varnothing$

Exercises 1.5 A, page 77

1. $8 = 8$, theorem 1

2. $2 = 2$, theorem 2

3. $4 = 4$, theorem 3

4. $10 = 10$, theorem 4

5. $5 = 5$, theorem 5

6. 2

7. 4

8. 2

9. 3

10. 2

11. $\frac{3}{5}$

12. $\frac{3}{4}$

13. 16

14. 512

15. $\frac{1}{4}$

16. $\frac{1}{125}$

17. $\frac{1}{1000} = 0.001$

18. 9

19. 25

20. $\frac{36}{49}$

21. $\frac{1}{8}$

22. 8

23. 4

24. -4

25. $\frac{1}{4}$

26. $-\frac{1}{4}$

27. 15,625

28. $\frac{1}{8}$

29. $\sqrt{10}$

30. $\sqrt[3]{4}$

31. $\sqrt[4]{14}$

32. 4

33. $\sqrt[3]{5}$

34. 100,000

35. $\sqrt{10}$

36. $\frac{1}{10} = 0.1$

37. $10\sqrt[4]{10}$

38. $\sqrt{5}$

39. 2

40. $\sqrt{10}$

41. 4

42. $y\sqrt{x}$

43. $1/64x^2$

44. $4/x^2$

45. $3y/x^2$

46. $1/1,000,000xy^5$

47. $x + y + 2\sqrt{xy}$

48. $x - 3$

49. $2x + 5$

50. 5

Exercises 1.6 A, page 82

1. 10

2. 10

3. $6x$

4. y^2

5. $2\sqrt{3}$

6. $5\sqrt{5}$

7. $9\sqrt{5}$

8. $3\sqrt{x}$

9. $2x\sqrt{6}$

10. $2y\sqrt{10y}$

11. x^2

12. $y^3\sqrt{y}$

13. $6x\sqrt{x}$

14. $2x^2\sqrt{10x}$

15. $4\sqrt{2x}$

16. $22\sqrt{3}$

17. $7x\sqrt{7x}$

18. $30\sqrt{2}$

19. $6x^3\sqrt{6}$

20. $200x^2y\sqrt{2y}$

21. 3

22. 13

23. 2

24. -1

25. 6

26. $3\sqrt[3]{2}$

27. $-3\sqrt[3]{2}$

28. $2\sqrt[3]{9}$

29. $2\sqrt[3]{2}$

30. 2

31. 9

32. $10x$

33. $2\sqrt[3]{4x}$

34. $3\sqrt[3]{4y^2}$

35. $2y\sqrt[3]{5y}$

36. x

37. $2y^2\sqrt[3]{2}$

38. $6y\sqrt[3]{2y^2}$

39. $6x$

40. $6\sqrt[3]{3}$

Exercises I.7 A, page 85

1. $\sqrt{2}/2$

2. $\sqrt{6}/6$

3. $\sqrt{15}/6$

4. $\sqrt{5}/5$

5. $\sqrt{5}/10$

6. $\sqrt{2}/5$

7. $\sqrt{6}/10$

8. $4\sqrt{3}/3$

9. $2\sqrt{7}$

10. $\sqrt{2}/2x$

11. $\sqrt{2x}/x$

12. $2\sqrt{2y}/y$

13. $\sqrt{2x}/2x^2$

14. $3\sqrt{5}$

15. $\sqrt{2x}/2x$

16. $\sqrt{2}+1$

17. $\sqrt{7}-\sqrt{3}$

18. $-(2\sqrt{5}+5)$

19. $(\sqrt{11}-1)/2$

20. $[2(\sqrt{x}+1)]/(x-1)$

21. $\sqrt{2}/4$

22. $\frac{5}{7}$

23. 1

24. $\sqrt{x}+\sqrt{y}$

25. $\sqrt{x}-\sqrt{2y}$

Exercises I.8 A, page 88

1. $8\sqrt{5}$

2. $3\sqrt{2}$

3. $6\sqrt{6}$

4. $\sqrt{10}$

5. $3\sqrt{5}+5\sqrt{3}$

6. $3\sqrt{3}$

7. $2\sqrt{14x}$

8. $2\sqrt{2x}+3\sqrt{3x}$

9. $4\sqrt{3}$

10. $\sqrt{10}$

11. $12\sqrt{6}$

12. $11\sqrt{2x}$

13. $\sqrt{13x}$

14. $5x\sqrt{7}$

15. $6x$

16. $6y$

17. $3\sqrt{10}/10$

18. $17\sqrt{2}/6$

19. $5\sqrt{5}+2\sqrt{3}$

20. $4\sqrt{x}-5\sqrt{y}$

21. $9\sqrt{42}/20$

22. $3\sqrt{3x}-3\sqrt{2x}$

23. $4\sqrt{14}-\sqrt{2}/2$

24. $5+2\sqrt{6}$

25. $30 - 12\sqrt{6}$

26. $9x + 6x\sqrt{2}$

27. $6 + 9\sqrt{2}$

28. $-5\sqrt{3} - 5\sqrt{2}$

29. 0

30. $4 - \sqrt{5}$

Exercises 1.9 A, page 91

1. $\sqrt{5}$

2. $\sqrt[3]{10}$

3. $\sqrt{7}$

4. $\sqrt{11x}$

5. $\sqrt{3t}$

6. $\sqrt[3]{4t^2}$

7. $2\sqrt{5}$

8. $3\sqrt{2a}$

9. $\sqrt[3]{2c}$

10. $5\sqrt[6]{5}$

11. $\sqrt[4]{216}$

12. $2y$

13. $\sqrt[3]{25}$

14. $\sqrt[3]{10}$

15. 3

16. $\sqrt[4]{7}$

17. $\sqrt[4]{2}/2$

18. $\sqrt[3]{3x}$

19. $\sqrt[6]{40}$

20. $\sqrt{5}$

21. $\sqrt[3]{4}$

22. 6

23. 3

24. $\sqrt[3]{2}$

25. 27

26. $\sqrt{3} + 3\sqrt{2}$

27. $6\sqrt{2x} + 2\sqrt{5x}$

28. $\sqrt[3]{49}/7$

29. $\sqrt[3]{100y^2} + \sqrt{10y}$

30. x

Exercises 1.10 A, page 94

1. 9.29×10^7

2. 6.6×10^{21}

3. 1.14×10^7

4. 9.5×10^{-8}

5. 2.4×10^{-9}

6. 2.21×10^9

7. 6.1×10^{-4}

8. 1.2×10^5

9. 2.205×10^{-3}

10. 6.67×10^{-8}

11. 2300

12. -460

13. $0.000\ 0180$

14. $864,000$

15. $0.000\ 000\ 0303$

16. 0.01745

17. 0.16667

18. $0.000\ 000\ 000\ 480$

19. $1,870,000,000$

20. $6,300,000,000,000,000,000$

21. 1080

22. 0.01

23. 0.03

24. $300,000$

25. 0.048

26. 1.047

27. 42

28. 8.9 hours, approx.

29. 2.88×10^{-10}

30. 1.33×10^{-9}

Chapter I Review Exercises, page 99

1. a. $2\sqrt{3}$ b. 10

2. a. 8.54   b. 0.02   c. 2.92   d. 0.03

3. a. $-3\sqrt[3]{3}$ b. 7   c. $\frac{2}{3}$ d. 0.0013

4. a. $8\sqrt{3}$ b. $6\sqrt{5}$ c. $2\sqrt[3]{18}$ d. $12\sqrt{15}$

5. a. $3\sqrt{5}/5$ b. $\sqrt{15}/5$ c. $6\sqrt{3}$ d. $2\sqrt{2}-5$ e. $-(\sqrt{3}+\sqrt{6})$

6. a. $5\sqrt{6}$ b. $5\sqrt{2}-20$ c. $8\sqrt{7}-3\sqrt{2}$ d. 288   e. $-\sqrt{6}/2$
   f. $31\sqrt{30}/30+3\sqrt{3}$

7. a. a^8b^{13} b. $-18a^8b^7/25m^8n^3$ c. $a^8b^4/192$ d. $-y^4$

8. a. $1/m^4$ b. p^2 c. x^2z/y d. $(a+b)/ab$

9. a. $\sqrt[4]{x}$ b. $2\sqrt{a}$ c. $\sqrt[4]{x^3}$ d. $\sqrt[3]{xy}/xy$

10. a. $y^{1/3}$ b. $x^{1/2}y^{5/6}$ c. $xy^{1/2}$ d. $(xy)^{2/3}$

11. a. 1   b. 5   c. $\frac{1}{8}$ d. $\frac{29}{6}$

12. a. 0   b. 0   c. 18   **13.** a. 7   b. $3\sqrt{5}$ c. $4\sqrt{6}$

14. a. 64   b. $\frac{399}{4}$ **15.** a. 0.09   b. 80,000

16. a. $5x\sqrt{2x}$ b. $2\sqrt{5}$ c. $\sqrt{3x}$

Exercises 2.I A, page I07

1. $2i$ **2.** $9i$

3. xi **4.** $\frac{1}{2}i$

5. $i\sqrt{2}$ **6.** $i2\sqrt{2}$

7. $40i$ **8.** $-40i$

9. $i6\sqrt{2}$ **10.** $-i6\sqrt{2}$

11. $4+5i$ **12.** $6-7i$

13. $7+6i$ **14.** $8-i2\sqrt{3}$

15. $5+0i$ **16.** $0+0i$

17. $0+2i$ **18.** $0+13i$

19. $x=4, y=2$ **20.** $x=4, y=-5$

21. $x=0, y=-5$ **22.** $x=4, y=0$

23. $x=5, y=-1$ **24.** $x=1, y=2$

25. $7+2i$ **26.** $5+7i$

27. $-2-3i$ **28.** $1-5i$

29. $5+3i$ **30.** $-3+5i$

31. $9-i\sqrt{3}$ **32.** $-4\sqrt{2}-i\sqrt{6}$

33. $-1-4i$ **34.** $0+0i$

35. $0+9i$ **36.** $x\geq 4$

37. $x\leq 9$ **38.** $x\leq -4$ or $x\geq 4$

39. $-3 \le x \le 3$

40. $x < -5$ or $x > 5$

41. $b = 0$

Exercises 2.2 A, page 111

1. -10

2. $-6\sqrt{2}$

3. $6i$

4. $i\sqrt{6}$

5. $-6i$

6. $-6 + 8i$

7. $6 + 15i$

8. $6 + 6i$

9. $12 - 5i$

10. $22 + 7i$

11. 20

12. 1

13. $5 + 5i$

14. $15 - 90i$

15. $5 + 12i$

16. $-4 + 78i$

17. $17 + i$

18. $22 - 7i$

19. $3 + i$

20. $5 - 15i$

21. $5 - 15i$

22. $9 - 40i$

23. 0

24. 0

25. 5

26. $(3 + i)^2 - 6(3 + i) + 10 = 8 + 6i - 18 - 6i + 10 = 0$

$(3 - i)^2 - 6(3 - i) + 10 = 8 - 6i - 18 + 6i + 10 = 0$

27. $4i^2 + 2i^2 + 6 = -4 - 2 + 6 = 0$ **28.** $x = 0, y = 0$

29. $x = 3$ or $x = -3$

30. $x = -1$

31. i

32. -1

33. $-i$

34. 1

35. $-i$

36. -1

37. 0

38. $-i$

Exercises 2.3 A, page 115

1. $5 - 7i$

2. $6 + 4i$

3. $2 - i$

4. $4 + 3i$

5. $-i$

6. i

7. 3

8. $-1 - i$

9. $\frac{3}{13} - \frac{2}{13}i$

10. $-\frac{5}{34} - \frac{3}{34}i$

11. i

12. $2 - \frac{3}{2}i$

13. $\frac{3}{5} + \frac{3}{10}i$

14. $-\frac{1}{2} - (\sqrt{3}/2)i$

15. $-2 - \sqrt{3}$

16. $4\sqrt{2}/3 + \frac{1}{3}i$

17. $(a^2 - b^2)/(a^2 + b^2) + 2abi/(a^2 + b^2)$

18. $\frac{3}{2} - \frac{1}{2}i$

19. $\frac{8}{5} + 0i$

20. $0.3 + 0.9i$

Exercises 2.4 A, page 120

1–10.

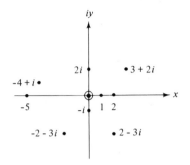

11. a. $8 + 6i$ b. 10 mph   c. 3 miles
12. $-8 + 16i$; $8\sqrt{5}$; -2

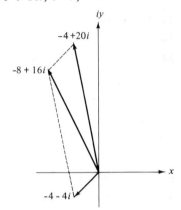

Exercises 2.5 A, page 126

1. $-5, 1$ **2.** $0, 4$
3. $-c, 2c$ **4.** $-3, -2$
5. $\frac{5}{2}, -\frac{5}{2}$ **6.** $0, -3$
7. $-2, -1$ **8.** $\frac{2}{3}, -1$
9. $2i, -2i$ **10.** $-3, 2$
11. $-5, 4$ **12.** $0, 4$
13. -5 **14.** $0, k$
15. $\sqrt{5}, -\sqrt{5}$ **16.** $a/2, -a/2$
17. $3, 1$ **18.** $2 + n, 2 - n$
19. $a + n, a - n$ **20.** $(-b + 5)/a, (-b - 5)/a$
21. d/c **22.** $8a, -2a$

23. 0, 1, 4

24. $-2b$

25. 5, -5

26. 4, -4

27. $\frac{1}{2}, -1$

28. c, d

29. 2

30. -5

31. $t = \pm \sqrt{2s/g}$

32. $r = \pm \sqrt{(\pi R^2 - A)/\pi}$

33. $n_1 = \pm \sqrt{kn_2^2/(1 - k)} = \pm n_2 \sqrt{k/(1 - k)}$

34. $c = 0$ or $c = d$

35. $d = L$ or $d = L/4$

Exercises 2.6 A, page 131

1. $5 + \sqrt{2}, 5 - \sqrt{2}$

2. $-3 + 2i, -3 - 2i$

3. $2 + \sqrt{3}, 2 - \sqrt{3}$

4. $1 + i\sqrt{2}, 1 - i\sqrt{2}$

5. $3, -1$

6. $5, 3$

7. $4 + i, 4 - i$

8. $1 + 2\sqrt{5}, 1 - 2\sqrt{5}$

9. $(1 + \sqrt{5})/2, (1 - \sqrt{5})/2$

10. $1, -\frac{2}{3}$

11. $1 + i, 1 - i$

12. $3, 1$

13. $0, -4$

14. $(3 + \sqrt{2})/2, (3 - \sqrt{2})/2$

15. $7 + 2i, 7 - 2i$

16. $\frac{4}{3}, -\frac{1}{2}$

17. $-3 + \sqrt{7}, -3 - \sqrt{7}$

18. $-3 + 4i, -3 - 4i$

19. $5 + i\sqrt{70}, 5 - i\sqrt{70}$

20. $4, -1$

21. $5 + i, 5 - i$

22. $2 + \sqrt{2}, 2 - \sqrt{2}$

23. $5y, y$

24. $2\sqrt{y^2 + 1}, -2\sqrt{y^2 + 1}$

25. $2\sqrt{25 - x^2}/5, -2\sqrt{25 - x^2}/5$

26. $-x/2 + ix\sqrt{3}/2, -x/2 - ix\sqrt{3}/2$

27. $(v \pm \sqrt{v^2 - 64s})/32$

28. $(-E \pm \sqrt{E^2 + 4RP})/2R$

29. $b \pm \sqrt{b^2 - c}$

30. $(-b \pm \sqrt{b^2 - 4ac})/2a$

Exercises 2.7 A, page 137

1. a. $2x^2 - 3x - 1 = 0; a = 2, b = -3, c = 1$
  b. 17, real and unequal

2. a. $4x^2 - 12x + 9 = 0; a = 4, b = -12, c = 9$
  b. 0, real and equal

3. a. $x^2 - 5 = 0; a = 1, b = 0, c = -5$
  b. 20, real and unequal

4. a. $x^2 + 5x = 0; a = 1, b = 5, c = 0$
  b. 25, real and unequal

5. a. $x^2 - 4x + 8 = 0; a = 1, b = -4, c = 8$
  b. -16, imaginary

6. $-\frac{2}{3}, -1$

7. $2, -\frac{1}{3}$

8. $(3 \pm \sqrt{5})/2$

9. $(-3 \pm \sqrt{13})/2$

10. $(3 \pm i\sqrt{3})/2$

11. $(-1 \pm i\sqrt{3})/2$

12. $(2 \pm \sqrt{7})/2$

13. $(1 \pm i\sqrt{5})/3$

14. $(-1 \pm \sqrt{7})/3$

15. $4i, -4i$

16. $0, \frac{5}{3}$

17. $\frac{1}{18}$

18. $(5 \pm \sqrt{5})/2$

19. $6, \frac{3}{2}$

20. $5 \pm 5\sqrt{5}$

21. $-p \pm \sqrt{p^2 - q}$

22. $(1 \pm \sqrt{5})n/2$

23. $y/3, -3y$

24. $2 \pm \sqrt{10 - 2x}$

25. $(-v \pm \sqrt{v^2 + 2sg})/g$

26. $\frac{1}{2}(2k + 1 \pm \sqrt{4k + 1})$

27. $\frac{1}{2}(a \pm \sqrt{a^2 - 4ab})$

28. $(-\pi s \pm \sqrt{\pi^2 s^2 + 4A\pi})/2\pi$

Exercises 2.8 A, page 142

1. $r + s = 8, rs = 5$

2. $r + s = -5, rs = -2$

3. $r + s = \frac{3}{2}, rs = 2$

4. $r + s = -2, rs = -\frac{1}{3}$

5. $r + s = \frac{5}{4}, rs = -\frac{7}{4}$

6. $r + s = 3$

7. $r + s = -2$

8. $r + s = -4$

9. $r + s = 9$

10. $r + s = 0$

11. $rs = 4$

12. $rs = -14$

13. $rs = 0$

14. $rs = -6$

15. $rs = 7$

16. $5 + \sqrt{7} + 5 - \sqrt{7} = 10$

$(5 + \sqrt{7})(5 - \sqrt{7}) = 25 - 7 = 18$

Yes

17. $1 - \sqrt{2} + 1 + \sqrt{2} = 2$

But $-b/a = -2$

No

18. $(2 + \sqrt{5})/2 + (2 - \sqrt{5})/2 = 2$

$[(2 + \sqrt{5})/2][(2 - \sqrt{5})/2] = -\frac{1}{4}$

But $c/a = -\frac{1}{2}$

No

19. $(-2 + \sqrt{5})/2 + (-2 - \sqrt{5})/2 = -2$

$[(-2 + \sqrt{5})/2][(-2 - \sqrt{5})/2] = -\frac{1}{4}$

Yes

20. $-3 + i - 3 - i = -6$

$(-3 + i)(-3 - i) = 10$

Yes

21. $x^2 - 6x + 2 = 0$

22. $x^2 + 3x = 0$

23. $3x^2 - 17x + 10 = 0$

24. $25x^2 + 10x + 1 = 0$

25. $4x^2 + 6x - 5 = 0$

26. $x^2 - x - 6 = 0$

27. $x^2 - 3 = 0$

28. $x^2 - 6x + 7 = 0$

29. $x^2 - 5x = 0$

30. $x^2 - 2x + 5 = 0$

31. $k = 1$

32. $k = 23$

33. $k = 3$

34. $k = \pm 8$

35. $k = 0$

36. $k = 0$

37. $\{2 + \sqrt{6}, 2 - \sqrt{6}\}$

38. $\{2 + 2i, 2 - 2i\}$

39. $\{(5 + \sqrt{17})/4, (5 - \sqrt{17})/4\}$

40. $\{(4 + 2i)/5, (4 - 2i)/5\}$

Exercises 2.9 A, page 147

1. $\{3, -3, \sqrt{5}, -\sqrt{5}\}$

2. $\{\sqrt{2}/2, -\sqrt{2}/2, i, -i\}$

3. $\{0, 1, 5, 6\}$

4. $\{\frac{1}{3}, -\frac{1}{4}\}$

5. $\{\frac{3}{2}, -\frac{3}{2}, i, -i\}$

6. $\{-64, 8\}$

7. $\{1, -1, \sqrt{2}/2, -\sqrt{2}/2\}$

8. $\{-2, 1, 3, 6\}$

9. $\{81, 256\}$

10. $\{\frac{1}{2}, -1, 2\}$

11. $\{2, -1 + i\sqrt{3}, -1 - i\sqrt{3}\}$

12. $\{1, -9, -4 + 2\sqrt{3}, -4 - 2\sqrt{3}\}$

13. $\{3, -\frac{1}{2}, -\frac{3}{7}\}$

14. $\{3, 3 + \sqrt{35}, 3 - \sqrt{35}\}$

Exercises 2.10 A, page 151

1. $\{11\}$

2. $\{2\}$

3. $\{-4\}$

4. $\{1\}$

5. $\{5\}$

6. $\varnothing$

7. $\{24\}$

8. $\{2\}$

9. $\{-1\}$

10. $\{-1, -5\}$

11. $\{8, -2\}$

12. $\varnothing$

13. $\{-1\}$

14. $\{1\}$

15. $\{1, 2\}$

16. $\{3, 7\}$

17. $\{20\}$

18. $\{0\}$

19. $\{13\}$

20. $\{\frac{1}{2}\}$

Exercises 2.11 A, page 157

1. $3\sqrt{2}/2$

2. Width $= 4$ inches, length $= 13$ inches

3. 4 or $-\frac{1}{4}$

4. 6 hours

5. 12 inches by 24 inches

6. (3 and 4) or (28 and -21)

7. 13 feet, 5 inches

8. $4\frac{1}{2}$ mph

9. 20 feet

10. 2.07 inches approx.

11. 5.2×10^{-5} (*Note:* x must be positive.)

12. 30 mph

13. Faster, 4 days @ \$50 = \$200
Slower, 12 days @ \$20 = \$240
Together, \$210
Answer: faster man alone

14. 60 inches = 5 feet

15. 2 seconds, 544 feet per second = 371 mph approx.

16. 2500 amperes or 250 amperes
(Actually, only 2500 amperes is an acceptable answer.)

Chapter 2 Review Exercises, page 162

1. a. $2 + 4i$ **b.** $1 + 8i$ **c.** $2 + i\sqrt{3}$ **d.** $18 + i$ **e.** $-48 - 14i$
   f. $-\frac{1}{13} + (-\frac{5}{13})i$

2. a. $x = 3, y = 2$ **b.** $x = \frac{2}{3}, y = -\frac{4}{3}$

3. Resultant: $12 - 5i$, magnitude: 13, direction: $-\frac{5}{12}$

4. $(3\sqrt{3}/2 + 3i/2)^2 = 27/4 + 18\sqrt{3}i/4 + 9i^2/4 = 9/2 + 9\sqrt{3}i/2$

5. a. $\{10, -4\}$ **b.** $\{-\frac{1}{6}, -\frac{5}{2}\}$ **c.** $\{0, \frac{4}{3}\}$ **d.** $\{3i, -3i\}$

6. a. $\{-5, -3\}$ **b.** $\{1 + i\sqrt{2}, 1 - i\sqrt{2}\}$ **c.** $\{2 + \sqrt{5}, 2 - \sqrt{5}\}$
   d. $\{-5 + 2\sqrt{2}, -5 - 2\sqrt{2}\}$

7. a. $\{(-5 + \sqrt{33})/4, (-5 - \sqrt{33})/4\}$ **b.** $\{(1 + \sqrt{17})/4, (1 - \sqrt{17})/4\}$
   c. $\{\frac{5}{2}, -\frac{4}{3}\}$ **d.** $\{(8 + 2\sqrt{10})/3, (8 - 2\sqrt{10})/3\}$

8. a. $\{\sqrt{42}/6, -\sqrt{42}/6\}$ **b.** $\{-2 + 2i, -2 - 2i\}$ **c.** $\{-2, \frac{5}{2}\}$
   d. $\{(-5 + \sqrt{13})/6, (-5 - \sqrt{13})/6\}$

9. a. 9   **b.** ± 2 **c.** $\frac{5}{4}$ **d.** ± 4

10. a. $x^2 + 2x + 4 = 0$ **b.** $4x^2 - 2x + 3 = 0$

11. a. $y = 5x$ or $y = -2x$ **b.** $y = (-1 \pm \sqrt{2})x^2$

12. a. 6   **b.** -2 **c.** 0

13. a. 4   **b.** -2 **c.** 3   **d.** 0

14. a. $\{1, 125\}$ **b.** $\{79\}$ **c.** $\{5, -\frac{1}{2}\}$ **d.** $\{10\}$ **e.** $\{4\}$ **f.** $\{10\}$

15. 30 minutes   **16.** 180 mph

17. 15 feet by 30 feet   **18.** 27 inches, 36 inches

19. $t = (cv + \sqrt{c^2v^2 - 2qH})/g$ **20.** $r = (-\pi h + \sqrt{\pi^2h^2 + 2\pi T})/2\pi$

21. $L/W = (1 + \sqrt{5})/2$ and $W/L = (\sqrt{5} - 1)/2$

22. \$20

Exercises 3.1 A, page 171

1. $(x + 5)(x + 4)$ **2.** $(y - 1)(y - 2)$

3. $(x + 6)(x + 2)$ **4.** $(3y - 13)(y + 1)$

5. $(t + 1)(t + 5)(t^2 + 6t + 10)$ **6.** $(x^2 + y + 6)(x^2 - y + 6)$

7. $(x^2 + y - 9)(x^2 - y + 9)$ **8.** $(x + y + 2)(x - y - 10)$

9. $(x + 2)(x^2 + 3)$ **10.** $(x - 8)(2x + 1)(2x - 1)$

11. $(2x + 5)(x - 2)(x^2 + 2x + 4)$ **12.** $(x + 6)(x + 3)(x - 3)$

13. $(x - 1)(x + 3)(x^2 - 3x + 9)$ **14.** $(x + 2y + 6)(x - 2y + 6)$

15. $(x + y + 8)(x - y - 8)$ **16.** $(x^2 + x - 9)(x^2 - x + 9)$

17. $(y^2 + 5x + 2)(y^2 - 5x + 2)$ **18.** $2(3x^2 + 1)$

19. $(x + 5)(x - 5)(x^2 + 10)$ **20.** $(t - 7)(t + 5)(t^2 - 2t + 2)$

21. $(c + 3 - 4x)(c + 3 + 4x)$ **22.** $(x - y)(x + y)(3x + 2)$

23. $(a - 3)(2a - 1)(2a + 1)$ **24.** $9(2x - 5)$

25. $(x - 3)^2(a + b)(a^2 - ab + b^2)$ **26.** $x^2(x^4 - 3x^2 + 3)$

27. $-(x - 5)^2$ **28.** $(2p - 3)(p^2 + 1)$

29. $(x - y)(x + y)(x^2 + y^2)(x^4 + y^4)$

30. $(t - 2)(t + 2)(t^2 + 2t + 4)(t^2 - 2t + 4)$

31. $p^3(r - s + 1)(r^2 - 2rs + s^2 - r + s + 1)$

32. $2(x - 2)(2x + 3)$ **33.** $(x + 3)(2x - 1)(2x + 5)$

34. $(r + 2s)(r + 2s - 1)$ **35.** $(x + 3y - 3)(x + 3y + 3)$

36. $(r^2 - 2rt + 2t^2)(r^2 + 2rt + 2t^2)$ **37.** $(a^2 - a + 1)(a^2 + a + 1)$

38. $(x - 3y + 1)(x + 3y + 1)$ **39.** $(x - y)^3(x - y - 2)(x - y + 2)$

40. $(x - y)(x + y)(x + y + 1)$

41. a. $(x^2 - 3)(x^2 + 1)$ b. $(x - \sqrt{3})(x + \sqrt{3})(x^2 + 1)$

   c. $(x - \sqrt{3})(x + \sqrt{3})(x - i)(x + i)$

42. a. $2x^2(x^2 - 5)^2$ b. and c. $2x^2(x - \sqrt{5})^2(x + \sqrt{5})^2$

43. a. $x^2 - 8x + 10$ b. and c. $(x - 4 + \sqrt{6})(x - 4 - \sqrt{6})$

44. a. and b. $y^2 - y + 1$ c. $(y - \frac{1}{2} + i\sqrt{3}/2)(y - \frac{1}{2} - i\sqrt{3}/2)$

45. a., b., and c. $(t + 1)(2t - 5)(2t + 5)$

Exercises 3.2 A, page 176

1. $2x^2 + x + 6; R, 7$ **2.** $x^3 - x^2 + 4x - 2; R, -1$

3. $x^3 - x^2 + x - 3; R, 8$ **4.** $x^2 + 2x - 5; R, 0$

5. $x^3 - 3x^2 + 10x - 30; R, 84$ **6.** $2x^3 + 10x^2 + 50x + 50; R, 256$

7. $x^2 + 8; R, 0$ **8.** $5y^3 - 8y^2 + 2y - 8; R, 0$

9. $x^2 + 3x + 9; R, 0$ **10.** $x^2 - 2x + 4; R, 0$

11. $3x^2 + 7x + 14; R, 21$ **12.** $3y^2 - 5y + 8; R, -38$

13. $x^4 - 2x^3 + 4x^2 - 8x + 16; R, 0$

14. $y^5 + 2y^4 + 4y^3 + 8y^2 + 16y + 32; R, 0$

15. $4x^2 - 8x - 3; R, -\frac{5}{2}$ **16.** $3x^2 + 3x - 6; R, 0$

17. $5x^3 - 2x + \frac{33}{5}; R, \frac{8}{25}$ **18.** $2x^2 - 4x + 3; R, -2$

19. $2x^2 - 4x + 3; R, 0$ **20.** $x^2 - 2x - 1; R, 0$

21. $x^3 + 2x^2 + 2; R, -2$

22. $x^2 + (2a + 1)x + 2a; R, 0$

23. $x^2 - 2bx - 2b^2; R, 0$

24. $x^2 - (1 + 2i)x + 2i; R, 0$

25. $x^2 + (2 + 3i)x + 6i; R, 0$

26. $x^2 + 5x + 5$

27. $x^2 + 5$

28. 56

29. $P(4) = 0$

30. $R = P(2) = 8$

Exercises 3.3 A, page 182

1. $P(2) = 0$, yes

2. $P(3) = 0$, yes

3. $P(-2) = 14$, no

4. $P(-3) = 0$, yes

5. $P(1) = 0$, yes

6. $P(1) = R = -5$

7. $P(-1) = R = 5$

8. $P(1) = R = 0$

9. $P(-1) = R = 0$

10. $P(2) = R = 0$

11. $P(1) = -18$

12. $P(-1) = -4$

13. $P(2) = -28$

14. $P(-2) = 0$

15. $P(3) = 0$

16. $P(-3) = 42$

17. $P(x) = (x + 2)(x - 3)(x^2 + x + 1)$

18. $P(1) = -30$

19. $P(-1) = -60$

20. $P(2) = 0$

21. $P(-2) = -96$

22. $P(5) = 870$

23. $P(-5) = 0$

24. $P(2i) = 0$

25. $P(-2i) = 0$

26. $(x + 5)(x - 2)(x + 2i)(x - 2i)$

27. $P(1) = -2,$ $P(-1) = 0$

    $P(2) = 0$ $\quad P(-2) = -20$

    $P(\frac{1}{2}) = 0$ $\quad P(-\frac{1}{2}) = \frac{5}{2}$

    $P(x) = (x + 1)(x - 2)(2x - 1)$

28. $k = 33$

29. $k = 39$

30. $k = 8$

31. $k = 2$

32. $k = -30$

33. $n = 0$ or $m = -\frac{1}{3}$

34. $k = 1$

35. a.

$$
\begin{array}{r}
2x^3 + 5x^2 + 20x + 85 \\
\hline
x - 4 \overline{)2x^4 - 3x^3 + \qquad\qquad 5x - 1} \\
2x^4 - 8x^3 \\
\hline
5x^3 \\
5x^3 - 20x^2 \\
\hline
20x^2 + 5x \\
20x^2 - 80x \\
\hline
85x - 1 \\
85x - 340 \\
\hline
\boxed{339}
\end{array}
$$

b. 4| 2   −3   0   5   −1
           8   20   80   340

     2   5   20   85   (339)

c. $P(4) = 2(4^4) − 3(4^3) + 5(4) − 1 = 512 − 192 + 20 − 1 = $ (339)

36. Yes, $P(2 + 4i) = 0$ **37.** $P(2i) = 0$

Exercises 3.4 A, page 188

1. $\{3, −2, −\frac{1}{2}\}$ **2.** $\{−4, 1 \pm i\sqrt{2}\}$

3. $\{\frac{4}{3}, (−1 \pm \sqrt{5})/2\}$ **4.** $\{−2, 3, i, −i\}$

5. $\{−\frac{1}{2}, 1, −1, 5\}$ **6.** $\{−3, −1, 1\}$

7. $\{5, (−5 \pm \sqrt{29})/2\}$ **8.** $\{\frac{2}{3}, −\frac{2}{3}, 1, −1\}$

9. $\{0, −3, \frac{1}{3}, 3\}$ **10.** $\{−2, 3, −3, 1 \pm i\sqrt{3}\}$

11. $\{1, −3, −\frac{3}{2}\}$ **12.** $\{1, 1, −2, −\frac{2}{3}\}$

13. $\{−4, 2 \pm 2i\sqrt{3}\}$ **14.** $\{1, (−1 \pm i\sqrt{3})/2\}$

15. $\{1, −1, −1, 5\}$ **16.** $\{4, (−1 \pm i\sqrt{15})/2\}$

17. $\{\sqrt{7}, \sqrt{7}, −\sqrt{7}, −\sqrt{7}\}$ **18.** $\{3, −2, i\sqrt{6}, −i\sqrt{6}\}$

19. $\{2, (−3 \pm i\sqrt{15})/2\}$ **20.** $\{2, 3, −5\}$

21. a. 3 inches   b. 4 inches   c.   2 inches

22. 2 centimeters   **23.** 6

24. 9 inches

Exercises 3.5 A, page 194

1.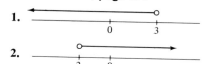

2.

3.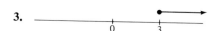

4.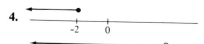

5.

6.

7.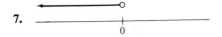

8.

9.

10.

11.

12.

13.

14.

15.

16.

17.

18.

19.

20.

Exercises 3.6 A, page 198

1. $\{x \mid x > \frac{4}{3}\}$

2. $\{x \mid x > 3\}$

3. $\{x \mid x < -3\}$

4. $\{x \mid x \geq 2\}$

5. $\{x \mid x > 2\}$

6. $\{x \mid x > 0\}$

7. $\{x \mid x \leq -1\}$

8. R

9. $\{x \mid x < 2\}$

10. $\{x \mid x \geq 0\}$

11. $\{x \mid x < 2\}$

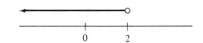

12. $\{x \mid 0 < x < \frac{4}{3}\}$

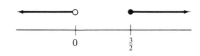

13. $\{x \mid x < 0 \text{ or } x \geq \frac{3}{2}\}$

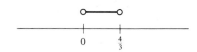

14. $\{x \mid 0 < x \leq 1\}$

15. $\{x \mid x < 1 \text{ or } x > 2\}$

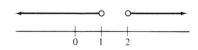

16. $\{x \mid -5 \leq x < -3\}$

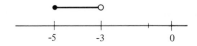

17. $\{x \mid -10 \leq x < -5\}$

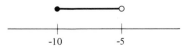

18. $\{x \mid x < 0\}$

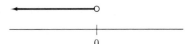

19. $\{x \mid x < -1 \text{ or } x > \frac{1}{2}\}$

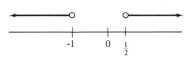

20. $\{x \mid x < -2 \text{ or } x \geq -1\}$

21. $\{x \mid x < -2\}$ **22.** $93 \leq x \leq 100$

23. 14   **24.** $10 \leq c \leq 25$

25. $13.2 \leq M \leq 16.8$

26. If $a > b$, then $a - b > 0$

If $b > c$, then $b - c > 0$

Adding, $a - c > 0$ or $a > c$

27. If $a > b$ and $ab > 0$, then $1/ab > 0$ and $a/ab > b/ab$ and $1/b > 1/a$ and $1/a < 1/b$

Exercises 3.7 A, page 207

1. $\{x \mid x < -3 \text{ or } x > -\frac{2}{3}\}$

2. $\{x \mid -\frac{3}{2} \leq x \leq 2\}$

3. $\{x \mid x \leq -4 \text{ or } x \geq 0\}$

4. $\{x \mid x \leq -2 \text{ or } x \geq 2\}$

5. $\{x \mid x < 0 \text{ or } x > 2\}$

6. $\{x \mid -1 < x < 4\}$

7. R

8. R

9. $\varnothing$

10. $\{-1\}$

11. R

12. $R - \{2\}$

13. $\{x \mid -2 < x < 6\}$

14. $\{x \mid x < -\frac{5}{2} \text{ or } x > 7\}$

15. $\{x \mid 4 - \sqrt{5} \leq x \leq 4 + \sqrt{5}\}$

16. $\{x \mid x < 1 - \sqrt{2} \text{ or } x > 1 + \sqrt{2}\}$

17. $\varnothing$

18. $\{x \mid x \leq k - 2 \text{ or } x \geq k\}$

19. $\{x \mid a - b \leq x \leq a + b\}$

20. $\{x \mid x > 0 \text{ and } x \neq 1\}$

21. $\{x \mid 1 < x < 4\}$

22. $\{x \mid 0 < x \leq 6\}$

23. $\{x \mid x < -\frac{1}{3} \text{ or } x > \frac{5}{4}\}$

24. $\{x \mid x < -4 \text{ or } x > -3\}$

25. $\{x \mid x < 0 \text{ or } x > 5\}$

26. $x \geq -2$

27. $x \leq -5 \text{ or } x \geq 5$

28. $-5 \leq x \leq 5$

29. $(a - b)^2 > 0$, $(a^2 - 2ab + b^2) > 0$, $a^2 + 2ab + b^2 > 4ab$, $(a + b)^2 > 4ab$, $a + b > 4ab/(a + b)$

30. $(a - 1)^2 > 0$; $a^2 - 2a + 1 > 0$; $a^2 + 1 > 2a$; $a + (1/a) > 2$

31. *a.* $0 < v < 66$ feet per second   *b.* $0 < v < 45$ mph

32. $\frac{1}{3} < t < \frac{1}{2}$

Exercises 3.8 A, page 213

1. $2, -2$

2. $5, -5$

3. $-3, -1$

4. $1, -\frac{5}{3}$

5. $5, -1$

6. $\frac{7}{2}, -\frac{1}{2}$

7. $7, 1$

8. $7, 1$

9. $\varnothing$

10. $\frac{9}{2}, -\frac{7}{2}$

11. $\varnothing$

12. $14, 2$

13. $-\frac{1}{4}, -\frac{7}{4}$

14. $4, 0$

15. $x > 0$

16. $x < 0$

17. $a + b, a - b$

18. $a/c, -a/c$

19. $(c - b)/a, (-c - b)/a$

20. $2a + 2b, 2b - 2a$

21. 4

22. 4

23. 5

24. 5

25. 7

26. 7

27. $|x - 4| = 3$

28. $|x + 2| = 5$

29. $3, -3$

30. $x \geq 0$

31. $\varnothing$

32. $x \leq 2$

33. $\varnothing$

34. R

35. $x \geq 2$

Exercises 3.9 A, page 217

1. $x < -2$ or $x > 2$

2. $-3 \leq x \leq 3$

3. $-\frac{1}{2} < x < \frac{1}{2}$

4. $1 < x < 7$

5. $-4 \leq x \leq 1$

6. $-1 \leq x \leq 2$

7. $x \leq -1$ or $x \geq 2$

8. $-\frac{1}{2} < x < \frac{9}{2}$

9. $x < -\frac{1}{2}$ or $x > \frac{1}{2}$

10. $x < -6$ or $x > 6$

11. $x < \frac{4}{3}$ or $x > \frac{8}{3}$

12. $-2 < x < 0$ or $0 < x < 2$

13. $\varnothing$

14. R

15. R

16. $\varnothing$

17. $x < \frac{5}{2}$ or $x > \frac{11}{2}$

18. $\varnothing$

19. $-\frac{4}{3} \leq x < 0$ or $0 < x \leq \frac{4}{3}$

20. $x < -3$ or $0 < x < 1$ or $x > 4$

Chapter 3 Review Exercises, page 219

1. $y^2(x + 1)(x^2 - x + 1)$

2. $(7x + 7a + 3y)(x + a - 2y)$

3. $(r - 5s)(r + 5s + 1)$

4. $(a + c)(x + ab)$

5. $(x - 3)(x^2 + 3x + 9)$

6. $(x^2 + x + 1)(x^2 - x + 1)$

7. $(x - 1)(x - 3)(x + 2)$

8. $(b + 1)(b^2 - 4b + 7)$

9. $(x + 1)(x + 5)(x^2 - 5x + 25)$

10. $(m - 1)(m + 3)(m^2 + 2m + 5)$

11. a. $2x^4 - 8x^3 + 2x^2 + x + 1, R, -7$ b. $3x^2 + 11x + 32, R, 100$

12. a. $P(-4) = 2(-4)^5 - 30(-4)^3 + 9(-4)^2 + 5(-4) - 3 = -7$

    b. $P(3) = 3(3)^3 + 2(3)^2 - 3 + 4 = 100$

13. -9

14. -30

15. $\{-2, 1, (1 \pm i\sqrt{7})/2\}$

16. $\{-2, \pm\sqrt{3}, 1 \pm i\sqrt{3}\}$

17. $\{-5, -1, 4\}$

18. $\{-1, 2, \pm i\sqrt{2}\}$

19. $x > 4$

20. $x < -3$ or $x > 0$

21. $x \geq -6$

22. $x < 0$ or $x > 2$

23. $-\frac{3}{2} < x < 2$

24. $0 < x \leq 3$

25. $x < -3$ or $x > 4$

26. $-3 \leq x \leq 5$

27. $x \leq -4$ or $x > -3$

28. $-2, -3$

29. $\frac{7}{3}, 1$

30. $x < -9$ or $x > 1$

31. $-\frac{3}{2} \leq x \leq \frac{9}{2}$

32. $x \leq 1$ or $x \geq 5$

33. $x \le -\frac{13}{5}$ or $x \ge -\frac{7}{5}$ **34.** $-1 < x < 7$

35. R **36.** $\varnothing$

Exercises 4.1 A, page 226

1. $\{(1, 2), (1, 4), (2, 2), (2, 4), (3, 2), (3, 4)\}$

2. $\{(2, 1), (4, 1), (2, 2), (4, 2), (2, 3), (4, 3)\}$

3. $\{(1, 1), (1, 2), (1, 3), (2, 1), (2, 2), (2, 3), (3, 1), (3, 2), (3, 3)\}$

4. $\{(2, 2), (2, 4), (4, 2), (4, 4)\}$

5. Domain $= \{1, 2, 3, 5\}$, range $= \{1, 2, 6\}$

6. Domain $= \{-1, 0, 3\}$, range $= \{-2, -1, 0\}$

7. Domain $= \{1, 4, 0\}$, range $= \{1, -1, 2, -2, 0\}$

8. Domain $= \{10, 12, 14, 15\}$, range $= \{20, 30\}$

9. Yes   **10.** No

11. No   **12.** No

13. Yes   **14.** Yes

15.

x	1	2	3	4	5
y	3	6	9	12	15

16.

x	3	4	5
y	$\frac{1}{3}$	$\frac{1}{4}$	$\frac{1}{5}$

17.

r	1	2	3	4
s	4	3	2	1

18.

a	1	2	3	4	5	...
b	1	2	3	4	5	...

19. $\{(x, y) \mid y = x, 0 < x < 6, x \text{ is a natural number}\}$

20. $\{(x, y) \mid y = x^2, 0 < x < 6, x \text{ is a natural number}\}$

21. $\{(x, y) \mid y = -2x, -6 \le x \le -1, x \text{ is an integer}\}$

22. $\{(x, y) \mid y = 0, 1 \le x \le 5, x \text{ is a natural number}\}$

23. $\{x \mid x \ne 3\}$ **24.** $\{x \mid x \ne 0 \text{ and } x \ne 3\}$

25. $\{x \mid x \le 16\}$ **26.** $\{x \mid x \ge -4\}$

27. $\{(-1, -2), (0, 0), (1, 2)\}$

28. $\{(-1, -2), (0, -1), (0, -2), (1, 0), (1, -1), (1, -2), (2, 1), (2, 0), (2, -1), (2, -2)\}$

29. {(−2, −1), (−2, 0), (−2, 1), (−2, 2), (−2, −2), (−1, −1), (−1, 0), (−1, 1), (−1, 2), (0, 0), (0, 1), (0, 2), (1, 1), (1, 2), (2, 2)}

30. {(−2, 2), (−1, 1), (0, 0), (1, 1), (2, 2)}

31. a.

t	0	1	2	3	4	5	6
s	0	16	64	144	256	400	576

b. and c.

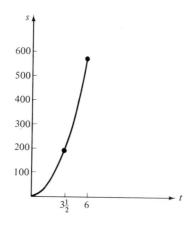

d. 196 feet

e. $4\frac{1}{2}$ seconds

32. a. {(1898, 39), (1904, 104), (1927, 204), (1935, 301), (1963, 407), (1964, 526), (1965, 601), (1970, 622)}

b.

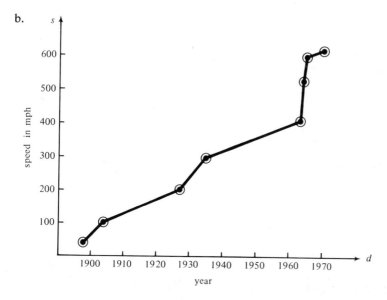

33. a. $\{(0, 0), (1, 9), (2, 16), (3, 21), (4, 24), (5, 25), (6, 24), (7, 21), (8, 16), (9, 9),$
$(10, 0)\}$
  b. $x = 5, P = \$25$

Exercises 4.2 A, page 235

1. a. $\{(-1, -4), (0, -3), (1, -2), (2, -1), (3, 0)\}$
  b. $\{-4, -3, -2, -1, 0\}$
2. a. $\{(-1, 6), (0, 5), (1, 4), (2, 3), (3, 2)\}$
  b. $\{6, 5, 4, 3, 2\}$
3. a. $\{(-1, 1), (0, \sqrt{2}), (1, \sqrt{3}), (2, 2), (3, \sqrt{5})\}$
  b. $\{1, \sqrt{2}, \sqrt{3}, 2, \sqrt{5}\}$
4. a. $\{(-1, 0), (0, 1), (0, -1), (1, \sqrt{2}), (1, -\sqrt{2}), (2, \sqrt{3}), (2, -\sqrt{3}), (3, 2),$
$(3, -2)\}$
  b. $\{0, 1, -1, \sqrt{2}, -\sqrt{2}, \sqrt{3}, -\sqrt{3}, 2, -2\}$
5. a. $\{(-1, 2), (0, 5), (1, 8), (2, 11), (3, 14)\}$
  b. $\{2, 5, 8, 11, 14\}$
6. 1, 2, 3, 5; exactly one second component for each first component

7. 22	**8.** 1
9. 2	**10.** $3a^2 + 4a + 2$
11. $3a^2 + 10a + 9$	**12.** $(3/a^2) + (4/a) + 2$
13. -33	**14.** -3
15. -2	**16.** -6
17. $b^3 + b - 3$	**18.** $(b + h)^3 + b + h - 3$

19. $b^3 + h^3 + b + h - 6$
20. a. $3x^2 + 4$
  b. $9x^2 + 24x + 16$

21. a.

1	2	4	5	8	10	50	75	100
0.50	22	68	92.50	172	230	2230	4292.50	6980

  b. $92.50, \$2230, \$6980$
22. a. $\{(1, \pi), (2, 4\pi), (3, 9\pi)\}$
  b. Yes, exactly one value of A for each value of r
23. $\{(1, 50), (2, 100), (5, 250), (7, 350)\}$
24. a. $\frac{88}{21}$ **b.** $\frac{792}{7}$ **c.** $\frac{11000}{21}$
25. $A = f(b) = 5b/2$
26. a. $\frac{15}{2}$ **b.** $[5(b + h)]/2$ **c.** $5h/2$ **d.** $\frac{5}{2}$

Exercises 4.3 A, page 244

1. $m = 2, b = -1$

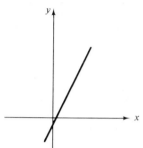

2. $m = 3, b = 2$

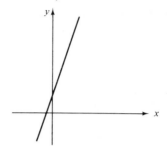

3. $m = \frac{1}{2}, b = -4$

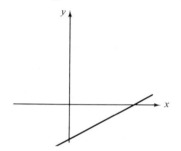

4. $m = -1, b = -1$

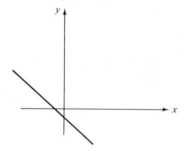

5. $m = 0, b = -2$

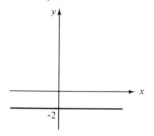

6. $m = -1, b = 0$

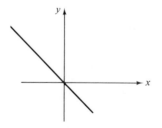

7. $m = \frac{1}{2}, b = 0$

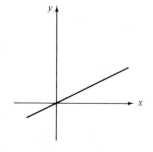

8. $m = 0$, $b = 0$; the graph is the x-axis

9. $m = \frac{3}{5}$, $b = \frac{1}{2}$

10. $m = -\frac{1}{4}$, $b = -1$

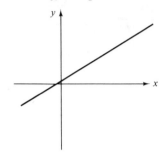

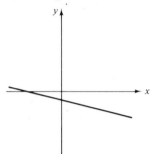

11. 1

12. $-\frac{1}{3}$

13. $\frac{5}{6}$

14. $-\frac{1}{2}$

15. 5

16. $y = \frac{2}{3}x + \frac{5}{3}$, $m = \frac{2}{3}$, $b = \frac{5}{3}$

17. $y = -\frac{1}{4}x + \frac{1}{4}$, $m = -\frac{1}{4}$, $b = \frac{1}{4}$

18. $y = \frac{1}{2}x - \frac{1}{2}$, $m = \frac{1}{2}$, $b = -\frac{1}{2}$

19. $y = 4x - 6$, $m = 4$, $b = -6$

20. $y = x$, $m = 1$, $b = 0$

21. a. -8 **b.** 0

22. a.

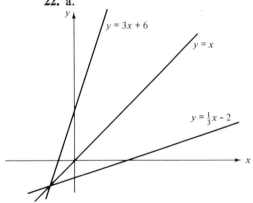

23. a. 132 feet   **b.** 85.8 feet

24. a. and b.

Exercises 4.4 A, page 250

1. 6

2. 9

3. 4

4. $2p^2 - 3p + 4$

5. $2(p + h)^2 - 3(p + h) + 4$

6. $2p^2 - 3p + 4 + h$

7. 6

8. $8x^2 - 6x + 4$

9. $(2/x^2) - (3/x) + 4$

10. $2x^2 + 5x + 6$

11. 3

12. -6

13. -6

14. $-2x - 1$

15. $3 - x^4$

16. $2 + 2x - x^2$

17. $\frac{26}{9}$ **18.** $3 - (1/x^2)$

19. a.

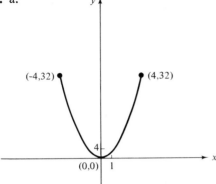

   b. $(0, 0)$ c. $y \geq 0$

20. a.

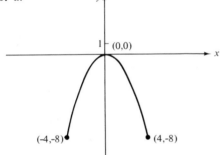

   b. $(0, 0)$ c. $y \leq 0$

21. a.

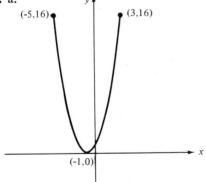

   b. $(-1, 0)$ c. $y \geq 0$

22. a.

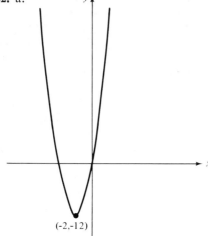

(-2,-12)

   b. $(-2, -12)$ c. $y \geq -12$

23. a.

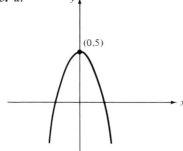

(0,5)

   b. $(0, 5)$ c. $y \leq 5$

24. a.

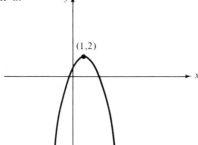

(1,2)

   b. $(1, 2)$ c. $y \leq 2$

25. a.

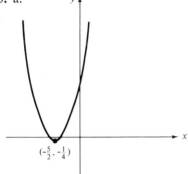

$(-\frac{5}{2}, -\frac{1}{4})$

b. $(\frac{5}{2}, -\frac{1}{4})$ c. $y \geq -\frac{1}{4}$

26. a.

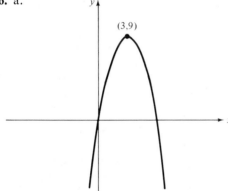

(3,9)

b. $(3, 9)$ c. $y \leq 9$

27. a.

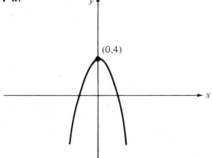

(0,4)

b. $(0, 4)$ c. $y \leq 4$

28. a.

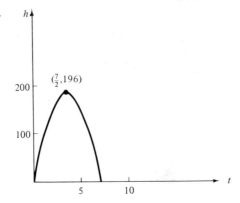

b. 196 feet, $3\frac{1}{2}$ seconds

Exercises 4.5 A, page 256

1. 10

2. 25

3. $3\sqrt{2}$

4. $\frac{13}{2}$

5. 5

6. $(x - 2)^2 + (y - 3)^2 = 9$

7. $(x + 1)^2 + (y - 2)^2 = 36$

8. $(x - 3)^2 + (y + 5)^2 = 25$

9. $x^2 + y^2 = 100$

10. $(x + 6)^2 + (y - 2)^2 = 3$

11. $(x - 1)^2 + (y - 5)^2 = 10$

12. $(x + 2)^2 + (y + 3)^2 = 73$

13. $(x + 4)^2 + (y - 1)^2 = 50$

14. $x^2 + y^2 = 25$

15. $(x - 3)^2 + (y - 4)^2 = 25$

16. $(x - 2)^2 + (y + 3)^2 = 16$; $C:(2, -3)$, $r = 4$

17. $(x + 5)^2 + (y - 3)^2 = 25$; $C:(-5, 3)$, $r = 5$

18. $(x - \frac{7}{2})^2 + (y - 1)^2 = 4$; $C:(\frac{7}{2}, 1)$, $r = 2$

19. $x^2 + (y - 7)^2 = 49$; $C:(0, 7)$, $r = 7$

20. $(x - 8)^2 + y^2 = 64$; $C:(8, 0)$, $r = 8$

21.

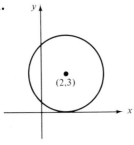

(2,3)

22.

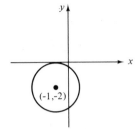

23.

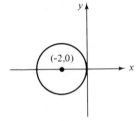

24.

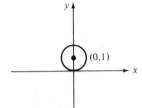

25.

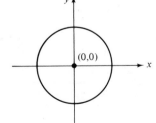

Exercises 4.6 A, page 264

1.

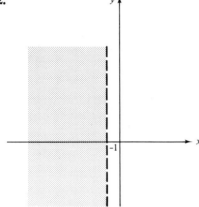

2.

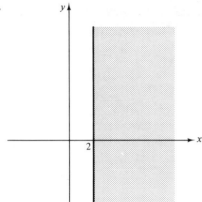

3.

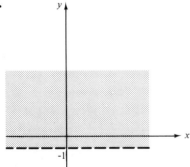

4.

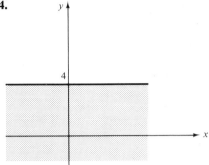

5.

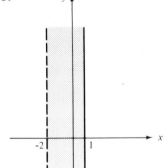

6.

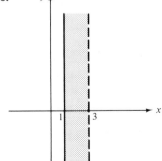

7.

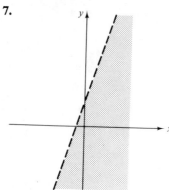

8.

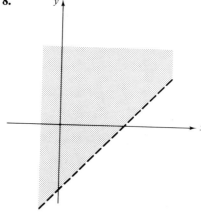

9.

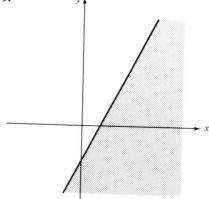

10.

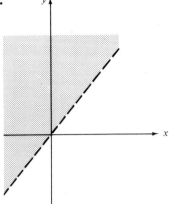

11.

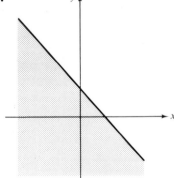

12.

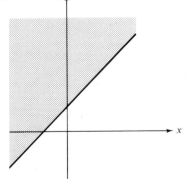

13.

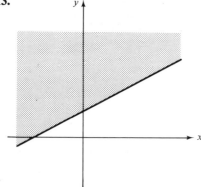

14.

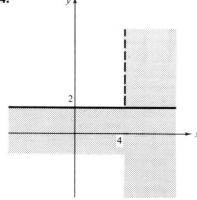

15.

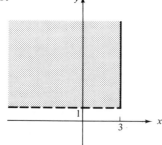

16.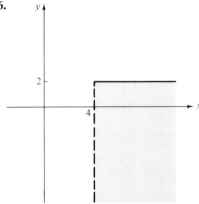

17.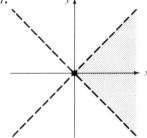

18.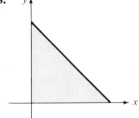

19.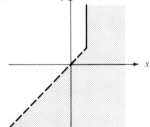

20.

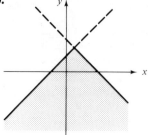

Exercises 4.7 A, page 269

1. $x < -3$ or $x > 3$
3. All real numbers
5. $0 < x < 6$
7. $-3 \le x \le \frac{1}{2}$
9. $-3.7 < x < -0.3$

2. $x \le -2$ or $x \ge 2$
4. $0 < x < 2$
6. $x < -3$ or $x > \frac{1}{2}$
8. All real numbers
10. $x \le -4.8$ or $x \ge 0.8$

11.

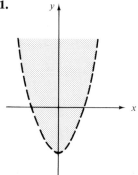

12.

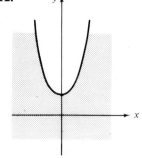

13.

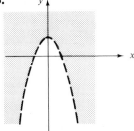

14.

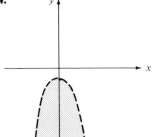

15.

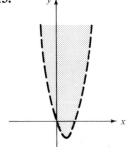

16.

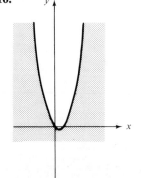

17.

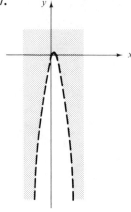

18.

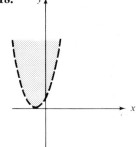

19.

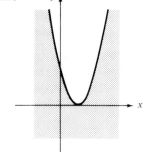

20.

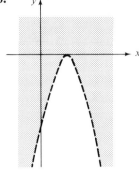

21.

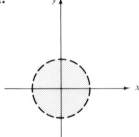

22.

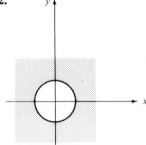

23.

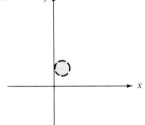

24.

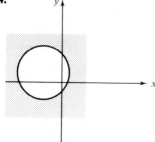

25.

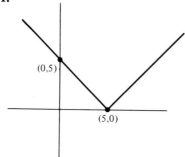

Exercises 4.8 A, page 274

1.

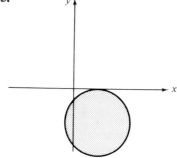

2.

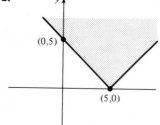

3.

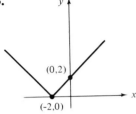

4.

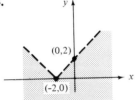

5.

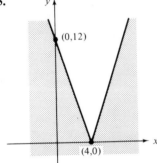

6.

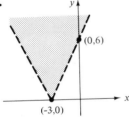

7.

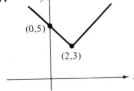

8.

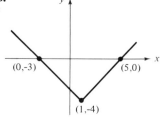

9.

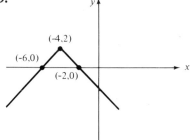

10.

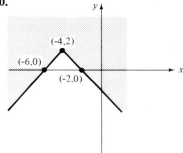

11.

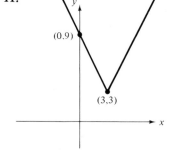

12.

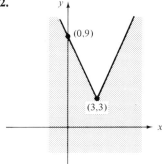

13.

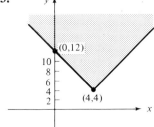

14.

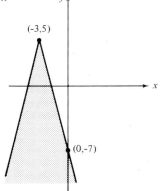

15.

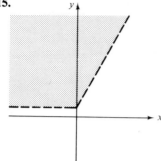

16.

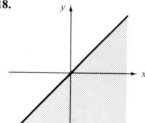

17. (0, 0)

18.

19.

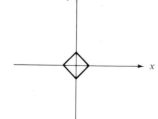

20.

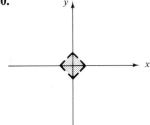

Exercises 4.9 A, page 279

1. a. $33 \leq x \leq 41$
   b. $1.22 \leq x \leq 1.32$
   c. $3.138 \leq x \leq 3.142$
   d. $3.33 \times 10^{18} \leq x \leq 3.39 \times 10^{18}$
   e. $7.9 - h \leq x \leq 7.9 + h$

2. $\$22.50 \leq C \leq \$25$

3.

 a.
b.

c.
d.

4. $0 < y < 60$ **5.** $x \leq 15$

6. $(0, 0)$, $(24, 24)$, $(15, 30)$ **7.** $(2, 0)$, $(6, 0)$, $(6, 6)$, $(2, 3)$, $(4, 6)$

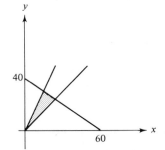

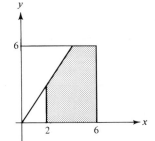

8. 32 10-cent stamps

9. 200 boxes oranges, 500 boxes lemons, 100 boxes limes

10. 48 literary pages, 12 art pages, 20 advertising pages; $C = \$2600$

Chapter 4 Review Exercises, page 282

1. a. $\{(1, 0), (1, -1), (1, 2), (2, 0), (2, -1), (2, 2), (3, 0), (3, -1), (3, 2)\}$
   b. $\{(0, 1), (-1, 1), (2, 1), (0, 2), (-1, 2), (2, 2), (0, 3), (-1, 3), (2, 3)\}$
   c. $\{(1, 1), (1, 2), (1, 3), (2, 1), (2, 2), (2, 3), (3, 1), (3, 2), (3, 3)\}$
   d. $\{(0, 0), (0, -1), (0, 2), (-1, 0), (-1, -1), (-1, 2), (2, 0), (2, -1), (2, 2)\}$

2. a. Domain $= \{1, 2, 3, 5\}$, range $= \{1, 2, 3\}$
   b. Function

3. a. Domain $= \{0, 1, 4\}$, range $= \{-2, 2, 3, 5\}$
   b. Not a function; 4 is paired with -2 and 3

4. a. Domain $= \{-1, 0, 1, 2, 3, 4\}$, range $= \{-1, 2, 5, 8, 11, 14\}$
   b. Function, one y for each x

5. a. Domain $= \{-3, -2, -1, 0, 1, 2, 3\}$, range $= \{0, 1, 4, 9\}$
   b. Function, one y for each x

6. a. $\{x \mid x \geq 0\}, \{y \mid y \geq 0\}$
   b. Function, one y for each x

7. a. $\{1, 2, 3\}, \{1, 2\}$
   b. The relation $= \{(1, 2), (1, 3), (2, 3)\}$
   Not a function; 1 is paired with 2 and 3

8. a. $\{x \mid -2 \leq x \leq 2\}, \{y \mid -2 \leq y \leq 2\}$
   b. Not a function, $y = \pm\sqrt{4 - x^2}$

9. a. $\{x \mid x \text{ is any real number}\}, \{y \mid y \geq 0\}$
   b. Function, one y for each x

10. a. $\{x \mid x \text{ is any real number}\}, \{y \mid y > 0\}$
   b. Not a function. For example, many values are paired with each x, $(1, 2)$, $(1, 3), (1, 4), \ldots$

11. a, d

12. 7

13. 5

14. 10

15. $8x^2 - 6x + 5$

16. $2x^2 + 5x + 7$

17. $2x^2 - 3x + 7$

18. $2(x + h)^2 - 3(x + h) + 5$

19. $4xh + 2h^2 - 3h$

20. -3

21. $\frac{9}{7}$

22. 1

23. $-\frac{5}{8}$

24. $m = 2, b = 3$

25. $m = -\frac{1}{2}, b = 4$

26. $m = 0, b = 4$

27. $m = -3, b = -1$

28. m is undefined, no y-intercept

29. $m = -\frac{3}{2}, b = 3$

30. a. and b.

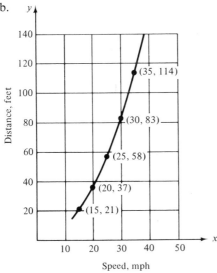

c. 27 mph   d. 37 mph

31. a., $x = 0$; $(0, 0)$ b., $x = 0$; $(0, -2)$ c., $x = -\frac{1}{2}$; $(-\frac{1}{2}, -\frac{5}{4})$

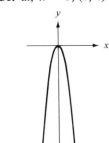 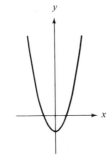 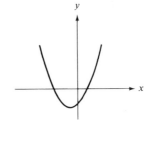

and d. $x = 2$; $(2, 7)$

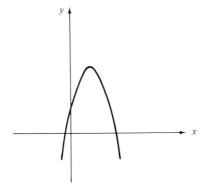

32. a.

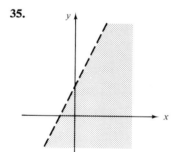

b. 100 feet   **c.** 5 seconds

33. $2\sqrt{10}$

34. $(x - 3)^2 + (y + 5)^2 = 25$; $C:(3, -5)$, $r = 5$

35.

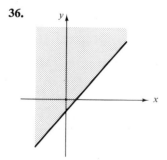

36.

37.

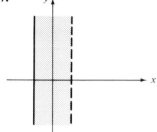

38.

39.

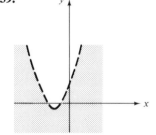

40.

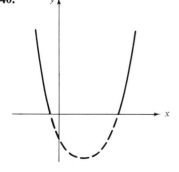

41.

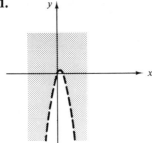

42.

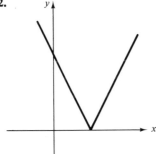

43.

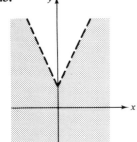

44.

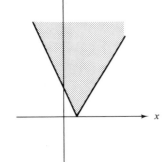

45.

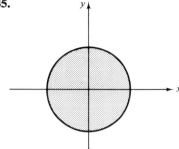

46.

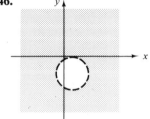

Exercises 5.1 A, page 291

1. a. $f(x) = x - 6$ b. $\{(x, y) \mid x = y - 6\}$ c. $f^{-1}(x) = x + 6$
2. a. $f(x) = x/2$ b. $\{(x, y) \mid x = y/2\}$ c. $f^{-1}(x) = 2x$
3. a. $f(x) = 8x$ b. $\{(x, y) \mid x = 8y\}$ c. $f^{-1}(x) = x/8$
4. a. $f(x) = x + 10$ b. $\{(x, y) \mid x = y + 10\}$ c. $f^{-1}(x) = x - 10$
5. a. $f(x) = 4x - 8$ b. $\{(x, y) \mid x = 4y - 8\}$ c. $f^{-1}(x) = (x + 8)/4$
6. a. $f(x) = 10 - 2x$ b. $\{(x, y) \mid 2y + x = 10\}$ c. $f^{-1}(x) = (10 - x)/2$
7. a. $f(x) = (3x - 6)/2$ b. $\{(x, y) \mid 3y - 2x = 6\}$
   c. $f^{-1}(x) = (2x + 6)/3$
8. a. $f(x) = 1/x$ b. $\{(x, y) \mid yx = 1\}$ c. $f^{-1}(x) = 1/x$
9. a. $f(x) = \sqrt{x + 3}$ b. $\{(x, y) \mid x = \sqrt{y + 3}\}$ c. $f^{-1}(x) = x^2 - 3$
10. a. $f(x) = 3$ b. $\{(x, y) \mid x = 3\}$ c. Not a function
11. a. $f(x) = x^2 - 4$ b. $\{(x, y) \mid x = y^2 - 4\}$ c. Not a function
12. a. $f(x) = x^3$ b. $\{(x, y) \mid x = y^3\}$ c. $f^{-1}(x) = \sqrt[3]{x}$
13. a. $f(x) = |x|$ b. $\{(x, y) \mid x = |y|\}$ c. Not a function
14. a. $f(x) = \sqrt[3]{x + 1}$ b. $\{(x, y) \mid x = \sqrt[3]{y + 1}\}$ c. $f^{-1}(x) = x^3 - 1$
15. a. $f(x) = -4x^2$ b. $\{(x, y) \mid 4y^2 + x = 0\}$ c. Not a function
16. a. $f^{-1}(x) = (x + 6)/2$ b. $[(2x - 6) + 6]/2 = 2x/x = x$
17. a. $f^{-1}(x) = x^2 - 3$ b. $(\sqrt{x + 3})^2 - 3 = x + 3 - 3 = x$
18. a. $f^{-1}(x) = 6/x$ b. $6/(6/x) = 6(x/6) = x$

19. a. $f^{-1}(x) = 2\sqrt[3]{x}$ b. $2\sqrt[3]{x^3/8} = 2(x/2) = x$

20. a. $f^{-1}(x) = \sqrt{x-1}$ b. $\sqrt{(x^2+1)-1} = \sqrt{x^2} = x$ since $x \geq 0$

21. a. $f^{-1}(x) = x^3$ b. $(\sqrt[3]{x})^3 = x$

22. a. $f^{-1}(x) = \sqrt{25-x^2}$

b. $\sqrt{25-(\sqrt{25-x^2})^2} = \sqrt{25-(25-x^2)} = \sqrt{x^2} = x$ since $x \geq 0$

23.

x	-3	-2	-1	$-\frac{1}{2}$	0	$\frac{1}{2}$	1	2	3
2^x	$\frac{1}{8}$	$\frac{1}{4}$	$\frac{1}{2}$	$\sqrt{2}/2$	1	$\sqrt{2}$	2	4	8

24.

x	$\frac{1}{8}$	$\frac{1}{4}$	$\frac{1}{2}$	$\sqrt{2}/2$	1	$\sqrt{2}$	2	4	8
$f^{-1}(x)$	-3	-2	-1	$-\frac{1}{2}$	0	$\frac{1}{2}$	1	2	3

25. 1

26. 3

27. 0

28. -2

29. $-\frac{1}{2}$

30. $\frac{1}{2}$

32. 100

33. 1

34. $\frac{1}{10}$

35. $\sqrt{10}$

36. $\sqrt{10}/100$

37. -2

38. 3

39. $\frac{1}{3}$

40. 0

Exercises 5.2 A, page 297

1. 3

2. -2

3. $\frac{3}{2}$

4. 4

5. $\frac{7}{2}$

6. -3

7. 5

8. 0

9. $-\frac{5}{3}$

10. $\frac{3}{4}$

Exercises 5.3 A, page 301

1. $\log_{10} 10,000 = 4$

2. $\log_5 1 = 0$

3. $\log_9 3 = \frac{1}{2}$

4. $\log_4 \left(\frac{1}{64}\right) = -3$

5. $\log_3 3 = 1$

6. $\log_{64} 16 = \frac{2}{3}$

7. $\log_9 \left(\frac{1}{27}\right) = -\frac{3}{2}$

8. $\log_{1/6} 36 = -2$

9. $3^2 = 9$

10. $10^0 = 1$

11. $10^1 = 10$

12. $2^{-3} = 0.125$

13. $36^{1/2} = 6$

14. $10^5 = 100,000$

15. $10^{-1} = 0.1$

16. $8^{-2/3} = 0.25$

17. 3

18. -2

19. -4

20. $\frac{5}{2}$

21. $-\frac{3}{2}$

22. -4

23. -2.5

24. 3

25. 6

26. $\frac{1}{81}$

27. 1

28. -2

29. 5

30. 64

31. 5

32. 0.001

33. 10,000

34. $10\sqrt{10}$

35. 3

36. $\frac{1}{4}$

37. 6

38. $\frac{1}{81}$

39. $\frac{1}{3}$

40. $\frac{1}{2}$

Exercises 5.4 A, page 304

1. $\log_b 3 + \log_b 5$

2. $\log_b 3 - \log_b 5$

3. $5 \log_b 3$

4. $\frac{1}{2} \log_b 3$

5. $\frac{1}{2}(\log_b 5 - \log_b 3)$

6. $-\frac{1}{3} \log_b 5$

7. $4 \log_b 3 - 2 \log_b 5$

8. $2 \log_b 5 + \frac{1}{4} \log_b 3$

9. $\frac{3}{2} \log_b 5 - \log_b 7$

10. $\frac{4}{3} \log_b 5 - \frac{2}{3} \log_b 3$

11. $\log_b 72$

12. $\log_b \frac{21}{8}$

13. $\log_b 3^5$

14. $\log_b \sqrt{17}$

15. $\log_b 2025$

16. $\log_b \sqrt[3]{7}/36$

17. $\log_b \sqrt[5]{75}$

18. $\log_b (\frac{1}{54})^3$

19. $\log_b 9\sqrt[4]{5}/7^3$

20. $\log_b \sqrt[3]{\frac{324}{85}}$

Exercises 5.5 A, page 308

1. a. and b. 0.6776

2. a. $0.9494 + 2$ b. 2.9494

3. a. $0.0899 - 1$ b. -0.9101

4. a. $0.6075 + 5$ b. 5.6075

5. a. $0.9768 - 2$ b. -1.0232

6. a. and b. 0.6990

7. a. $0.8451 - 3$ b. -2.1549

8. a. $0.8142 + 3$ b. 3.8142

9. a. $0.7513 - 1$ b. -0.2487

10. a. $0.4487 + 3$ b. 3.4487

11. a. $0.5647 + 4$ b. 4.5647

12. a. $0.9289 + 1$ b. 1.9289

13. a. $0.1790 - 4$ b. -3.8210

14. a. $0.5441 - 1$ b. -0.4559

15. a. and b. 0.4969

16. a. $0.9031 - 2$ b. -1.0969

17. a. 6.93×10^1 b. 69.3

18. a. 4.20×10^4 b. 42,000

19. a. 1.17×10^{-1} b. 0.117

20. a. 9.44×10^0 b. 9.44

21. a. 6.83×10^{-3} b. 0.00683

22. a. 8.00×10^6 b. 8,000,000

23. a. $1.95 + 10^{-2}$ b. 0.0195

24. a. 3.75×10^0 b. 3.75

25. a. 3.90×10^2 b. 390

26. a. 6.07×10^3 b. 6070

27. a. 1.75×10^{-1} b. 0.175

28. a. 3.73×10^{-4} b. 0.000373

29. a. 2.65×10^{-2} b. 0.0265

30. a. 6.31×10^1 b. 63.1

31. a. 1.09×10^0 b. 1.09

32. a. 6.31×10^{-2} b. 0.0631

Exercises 5.6 A, page 311

1. 0.4972

2. 1.8935

3. 0.6391 − 1

4. 0.9644 − 2

5. 2.3498

6. 3.0913

7. 3.7224

8. 0.8121 − 3

9. 0.9275 − 1

10. 0.7872

11. 1.4707

12. 0.0814 − 2

13. 3.356

14. 743.7

15. 0.4966

16. 0.002512

17. 59.84

18. 1507

19. 0.01903

20. 2.342

21. 0.1202

22. 0.05192

23. 0.001517

24. 0.0005063

Exercises 5.7 A, page 313

1. 27.8

2. 8.12

3. 34.9

4. 4.90

5. 24.7

6. 0.960

7. 3.62

8. 0.516

9. 0.000412

10. 2.37

11. 1.34

12. 0.0364

13. 2.07

14. 0.521

15. 0.768

16. 23.1

17. 1909

18. 373.7

19. − 1.27

20. 769

Exercises 5.8 A, page 316

1. 7

2. $-\frac{5}{2}$

3. $\log 2/\log 5 \approx 0.43$

4. -2

5. $3 + (\log 4.5/\log 6) \approx 3.84$

6. $\frac{2}{3}$

7. $-\frac{3}{2}$

8. $-\frac{5}{3}$

9. 5, 13

10. $-5 + 2\sqrt{2}$

11. $\frac{1}{5}$

12. 25

13. $\sqrt{3}/3$

14. $2 + (1/\log 3) \approx 4.10$

15. $x = \log 7/\log 2 \approx 2.81$

Exercises 5.9 A, page 318

1. 2.303

2. 4.606

3. 0.4447

4. 1.145

5. 3.227

6. $0.1313 - 1$

7. -2.495

8. 2.27

9. -2.10

10. 3.61

Exercises 5.10 A, page 323

1. a. 2.3
   b. 6.5
   c. 4.1

2. a. 1.0×10^{-7}
   b. 2.5×10^{-3}
   c. 2.0×10^{-9}

3. a. 3.05 minutes
   b. 1.09 grams

4. 0.000436

5. 666

6. $5353

7. $8\frac{3}{4}$ years

8. a. 65 amperes approx.
   b. 22 amperes approx.

9. 55.4 volts approx.

10. $2750

11. 217 feet approx.

12. a. 0.03924 approx.
   b. 70 minutes approx., infinitely long

Chapter 5 Review Exercises, page 327

1. 1

2. 0

3. Undefined

4. Undefined

5. -3

6. 2

7. 1.2

8. -0.4

9. 0.2

10. $-\frac{4}{15}$

11. -0.4

12. $2\sqrt{3}/15$

13. -4.6

14. $\sqrt{5}/5$

15. $\log_5 3$

16. $-1 + 0.79$

17. 1.7

18. 40

19. $\frac{1}{10}$

20. -2

21. a. $\log x = \frac{1}{3}(\log 42.21 + \log 23.50 - \log 5650)$
   b. 0.5600

22. 2.32 inches

23. $\frac{1}{2}$

24. 1.77

Exercises 6.1 A, page 336

1. $x + y = 5 + 3 = 8$ and $x - y = 5 - 3 = 2$

2. $3x + y = 3(-1) + 4 = 1$ and $2x + 6 = 2(-1) + 6 = 4 = y$

3. a. $x^2 + y^2 = 0^2 + (-3)^2 = 9$ and $x - 3 = 0 - 3 = -3 = y$
   b. $x^2 + y^2 = 3^2 + 0^2 = 9$ and $x - 3 = 3 - 3 = 0 = y$

4. a. $x^2 - 4x - 1 = 2^2 - 4(2) - 1 = -5$ and $4x + y = 8 - 5 = 3$
   b. $x^2 - 4x - 1 = 4 + 8 - 1 = 11$ and $4x + y = -8 + 11 = 3$

5. a. $x^2 + y^2 = 0^2 + 5^2 = 25$ and $5 - x^2 = 5 - 0^2 = 5 = y$
   b. $x^2 + y^2 = 3^2 + (-4)^2 = 25$ and $5 - x^2 = 5 - 9 = -4 = y$
   c. $x^2 + y^2 = (-3)^2 + (-4)^2 = 25$
      and $5 - x^2 = 5 - (-3)^2 = -4 = y$

6.

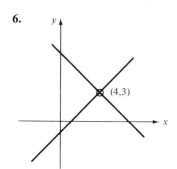

(4,3)

7.

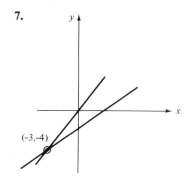

(-3,-4)

8.

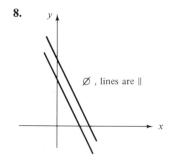

∅ , lines are ∥

9.

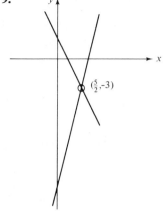

$(\frac{5}{2}, -3)$

10.

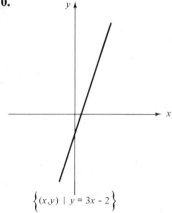

$\left\{ (x,y) \mid y = 3x - 2 \right\}$

11.

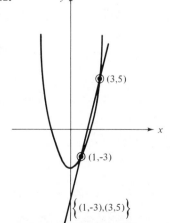

$(3,5)$

$(1,-3)$

$\left\{ (1,-3),(3,5) \right\}$

12.

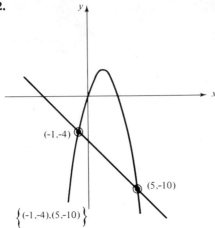

(-1,-4)

(5,-10)

$\left\{ (-1,-4),(5,-10) \right\}$

13.

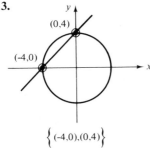

(0,4)

(-4,0)

$\left\{ (-4,0),(0,4) \right\}$

14.

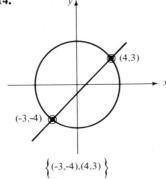

(4,3)

(-3,-4)

$\left\{ (-3,-4),(4,3) \right\}$

15.

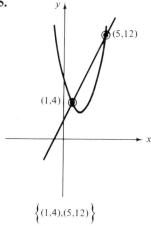

$$\left\{(1,4),(5,12)\right\}$$

16.

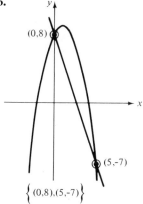

$$\left\{(0,8),(5,-7)\right\}$$

17.

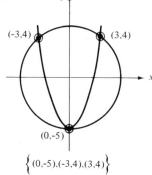

$$\left\{(0,-5),(-3,4),(3,4)\right\}$$

18.

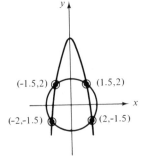

$$\left\{(-2,-1.5),(-1.5,2),(1.5,2),(2,-1.5)\right\}$$

19.

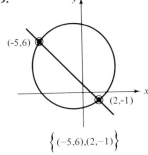

$$\left\{(-5,6),(2,-1)\right\}$$

20.

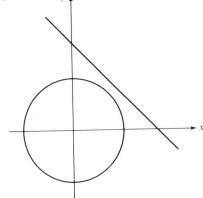

Solution set is empty

Exercises 6.2 A, page 341

1. $\{(-3, 1), (5, 3)\}$

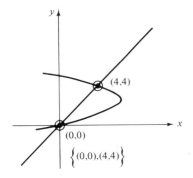

2. $\{(0, 0), (4, 4)\}$

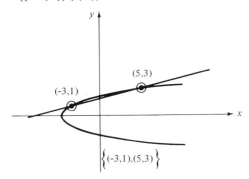

3. $\{(-5, -1), (-2, 2)\}$

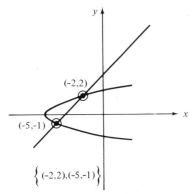

4. $\{(0, 0), (\frac{1}{4}, \frac{1}{2})\}$

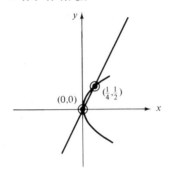

5. $\{(6, -5), (6, -3)\}$

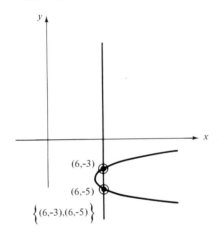

6. $\{(-5, 1), (3, 5)\}$

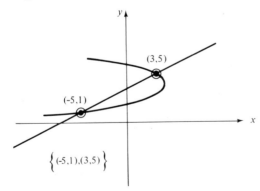

7. $\{(0, 0), (-1, 1)\}$

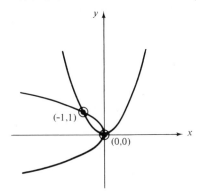

8. $\{(-2, 0), (0, 4)\}$

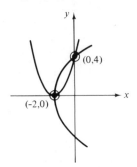

9. $\{(0, 8), (8, 0), (6, -4)\}$

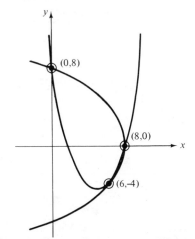

10. {(0, 0), (1, 2), (4, 4), (9, − 6)}

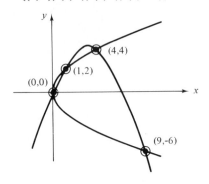

Exercises 6.3 A, page 349

1. Ellipse; domain: $|x| \leq 10$, range: $|y| \leq 2$

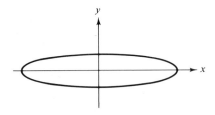

2. Ellipse; domain: $|x| \leq 10$, range: $|y| \leq 8$

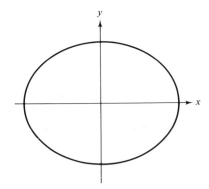

3. Ellipse; domain: $|x| \leq \sqrt{10}$, range: $|y| \leq 5$

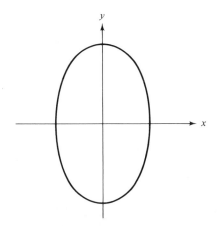

4. Hyperbola; domain: $|x| \geq 3$, range: all R, asymptotes: $y = \pm x$

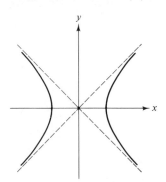

5. Hyperbola; domain: all R, range: $|y| \geq 3$, asymptotes: $y = \pm x$

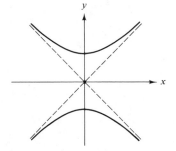

6. Hyperbola; domain: all R, range: $|y| \geq 6$, asymptotes: $y = \pm \frac{3}{4}x$

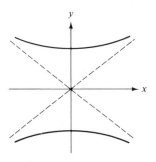

7. Hyperbola; domain: all $x \neq 0$, range: all $y \neq 0$, asymptotes: $x = 0$, $y = 0$

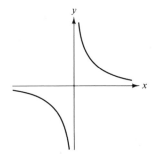

8. Hyperbola; domain: all $x \neq 0$, range: all $y \neq 0$, asymptotes: $y = 0$, $x = 0$

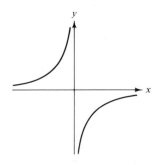

9. Hyperbola; domain: all R, range: $|y| \geq 8$, asymptotes: $y = \pm 2x$

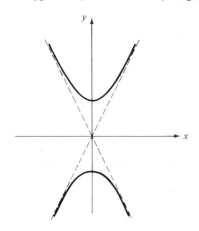

10. Two intersecting lines; domain = range = R

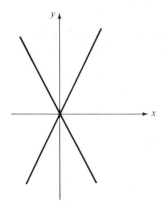

Exercises 6.4 A, page 355

1. $(-2, 3), (3, -2)$ **2.** $(-0.8, 1.8)$

3. $(4.2, 1.4), (-4.2, -1.4)$

4. $(4, 6), (4, -6), (0.5, 2.1), (0.5, -2.1)$

5. $(2.8, 2.1), (-2.8, -2.1)$ **6.** $(9, 0), (-5, 7.5), (-5, -7.5)$

Exercises 6.5 A, page 358

1. $(2, -3), (-2, -1)$ **2.** $(-4, 12), (3, 19)$

3. $(3 + i, -1 - 3i), (3 - i, -1 + 3i)$ **4.** $(2, 3), (-\frac{3}{2}, -4)$

5. $(6, -2), (-3, 4)$ **6.** $(1, 2), (1, -2), (3, \pm 2\sqrt{3})$

7. $(0, -4), (\pm\sqrt{3}, 2)$

8. $(\sqrt{5}, 2\sqrt{5}/5), (-\sqrt{5}, -2\sqrt{5}/5), (2i, -i), (-2i, i)$

9. $(\pm\frac{3}{2}, 2), (\pm 2, -\frac{3}{2})$ **10.** $(0, 0), (1, 2), (4, 4), (9, -6)$

Exercises 6.6 A, page 362

1. $(2, 4), (2, -4), (-2, 4), (-2, -4)$ **2.** $(\pm 5, \pm\sqrt{2})$

3. $(\sqrt{6}, 0), (-\sqrt{6}, 0), (\sqrt{5}, 2), (-\sqrt{5}, 2)$

4. $(2, 3), (2, -3), (-2, 3), (-2, -3)$ **5.** $(1, 4), (-1, -4), (4, -14), (-4, 14)$

6. $(-4, 2), (2, -4)$ **7.** $(1, 1 + 2\sqrt{3}), (1, 1 - 2\sqrt{3})$

8. $(1, 1), (-3, -\frac{1}{3})$ **9.** $(3, 16), (-8, -6)$

10. $(4, 2), (-4, -2), (3\sqrt{2}, \sqrt{2}), (-3\sqrt{2}, -\sqrt{2})$

Exercises 6.7 A, page 365

1. $(3, -5), (-5, 3), (-2 + \sqrt{6}, -2 - \sqrt{6}), (-2 - \sqrt{6}, -2 + \sqrt{6})$

2. $(2 + i\sqrt{2}, 2 - i\sqrt{2}), (2 - i\sqrt{2}, 2 + i\sqrt{2}), (-3 + i\sqrt{7}, -3 - i\sqrt{7}),$
$(-3 - i\sqrt{7}, -3 + i\sqrt{7})$

3. $(\frac{3}{2} + i\sqrt{3}/2, \frac{3}{2} - i\sqrt{3}/2), (\frac{3}{2} - i\sqrt{3}/2, \frac{3}{2} + i\sqrt{3}/2), (-1 + i\sqrt{2}, -1 - i\sqrt{2}),$
$(-1 - i\sqrt{2}, -1 + i\sqrt{2})$

4. $(-1 + \sqrt{2}, -1 - \sqrt{2}), (-1 - \sqrt{2}, -1 + \sqrt{2}), (4 + i, 4 - i),$
$(4 - i, 4 + i)$

5. $(-5 + \sqrt{5}, -5 - \sqrt{5}), (-5 - \sqrt{5}, -5 + \sqrt{5}), (\frac{1}{2} + \sqrt{5}, \frac{1}{2} - \sqrt{5}),$
$(\frac{1}{2} - \sqrt{5}, \frac{1}{2} + \sqrt{5})$

Exercises 6.8 A, page 366

1. 10 feet, 28 feet, 58 feet   **2.** 900 miles, 2400 miles

3. At about $(65, 40)$ **4.** 10 inches by 20 inches

5. 24 inches by 10 inches   **6.** 32 dresses, \$15 each

7. 80 persons, 90 cents each

8. b. 2925 feet   c. 7800 feet   d. 1300 feet and 6500 feet

9. 10 ohms, 15 ohms   **10.** $x = 9, y = 5$

Chapter 6 Review Exercises, page 371

1. $(-5, 0), (4, 3), (4, -3)$ **2.** $(\pm 2\sqrt{6}, -\frac{11}{2}), (\pm 2i, \frac{3}{2})$

3. $(\pm 2\sqrt{2}, \pm\sqrt{7})$ **4.** $(3, 8), (-3, -8), (4, 6), (-4, -6)$

5. $(3. 0), (-5, 2), (-5, -2)$ **6.** $(\pm 2, 3), (\pm i\sqrt{6}, -7)$

7. $(1, 2), (-1, -2), (2i, -i), (-2i, i)$
8. $(2, -1), (-2, 1), (9i, 3i), (-9i, -3i)$
9. $(2, 6), (-2, 6), (2\sqrt{7}, 18), (-2\sqrt{7}, 18)$
10. $(1, 1), (-3, -1)$
11. $(1, 1), (-1, 1), (\sqrt{21}/3, -\frac{5}{3}), (-\sqrt{21}/3, -\frac{5}{3})$
12. $(15, 32), (-10, -48)$
13. $xy = 750$ and $(x - 25)(y + 1) = 750$
    5 sets, \$150 each
14. $xy = 800$ and $(x - 8)(y + 5) = 800$
    40 mph and 20 hours
15. $x^2 + y^2 = (17)^2$ and $xy = 120$
    8 feet by 15 feet

Exercises 7.1 A, page 377

1. $(-1, 1, -2)$ **2.** $(5, -2, 1)$
3. $(\frac{1}{6}, \frac{5}{6}, \frac{2}{3})$ **4.** $\varnothing$
5. $(a, -4a + 13, 3a - 8)$ **6.** $(3, -2.5, -1.5)$
7. $(x, 2x, 5x/2)$ or $(2z/5, 4z/5, z)$ **8.** $(5, 3, 6)$
9. $\varnothing$ **10.** $(6, 1, 2)$

11. $(6, y, 3 - y)$ or $(6, 3 - z, z)$ **12.** $\left(\dfrac{a + b}{2}, \dfrac{a - c}{2}, \dfrac{b - c}{2}\right)$

13. $\left(x, \dfrac{4x}{5}, \dfrac{7x + 5}{5}\right)$ **14.** $(2, -1, 3)$

15. $(1, -2, -1, 3)$ **16.** $(5, 6, 3, 5)$

Exercises 7.2 A, page 385

1. $\begin{bmatrix} 4 & 5 \\ -7 & -1 \end{bmatrix}$ **2.** $\begin{bmatrix} 5 & 8 & 2 \\ 5 & 7 & 9 \end{bmatrix}$

3. Undefined, dimensions are different
4. Undefined, dimensions are different

5. $\begin{bmatrix} 5 & 3 & -4 \\ 2 & -3 & 0 \\ 3 & -1 & -1 \end{bmatrix}$

6. Undefined, dimensions are different

7. $[6 \quad 15 \quad 14]$ **8.** $\begin{bmatrix} -1 & -6 & -5 \\ -2 & 0 & 2 \\ -6 & 4 & -2 \end{bmatrix}$

9. $[6 \quad -9 \quad 12]$ **10.** $\begin{bmatrix} 10 & -15 & 25 \\ 5 & 20 & -10 \end{bmatrix}$

11. $\begin{bmatrix} -2 & 10 \\ 16 & 11 \end{bmatrix}$

12. $\begin{bmatrix} 2a + 12 & 2b + 6 & 2c + 9 \\ a - 8 & b - 4 & c - 6 \\ 4a + 12 & 4b + 6 & 4c + 9 \end{bmatrix}$

13. Undefined, dimensions not compatible

14. $\begin{bmatrix} x - y + z \\ 2x + y - 3z \\ 3x + 2y - z \end{bmatrix}$

15. $\begin{bmatrix} a - 2b + c + 3d & 7 \\ 2a & + 5c + 4d & -1 \end{bmatrix}$

16. $x = 0, y = 6, z = 9$

17. $x = -3, y = \frac{12}{5}$

18. $x = 2, y = -3, z = 4$

19. $x = \frac{9}{2}, y = -2, z = -\frac{15}{4}$

20. $x = -3, y = 2, z = 3$

21. $A + B = \begin{bmatrix} a + 1 & b + 1 \\ c - 1 & d + 1 \end{bmatrix}$, an element of S

$AB = \begin{bmatrix} a - b & a + b \\ c - d & c + d \end{bmatrix}$, an element of S

22. $A + B = \begin{bmatrix} a + 1 & b + 1 \\ c - 1 & d + 1 \end{bmatrix} = \begin{bmatrix} 1 + a & 1 + b \\ -1 + c & 1 + d \end{bmatrix} = B + A$

23. $BC = \begin{bmatrix} 2 & 0 \\ 0 & -2 \end{bmatrix}, CB = \begin{bmatrix} 0 & 2 \\ 2 & 0 \end{bmatrix}$

24. $(A + B) + C = \begin{bmatrix} a + 2 & b + 2 \\ c & d \end{bmatrix} = A + (B + C)$

$(AB)C = \begin{bmatrix} 2a & -2b \\ 2c & -2d \end{bmatrix} = A(BC)$

25. $0 = \begin{bmatrix} 0 & 0 \\ 0 & 0 \end{bmatrix}$

26. $I = \begin{bmatrix} 1 & 0 \\ 0 & 1 \end{bmatrix}$

27. $-A = \begin{bmatrix} -a & -b \\ -c & -d \end{bmatrix}$

28. $A^{-1} = \begin{bmatrix} \dfrac{d}{ad - bc} & \dfrac{-b}{ad - bc} \\ \dfrac{-c}{ad - bc} & \dfrac{a}{ad - bc} \end{bmatrix}$

Yes, $A^{-1}A = AA^{-1}$

Given $AA^{-1} = 1$. If X exists so that $XA = 1$

then $(XA)A^{-1} = A^{-1}$

$X(AA^{-1}) = A^{-1}$

$X(I) = A^{-1}$

$X = A^{-1}$

29. $X = A + (-B)$
    $A - B = A + (-B)$

30. $Y = A^{-1}B$ if $ad - bc \neq 0$
    $B/A = A^{-1}B$ if $ad - bc \neq 0$

Exercises 7.3 A, page 394

1. $(3, -2)$

2. $(-2 \quad 3)$

3. $(3, 6, 12)$

4. $(6, 3, 4)$

5. $(x, -x - 1, 2 - 3x)$

6. $\left(a, 1 - 5a, \dfrac{4 - 7a}{2}\right)$

7. $\varnothing$

8. $(\frac{1}{2}, \frac{1}{4}, \frac{3}{4})$

9. $\varnothing$

10. $(\frac{3}{4}, \frac{5}{28}, \frac{13}{28})$

Exercises 7.4 A, page 401

1. 7

2. -22

3. -2

4. 29

5. 17

6. 18

7. a. $2(4 - 2) - 1(3 - 5) + 3(6 - 20) = 4 + 2 - 42 = -36$

b. $2(4 - 2) - 3(1 - 6) + 5(1 - 12) = 4 + 15 - 55 = -36$

8. a. $7(0 - 3) - 1(0 + 5) + 2(9 - 10) = -21 - 5 - 2 = -28$

b. $5(-1 - 4) + 3(-7 + 6) + 0 = -25 - 3 = -28$

9. $\begin{vmatrix} -2 & 7 & 7 \\ 1 & 0 & 0 \\ 2 & -1 & -3 \end{vmatrix} = 14$

10. $\begin{vmatrix} 1 & 1 & 1 \\ 0 & 3 & 8 \\ 0 & 7 & 26 \end{vmatrix} = 22$

11. $\begin{vmatrix} 1 & 2 & 0 \\ 1 & 4 & 0 \\ 1 & -1 & 2 \end{vmatrix} = 4$

12. $\begin{vmatrix} -3 & -2 & 9 \\ 0 & 3 & 0 \\ 8 & 1 & -5 \end{vmatrix} = -171$

13. $\begin{vmatrix} 1 & 2 & 3 \\ 0 & 2 & 6 \\ 0 & 0 & 6 \end{vmatrix}$

14. $\begin{vmatrix} 1 & -5 & 6 \\ 0 & 13 & -8 \\ 0 & 0 & 17 \end{vmatrix}$

15. $\begin{vmatrix} 1 & 3 & 2 & -2 \\ 0 & -1 & -5 & 5 \\ 0 & 11 & 7 & 0 \\ 0 & 7 & 7 & 0 \end{vmatrix} = 5 \begin{vmatrix} 11 & 7 \\ 7 & 7 \end{vmatrix} = 140$

16. $\begin{vmatrix} 1 & 1 & 1 & 1 \\ 0 & 1 & 2 & 3 \\ 0 & 0 & 2 & 6 \\ 0 & 0 & 0 & 6 \end{vmatrix} = 12$

Exercises 7.5 A, page 409

1. $(4, -2)$

2. $(2, -1)$

3. $(-2, 1)$

4. $(-5, 18)$

5. $(1, 6, 3)$

6. $(-3, 0, -4)$

7. $(\frac{3}{4}, -\frac{1}{2}, -\frac{2}{3})$

8. $(\frac{5}{12}, \frac{1}{3}, \frac{3}{4})$

9. $(4 - 2y, y, 4y + 2)$

10. $\varnothing$

11. $(0, 5, 3, -1)$

12. $(-6, 7, -5, 4)$

Exercises 7.6 A, page 411

1. 15 nickels, 25 dimes, 7 quarters

2. 90 minutes, 45 minutes, 60 minutes

3. $I_1 = 8, I_2 = 5, I_3 = 3$

4. \$16,000 at 6 percent, \$14,000 at 5 percent, \$10,000 at 10 percent

5. 40 of P, 25 of Q, 6 of R **6.** $a = 1, b = -2, c = 6$

7. 4 ounces of I, 1 ounce of II, 2 ounces of III

8. $(\frac{1}{4}, \frac{1}{2}, \frac{1}{4})$

Chapter 7 Review Exercises, page 415

1. $\begin{bmatrix} 6 & -2 & 3 \\ 10 & -3 & -7 \\ 0 & 8 & 10 \end{bmatrix}$

2. $\begin{bmatrix} 16 & 2x + 5y + 3z & \\ 6 & 3x + y + 2z & - t \\ 21 & x + 7y & - 2t \end{bmatrix}$

3. $\begin{bmatrix} 1 & -3 & 2 & -5 \\ 0 & 1 & 0 & 2 \\ 0 & 0 & 7 & -7 \end{bmatrix}$

4. $\begin{vmatrix} 2 & -4 & 6 \\ 0 & -5 & 10 \\ 0 & 0 & 2 \end{vmatrix}$

5. $(-4, 2, 3)$

6. $(-\frac{11}{4}, \frac{3}{2}, -\frac{33}{4})$

7. $\{(10t, -24t, -13t)$ where t is any real or complex number$\}$

8. $P = 50, Q = 35, R = 40$

Exercises 8.1 A, page 420

1. $1, 3, 5, 7, 9$

2. $1, 4, 9, 16, 25$

3. $0.5, 0.05, 0.005, 0.0005, 0.00005$

4. $1, 2, 3, 4, 5$

5. $-1, \frac{1}{4}, -\frac{1}{9}, \frac{1}{16}, -\frac{1}{25}$

6. $-\frac{3}{2}, \frac{3}{4}, -\frac{3}{8}, \frac{3}{16}, -\frac{3}{32}$

7. $0, 1, 0, \frac{1}{2}, 0$

8. $0, -\frac{1}{3}, -\frac{1}{2}, -\frac{3}{5}, -\frac{2}{3}$

9. $\frac{1}{2}, \frac{1}{4}, \frac{1}{8}, \frac{1}{16}, \frac{1}{32}$

10. $\frac{1}{2}, \frac{1}{5}, \frac{1}{10}, \frac{1}{17}, \frac{1}{26}$

11. $c(n) = n/(n + 1); \frac{5}{6} + \frac{6}{7}$

12. $c(n) = 6n; 30 + 36$

13. $c(n) = 3n + 1; 16 + 19$

14. $c(n) = 27 - 4n; 7 + 3$

15. $c(n) = 5(-3)^{1-n}; \frac{5}{81} - \frac{5}{243}$

16. $c(n) = (-1)^{n-1}2^{-n}; \frac{1}{32} - \frac{1}{64}$

17. $c(n) = (n + 3)/n^2; \frac{8}{25} + \frac{9}{36}$

18. $c(n) = 1/(n^2 + 1); \frac{1}{26} + \frac{1}{37}$

19. $c(n) = (1 - n)/(1 + n); -\frac{4}{6} - \frac{5}{7}$

20. $c(n) = (-1)^{n-1}3^n; 243 - 729$

Exercises 8.2 A, page 423

1. $1 + 3 + 5 + 7$
2. $1 + 8 + 27 + 64 + 125$
3. $0.3 + 0.03 + 0.003$
4. $-2 + 4 - 8 + 16 - 32$
5. $\frac{1}{2} + \frac{1}{3} + \frac{1}{4} + \frac{1}{5} + \frac{1}{6} + \frac{1}{7}$
6. $-5 - 12 - 19 - 26 - 33$
7. $\frac{3}{4} + \frac{4}{5} + \frac{5}{6} + \frac{6}{7} + \frac{7}{8}$
8. $497 - 502 + 507$
9. $1 + \frac{1}{3} + \frac{1}{5} + \frac{1}{7} + \cdots$
10. $24 + 0.24 + 0.0024 + \cdots$

11. $\displaystyle\sum_{k=1}^{9} (2k - 1)$
12. $\displaystyle\sum_{k=1}^{5} (\tfrac{1}{5})^{k-1}$

13. $\displaystyle\sum_{k=1}^{\infty} k^2$
14. $\displaystyle\sum_{n=1}^{5} \frac{(-1)^{n-1}}{n^2}$

15. $\displaystyle\sum_{n=1}^{\infty} (-1)^{n-1} 7^n$
16. $\displaystyle\sum_{n=1}^{7} (n^2 + 1)$

17. $\displaystyle\sum_{n=1}^{\infty} 4(6 - n)$
18. $\displaystyle\sum_{n=2}^{18} \frac{1}{n}$

19. $\displaystyle\sum_{n=1}^{\infty} \left(\frac{1}{n}\right)^n$
20. $\displaystyle\sum_{n=1}^{3} (2n - 1)(2n + 1)$

21. $-2 + 0 + 2 + 4 = 4$
22. $3(-1) + 3(0) + 3(1) + 3(2) + 3(3) = 15$
23. $1 + 3 + 1 + 3 = 8$
24. $2 + 8 + 24 + 64 = 98$

Exercises 8.3 A, page 428

1. 98
2. 20.5
3. -213
4. $-22\frac{2}{3}$
5. $32\sqrt{5}$
6. 48
7. 120
8. 108
9. 75
10. 145
11. 20,100
12. 892.5
13. n^2
14. 245
15. $[n(n + 2)]/2$
16. 140
17. 950
18. 210
19. $31\frac{1}{2}$
20. 21
21. 82.5, 85, 87.5
22. 67.5
23. 14, 4
24. 1.2, 1.4, 1.6, 1.8
25. 4.5
26. 14
27. $12,000 and $16,500
28. $2105
29. $20,100
30. 2200

Exercises 8.4 A, page 434

1. $\frac{256}{243}$

2. 2401

3. -390, $\frac{625}{4096}$

4. 9×10^{-14}

5. 36, 24

6. 6, 18, 54 or -6, 18, -54

7. 6 or -6

8. $5\sqrt[3]{2}$, $5\sqrt[3]{4}$

9. 41.496

10. 19,608

11. $\frac{1820}{729}$

12. -252

13. 511

14. 171

15. $\frac{25}{2}$

16. Sum does not exist; $r = \frac{5}{4} > 1$

17. $\frac{2}{275}$

18. 300

19. 360 inches

20. 8190

21. 800 gallons

22. 5

Exercises 8.5 A, page 438

1. I. $n = 1$, $2 = 1(1 + 1) = 2$

   II. If $2 + 4 + \cdots + 2k \qquad\qquad = k(k + 1)$

   then $2 + 4 + \cdots + 2k + 2(k + 1) = k(k + 1) + 2(k + 1)$
   $$= (k + 1)(k + 2)$$

2. I. $1^2 = [1(1 + 1)(2 + 1)]/6 = 2(3)/6 = 1$

   II. $1^2 + 2^2 + \cdots + k^2 + (k + 1)^2 = k(k + 1)(2k + 1)/6 + (k + 1)^2$
   $$= (k + 1)[k(2k + 1) + 6(k + 1)]/6$$
   $$= (k + 1)(2k^2 + 7k + 6)/6$$
   $$= (k + 1)(k + 2)(2k + 3)/6$$

3. I. $1/1(2) = 1/(1 + 1) = \frac{1}{2}$

   II. $1/1(2) + 1/2(3) + \cdots + 1/[k(k + 1)] + 1/[(k + 1)(k + 2)]$
   $$= k/(k + 1) + 1/[(k + 1)(k + 2)]$$
   $$= (k^2 + 2k + 1)/[(k + 1)(k + 2)]$$
   $$= (k + 1)/(k + 2)$$

4. I. $1 < 2^1$

   II. If $k < 2^k$, then $2k < 2(2^k)$ or $2k < 2^{k+1}$

   For $1 < k$, $k + 1 < 2k$ and thus $k + 1 < 2^{k+1}$.

5. I. $x - y = x - y$

   II. $x^{k+1} - y^{k+1} = x^k(x - y) + y(x^k - y^k)$

   If $x - y$ is a factor of $x^k - y^k$, then it is a factor of each term of the sum and thus a factor of the sum.

6. I. $1(1 + 1)(1 + 2) = 2(3)$

   II. $(k + 1)(k + 2)(k + 3) = k(k + 1)(k + 2) + 3(k + 1)(k + 2)$

   Since each term on the right is divisible by 3, the sum is also divisible by 3.

Exercises 8.6 A, page 442

1. $x^3 + 3x^2 + 3x + 1$ **2.** $x^4 + 4x^3 + 6x^2 + 4x + 1$

3. $x^3 - 3x^2 + 3x - 1$ **4.** $x^4 - 4x^3 + 6x^2 - 4x + 1$

5. $x^5 + 10x^4 + 40x^3 + 80x^2 + 80x + 32$

6. $y^5 - 10y^4 + 40y^3 - 80y^2 + 80y - 32$

7. $x^7 + 14x^6 + 84x^5 + 280x^4 + 560x^3 + 672x^2 + 448x + 128$

8. $y^8 - 8y^7 + 28y^6 - 56y^5 + 70y^4 - 56y^3 + 28y^2 - 8y + 1$

9. $b^6 - 6b^5 + 15b^4 - 20b^3 + 15b^2 - 6b + 1$

10. $x^5 + 5x^4y + 10x^3y^2 + 10x^2y^3 + 5xy^4 + y^5$

11. $x^6 + 9x^4 + 27x^2 + 27$

12. $1 - 5y^2 + 10y^4 - 10y^6 + 5y^8 - y^{10}$

13 $x^4 + 2x^3 + \frac{3}{2}x^2 + \frac{1}{2}x + \frac{1}{16}$ **14.** $a^6 - 3a^4b^2 + 3a^2b^4 - b^6$

15. $8x^3 + 60x^2 + 150x + 125$

16. $81x^4 - 216x^3 + 216x^2 - 96x + 16$

17. $17010x^6$ **18.** $\frac{35}{512}x^{13}$

19. $-1792y^5$ **20.** 924

21. 1.268   **22.** 0.85

23. 16

Chapter 8 Review Exercises, page 444

1. G, $\frac{25}{4}, \frac{125}{8}$ **2.** A, $(x + 4)/x$, $(x + 5)/x$

3. A, $-24, -33$ **4.** G, $-24, 48$

5. B, $-\frac{35}{8}, \frac{35}{16}$; $(1 - \frac{1}{2})^7$ **6.** G, $1/(3 - 2\sqrt{2})$, $1/(7 - 5\sqrt{2})$

7. A, $\log 128$, $\log 512$ **8.** G, $\log \sqrt[4]{3}$, $\log \sqrt[8]{3}$

9. G, $4i, 4i - 4$ **10.** B, $540, 405\sqrt{2}$; $(\sqrt{2} + 3)^5$

11. $1 + 3 + 5 + 7 + 9 + 11$

12. $9 + 16 + 25 + 36 + 49 + 64 + 81$

13. $3 + 0.3 + 0.03 + 0.003 + 0.0003$

14. $\frac{1}{4} + \frac{2}{5} + \frac{3}{6} + \frac{4}{7} + \frac{5}{8} + \cdots$

15. $\frac{7}{9} + \frac{8}{10} + \frac{9}{11} + \frac{10}{12} + \frac{11}{13} + \frac{12}{14} + \frac{13}{15}$

16. 650   **17.** -145

18. $[3n(n + 1)]/2$ **19.** 1550

20. 662.5   **21.** 45

22. 24,570   **23.** $\frac{127}{128}$

24. 32   **25.** -1705

26. 498   **27.** 295

28. $r = 2 > 1$; sum does not exist   **29.** $\frac{50}{3} = 16\frac{2}{3}$

30. $-\frac{1}{7}$ **31.** $\pm \sqrt[4]{8}/2$, $\pm i\sqrt[4]{8}/2$

32. a. $1 - 3x + 3x^2 - x^3$

b. $16 - 32t + 24t^2 - 8t^3 + t^4$

c. $a^5 + 15a^4 + 90a^3 + 270a^2 + 405a + 243$

33. 0.735

34. $-147,000x^3$

35. $\frac{62}{11}$

36. 1.0617

37. a. $1/2(5) = 1/(6 + 4) = \frac{1}{10}$

b. $1/2(5) + \cdots + 1/[(3k - 1)(3k + 2)] + 1/[(3k + 2)(3k + 5)]$

$$= 1/(6k + 4) + 1/[(3k + 2)(3k + 5)]$$
$$= (3k^2 + 5k + 2)/[2(3k + 2)(3k + 5)]$$
$$= (k + 1)/[2(3k + 5)]$$
$$= (k + 1)/(6k + 10)$$
$$= (k + 1)/[6(k + 1) + 4]$$

38. $x + y - 2\sqrt{xy} = (\sqrt{x} - \sqrt{y})^2 \geq 0$

$x + y \geq 2\sqrt{xy}$ and $\sqrt{xy} \leq (x + y)/2$

39. $10,750

40. $8540

INDEX

Abscissa 37, 225
Absolute value 62, 209
  equalities 209
  inequalities, 215
  relations 271
Addition method
  linear systems 38, 39
  quadratic systems 359
Antilogarithm 307
Arithmetic means 426
Arithmetic progression 425
Ascending powers 24
Associative axiom 13
Asymptotes 346
Augmented matrices 388
Axis 37, 224
  of symmetry 247

Bar 3
Base 23, 48
Binomial 23, 24
Binomial theorem 438
Braces 3
Brackets 3

Cartesian coordinate system 37
Cartesian product 222
Center of circle 252
Characteristic 306
Circle
  definition 252
  equation 254
Closure 13, 14
Coefficient 23
Common logarithms 305

Commutative axion 13
Completeness axiom 13, 63
Completing the square 128
Complex numbers 102
Composite number 10
Conic section 350
Conjugate
  complex 113
  irrational 84
Constant 1
Constant function 242
Coordinate 12, 37, 225
Counting numbers 12
Cramer's rule 406
Cube 2
Cube root 2, 70

Degree
  of monomial 23
  of polynomial 24
Descending powers 24
Determinant 395
Diagonal, main (principal) 384
Difference 2
  of cubes 24
  of squares 24
Digits 8
Dimensions of matrix 381
Direction of vector 118
Discriminant 133, 134
Distance formula 252
Distance on number line 211
Distributive axiom 13, 17
Dividend 2
Divisible 9

Divisor   2, 9
Domain   222
Double root   125
Dummy variable   422

Eccentricity   351
Ellipse   342
Empty set   9
Equal sets   9
Equality axioms   17
Equation   4
Equivalence theorems   17
Equivalent equations   16
Equivalent inequalities   195
Equivalent matrices   389
Even number   9
Expansion of determinant   396
Exponent   23, 48, 74
Exponent theorems   51
Exponential equations   314
Exponential function   293
Extraneous root   150

Factor   2, 9
Factoring   25, 166
Factor theorem   180
Finite set   8
Function   232
Fractions   28

Gauss's reduction method   391
Geometric means   431
Geometric progression   431
Graph   12, 37, 238
Graphical method
   linear systems   38, 39
   quadratic systems   352

Hyperbola   344
Hypotenuse   64

Identity axiom   13
Identity function   242
Identity matrix   384
Imaginary number   102
Imaginary part of complex number   104
Improper subset   9
Index
   of radicand   71
   of summation   421
Induction, mathematical   436
Inequality   4, 190
Infinite set   8
Integers   12
Integral exponents   55
Integral roots theorem   186
Interpolation, linear   309
Intersection of sets   10
Inverse axiom   13
Inverse matrix   384
Inverse relation   285
Irrational equations   148
Irrational numbers   12

Legs of right triangle   64
Like radicals   87
Like terms   24
Linear equation   16, 373
Linear function   238
Linear inequalities   195, 258
Linear programming   275
Linear system   334, 372
Listing method   8

Logarithm   298
Logarithmic equations   314
Logarithmic function   298

Magnitude of vector   118
Mantissa   306
Mathematical induction   436
Matrix   380
Mean proportional   432
Mean
   arithmetic   426
   geometric   431
Minor   396
Monomial   23
Multiple   9
Multiplicity of root   185

Natural logarithms   317
Natural numbers   8, 12
Negative product theorem   201
nth power   23, 48
nth root   71
Null set   9
Number line   12
Numeral   1

Odd number   9
Open equation   16
Order of determinant   396
Ordered pair   37, 221
Ordinate   37, 225
Origin   12, 37, 224

Parabola   247, 337
Parentheses   3
Pascal's triangle   441
Polynomial   23
Polynomial equations   184
Positive product theorem   201
Prime number   9
Product   2
Progression
   arithmetic   425
   geometric   431
   infinite geometric   433
Proper subset   9
Pythagoras, theorem of   64

Quadrant   225
Quadratic equation   101, 121, 334
Quadratic form equations   145
Quadratic formula   133, 134
Quadratic function   246
Quadratic inequalities
   one variable   201
   two variables   264
Quadratic system   334
Quotient   2

Radical   47, 62, 70, 79, 87
Radical equations   148
Radicand   71
Radius of circle   252
Range   222
Rational numbers   12
Rationalizing   83
Real numbers   12
Real part of complex number   104
Reciprocal   13
Rectangular coordinate system   37, 224
Reducing the radicand   79
Relation   4, 222
Remainder   2

Remainder theorem   179
Restricted values   32
Root
  cube   2, 70
  of equation   16
  square   2, 61

Scalar   382
Scientific notation   93, 305
Sequence   417
Series   418
Set   8
Set-builder method   8
Sigma notation   421
Similar terms   24
Simultaneous equations   330
Slope   240
Slope-intercept form   242
Solution
  of equation   16, 37, 373
  of inequality   195
Solution set   16, 37, 195, 223, 331, 373, 374
Square   2
Square matrix   381
Square root   2, 62
Standard form of complex number   104
Subset   9
Substitution axiom   4

Substitution method
  linear systems   38, 40
  quadratic systems   356
Sum   1
  of cubes   24
  product of roots   139
Symmetric equations   363
Synthetic division   173
System of equations   330

Term   2
Trichotomy axiom   13
Trinomial   23, 24

Union of sets   10
Universal set   9

Variable   1
Vector   117
Vertex   247, 337
Vinculum   3

Well-defined   8

y-intercept   239

Zero-product theorem   122

TABLE OF COMMON LOGARITHMS

n	0	1	2	3	4	5	6	7	8	9
1.0	.0000	.0043	.0086	.0128	.0170	.0212	.0253	.0294	.0334	.0374
1.1	.0414	.0453	.0492	.0531	.0569	.0607	.0645	.0682	.0719	.0755
1.2	.0792	.0828	.0864	.0899	.0934	.0969	.1004	.1038	.1072	.1106
1.3	.1139	.1173	.1206	.1239	.1271	.1303	.1335	.1367	.1399	.1430
1.4	.1461	.1492	.1523	.1553	.1584	.1614	.1644	.1673	.1703	.1732
1.5	.1761	.1790	.1818	.1847	.1875	.1903	.1931	.1959	.1987	.2014
1.6	.2041	.2068	.2095	.2122	.2148	.2175	.2201	.2227	.2253	.2279
1.7	.2304	.2330	.2355	.2380	.2405	.2430	.2455	.2480	.2504	.2529
1.8	.2553	.2577	.2601	.2625	.2648	.2672	.2695	.2718	.2742	.2765
1.9	.2788	.2810	.2833	.2856	.2878	.2900	.2923	.2945	.2967	.2989
2.0	.3010	.3032	.3054	.3075	.3096	.3118	.3139	.3160	.3181	.3201
2.1	.3222	.3243	.3263	.3284	.3304	.3324	.3345	.3365	.3385	.3404
2.2	.3424	.3444	.3464	.3483	.3502	.3522	.3541	.3560	.3579	.3598
2.3	.3617	.3636	.3655	.3674	.3692	.3711	.3729	.3747	.3766	.3784
2.4	.3802	.3820	.3838	.3856	.3874	.3892	.3909	.3927	.3945	.3962
2.5	.3979	.3997	.4014	.4031	.4048	.4065	.4082	.4099	.4116	.4133
2.6	.4150	.4166	.4183	.4200	.4216	.4232	.4249	.4265	.4281	.4298
2.7	.4314	.4330	.4346	.4362	.4378	.4393	.4409	.4425	.4440	.4456
2.8	.4472	.4487	.4502	.4518	.4533	.4548	.4564	.4579	.4594	.4609
2.9	.4624	.4639	.4654	.4669	.4683	.4698	.4713	.4728	.4742	.4757
3.0	.4771	.4786	.4800	.4814	.4829	.4843	.4857	.4871	.4886	.4900
3.1	.4914	.4928	.4942	.4955	.4969	.4983	.4997	.5011	.5024	.5038
3.2	.5051	.5065	.5079	.5092	.5105	.5119	.5132	.5145	.5159	.5172
3.3	.5185	.5198	.5211	.5224	.5237	.5250	.5263	.5276	.5289	.5302
3.4	.5315	.5328	.5340	.5353	.5366	.5378	.5391	.5403	.5416	.5428
3.5	.5441	.5453	.5465	.5478	.5490	.5502	.5514	.5527	.5539	.5551
3.6	.5563	.5575	.5587	.5599	.5611	.5623	.5635	.5647	.5658	.5670
3.7	.5682	.5694	.5705	.5717	.5729	.5740	.5752	.5763	.5775	.5786
3.8	.5798	.5809	.5821	.5832	.5843	.5855	.5866	.5877	.5888	.5899
3.9	.5911	.5922	.5933	.5944	.5955	.5966	.5977	.5988	.5999	.6010
4.0	.6021	.6031	.6042	.6053	.6064	.6075	.6085	.6096	.6107	.6117
4.1	.6128	.6138	.6149	.6160	.6170	.6180	.6191	.6201	.6212	.6222
4.2	.6232	.6243	.6253	.6263	.6274	.6284	.6294	.6304	.6314	.6325
4.3	.6335	.6345	.6355	.6365	.6375	.6385	.6395	.6405	.6415	.6425
4.4	.6435	.6444	.6454	.6464	.6474	.6484	.6493	.6503	.6513	.6522
4.5	.6532	.6542	.6551	.6561	.6571	.6580	.6590	.6599	.6609	.6618
4.6	.6628	.6637	.6646	.6656	.6665	.6675	.6684	.6693	.6702	.6712
4.7	.6721	.6730	.6739	.6749	.6758	.6767	.6776	.6785	.6794	.6803
4.8	.6812	.6821	.6830	.6839	.6848	.6857	.6866	.6875	.6884	.6893
4.9	.6902	.6911	.6920	.6928	.6937	.6946	.6955	.6964	.6972	.6981
5.0	.6990	.6998	.7007	.7016	.7024	.7033	.7042	.7050	.7059	.7067
5.1	.7076	.7084	.7093	.7101	.7110	.7118	.7126	.7135	.7143	.7152
5.2	.7160	.7168	.7177	.7185	.7193	.7202	.7210	.7218	.7226	.7235
5.3	.7243	.7251	.7259	.7267	.7275	.7284	.7292	.7300	.7308	.7316
5.4	.7324	.7332	.7340	.7348	.7356	.7364	.7372	.7380	.7388	.7396